U.S. Department
of Transportation
**National Highway
Traffic Safety
Administration**

www.nhtsa.gov

DOT HS 811 766

May 2013

Injury Vulnerability and Effectiveness of Occupant Protection Technologies for Older Occupants and Women

DISCLAIMER

This publication is distributed by the U.S. Department of Transportation, National Highway Traffic Safety Administration, in the interest of information exchange. The opinions, findings, and conclusions expressed in this publication are those of the authors and not necessarily those of the Department of Transportation or the National Highway Traffic Safety Administration. The United States Government assumes no liability for its contents or use thereof. If trade names, manufacturers' names, or specific products are mentioned, it is because they are considered essential to the object of the publication and should not be construed as an endorsement. The United States Government does not endorse products or manufacturers.

Technical Report Documentation Page

1. Report No. DOT HS 811 766	2. Government Accession No.	3. Recipient's Catalog No.
4. Title and Subtitle Injury Vulnerability and Effectiveness of Occupant Protection Technologies for Older Occupants and Women		5. Report Date May 2013
		6. Performing Organization Code
7. Author(s) Charles J. Kahane, Ph.D.		8. Performing Organization Report No.
9. Performing Organization Name and Address Office of Vehicle Safety National Highway Traffic Safety Administration Washington, DC 20590		10. Work Unit No. (TRAIS)
		11. Contract or Grant No.
12. Sponsoring Agency Name and Address National Highway Traffic Safety Administration 1200 New Jersey Avenue SE. Washington, DC 20590		13. Type of Report and Period Covered NHTSA Technical Report
		14. Sponsoring Agency Code
15. Supplementary Notes		

16. Abstract

Aging increases a person's fragility (likelihood of injury given a physical insult) and frailty (chance of dying from a specific injury). Young adult females are more fragile than males of the same age, but later in life women are less frail than their male contemporaries. Double-pair-comparison and logistic-regression analyses of 1975-2010 FARS, 1987-2007 MCOD, and 1988-2010 NASS-CDS data allow quantifying the effects of aging and gender on fatality and injury risk and studying how trends have changed as vehicle-safety technologies developed. In crashes of cars and LTVs of the past 50 model years, fatality risk increases as occupants age, given similar physical insults, by an average of $3.11 \pm .08$ percent per year that they age. Fatality risk is, on average, 17.0 ± 1.5 percent higher for a female than for a male of the same age (but more so for young adults and much less so for elderly occupants). The relative risk increases for aging and females may have both intensified slightly from vehicles of the 1960s up to about 1990 (even while safety improvements greatly reduced the absolute risk for men and women of all age groups); since then, the added risk for females has substantially diminished, probably to less than half, while the increase for aging may also have diminished, but by a much smaller amount. AIS ≥ 2 nonfatal-injury risk increases only by $1.58 \pm .35$ percent per year of aging, but it is 28.8 ± 6.0 percent higher for a female than for a male. Older occupants are susceptible to thoracic injuries, especially multiple rib fractures. Females are susceptible to neck and abdominal injuries and, at lower severity levels, highly susceptible to arm and leg fractures. Female drivers are especially vulnerable to leg fractures from toe-pan intrusion. All of the major occupant protection technologies in vehicles of recent model years have at least some benefit for adults of all age groups and of either gender; none of them are harmful for a particular age group or gender. Nevertheless, seat belts have been historically somewhat less effective for older occupants and female passengers, but more effective for female drivers. Frontal air bags are about equally effective across all ages; side air bags may be even more effective for older occupants than for young adults. Air bags and other non-belt protection technologies are helping females just as much and quite possibly even more than they protect males; this may have contributed to shrinking the historical risk increase for females relative to males of the same age.

17. Key Words FARS, MCOD, NASS, CDS, fragility, frailty, fatality risk, injury risk, aging, gender, effectiveness	18. Distribution Statement Document is available to the public from the National Technical Information Service www.ntis.gov		
19. Security Classif. (Of this report) Unclassified	20. Security Classif. (Of this page) Unclassified	21. No. of Pages 349	22. Price

Form DOT F 1700.7 (8-72)   Reproduction of completed page authorized

TABLE OF CONTENTS

LIST OF ABBREVIATIONS

AIS	Abbreviated Injury Scale
AMC	American Motors Corporation
BMI	Body mass index
CAC	Certified advanced compliant air bags
CDC	Centers for Disease Control and Prevention, U.S. Department of Health and Human Services
CDS	Crashworthiness Data System of NASS
CI	Confidence interval
CIREN	Crash Injury Research and Engineering Network, a database of severe crashes and medical information since 1996
CUV	Crossover utility vehicle
CY	Calendar year
delta v	A vehicle's velocity change as a result of a crash impact
df	Degrees of freedom
ESC	Electronic stability control
FARS	Fatality Analysis Reporting System, a census of fatal crashes in the United States since 1975
FMVSS	Federal Motor Vehicle Safety Standard
GES	General Estimates System of NASS
GM	General Motors
GVWR	Gross vehicle weight rating, specified by the manufacturer, equals the vehicle's curb weight plus maximum recommended loading
ICD	International Classification of Diseases
ICD-9	International Classification of Diseases, 9th revision
ICD-10	International Classification of Diseases, 10th revision

LTV	Light trucks and vans, includes pickup trucks, SUVs, minivans, and full-size vans
MAIS	A person's maximum-severity injury on the abbreviated injury scale
MCOD	Multiple cause of death file, a part of FARS
mph	Miles per hour
MY	Model year
NASS	National Automotive Sampling System, a probability sample of police-reported crashes in the United States since 1979, investigated in detail
NCAP	New Car Assessment Program: ratings of new vehicles since 1979 based on performance in frontal impact tests
NCHS	National Center for Health Statistics
NHTSA	National Highway Traffic Safety Administration, U.S. Department of Transportation
NS	Not [statistically] significant
NTDB	National Trauma Data Bank
PSU	Primary Sampling Unit
RATWGT	[Inverse sampling] ratio weight
RF	Right-front seat
SAS	Statistical and database management software produced by SAS Institute, Inc.
SUV	Sport utility vehicle
TTI(d)	Thoracic trauma index measured on a side-impact dummy
UEF	Universal exaggeration factor for belt effectiveness estimates after buckle-up laws
UTS	Ultimate tensile strength
VIN	Vehicle Identification Number
WHO	World Health Organization

EXECUTIVE SUMMARY

Older occupants have much higher injury and fatality risk in crashes than young adults. That has long been evident to crash investigators, data analysts, and medical personnel who treat the victims. As people age, they become more fragile: more susceptible to injury, given similar physical insults or impacts. Later on, they also become frailer: more likely to die, given the same injury. The analyses of this report will show that occupants' fatality risk, given similar physical impacts, grows by 3 percent or slightly more for each year that they get older, starting at about 21. The analyses will also demonstrate that young-adult women up to age approximately 35 have 25 to 30-percent higher fatality risk, given similar physical insults, than men of the same age. Men's advantage, however, diminishes after age 35; by age 70, female and male drivers are about equally at risk. Evidently, young women are more fragile than young men, but eventually men's fragility catches up and/or men become frail sooner than women (as evidenced by men's lower life expectancy).

The increased risk for older occupants and women may to a large extent be a consequence of intrinsic human anatomy and physiology. But a vehicle's design and technology and the crash environment could also be influential. Specifically, safety technology that is even more effective for the elderly and women than it is for young males would shrink the relative risk increase for older occupants and women. However, another technology that is especially effective for young males would tend to augment the risk disparity – even if it is also effective to some extent for the older or female occupants, but just not as effective as for young males. . .

During the 1980s, Evans developed a statistical technique – double-pair comparison analysis – and applied it to the then-new Fatality Analysis Reporting System (FARS) to quantify how fatality risk from the same crash situation increases with age and is different for males and females.[1] With only the first nine years of FARS data available, Evans was limited to estimating the average, overall effect on fatality risk of aging and gender in all crashes for the vehicle fleet that was on the road in CY 1975-1983 (when 97% of occupant fatalities were unbelted). Now, there are 36 years of FARS (1975-2010) as well as other databases such as FARS-Multiple Cause of Death (MCOD, 1987-2007) and NHTSA's Crashworthiness Data System (CDS, 1988-2010) that identify specific injuries. With these databases and an additional statistical technique – logistic regression – this report can focus on the details: the relatively small changes over time in the relationship between age, gender, and fatality or injury risk, or their variation by type of vehicle, impact, or injury. Have technologies introduced during the past 50 years and the substantial increase in belt use shrunk or augmented the risk increase for older occupants and women? These technologies include major occupant protection such as air bags as well as small, year-to-year changes in materials such as the shift to high-strength steels or incremental revisions in design to improve performance on crash ratings.

The report includes many estimates of the fatality- or injury-risk increase for aging a year, or the risk for a female relative to a male of the same age, given similar physical insults: by vehicle type, model year, seat position, types of occupant protection available, belt use, impact location, body region of the injury, or severity (AIS) of the injury. In addition, for the major occupant

---

[1] Evans, L. (1991). *Traffic Safety and the Driver*. New York: Van Nostrand Reinhold, pp. 22-28.

protection technologies – seat belts, air bags, and energy-absorbing materials – analysis methods of past NHTSA evaluations will show if fatality-reducing effectiveness is significantly different for younger and older occupants or for men and women. Finally, the report identifies specific injuries and injury sources (vehicle components contacted) that are prevalent among older or female occupants.

The first purpose of the analyses is to flag any safety technologies or vehicle components that have little benefit or possibly even increase harm to older occupants or women. The analyses did not raise any alarm flags for recent-model vehicles. The major current safety technologies appear to be effective (but not necessarily equally effective) for male and female adults of all age groups. However, seat belts in the back seat are not as effective for passengers over age 70 as they are for young adults. Likewise, some obsolete safety technologies did not perform as well for older occupants as for young adults: 2-point automatic seat belts and some of the static energy-absorbing devices before they were supplemented with air bags; also, older occupants and women, when belted, are at high risk of neck injury in frontal impacts of vehicles without air bags.

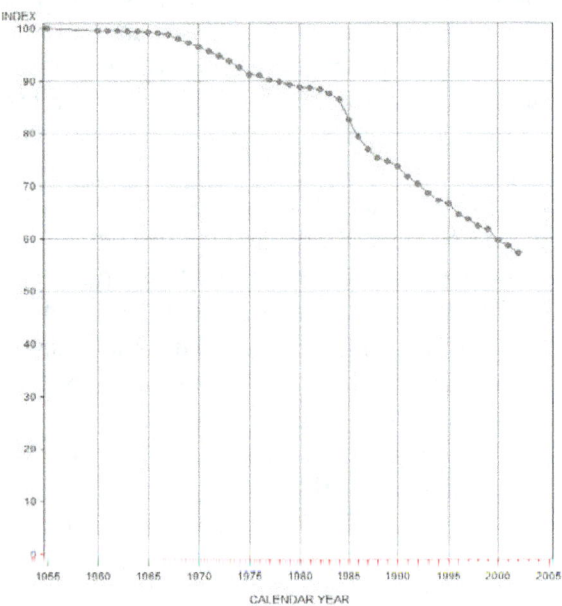

FIGURE 1: OCCUPANT FATALITY-RISK INDEX BY CALENDAR YEAR
BASED ON LIVES SAVED BY THE FMVSS (1955 = 100)

A second purpose is to build a narrative of how fatality and injury risk of older occupants and women, **relative** to young adult males, has changed over the past 50 model years. Of course, it is important to note that in **absolute** terms, vehicles have become much safer for everybody. Figure 1, based on a 2004 NHTSA report, shows that the overall, absolute occupant fatality risk decreased by 42 percent from CY 1955-1960 to CY 2002 due to increased belt use, air bags, and the other FMVSS.[2] This rising tide of safety benefited everyone: absolute risk decreased substantially for males and females of all ages, but not necessarily by exactly equal amounts. If safety improvements help young people even more than they help the elderly, the relative gap between the elderly and the young can increase even while absolute safety improves for both. The general impression from the analyses of this report is that the early safety technologies and design changes tended to benefit young adult males somewhat more than other groups, consequently augmenting slightly the relative risk increases for aging and for females. More recent technologies such as air bags, pretensioners, and load limiters, perhaps because they were developed with a deeper understanding of biomechanics and human tolerance limits, have substantially reduced the excess risk of females relative to male occupants and might perhaps also be diminishing the

[2] Based on Kahane, C. J. (2004). *Lives Saved by the Federal Motor Vehicle Safety Standards and Other Vehicle Safety Technologies, 1960-2002*. (Report No. DOT HS 809 833). Washington, DC: National Highway Traffic Safety Administration, Table 2 on p. xv. http://www-nrd.nhtsa.dot.gov/Pubs/809833.PDF.

relative risk increase associated with aging. However, these are mostly just tentative conclusions, because the changes over time in the trends are small, as may be seen in Figures 5 and 6 in the next section. Only the recent reduction in the fatality risk for women relative to men of the same age looks substantial.

The third purpose is to identify specific injuries and injury sources that are especially prevalent among older occupants and/or females. It should be understood that these vulnerabilities are primarily a consequence of the intrinsic physiological differences between young and old people, males and females. They may be difficult to mitigate. Nevertheless, those injury mechanisms might be priority candidates for additional study of potential safety improvements. This report, of course, confirms the vulnerability of older occupants to thoracic injuries, especially multiple fractures, which has already been widely reported in the literature. Another key finding is that females are especially vulnerable to neck and abdominal injuries. At a lower injury level, women are quite vulnerable to leg fractures from the floor, toe pan, and pedals and to arm fractures from numerous sources.

PRINCIPAL FINDINGS AND CONCLUSIONS

OVERALL AVERAGE EFFECT OF AGING AND GENDER ON FATALITY RISK

AGING

- For drivers and right-front (RF) passengers 21 to 96 of MY 1960-2011 cars and LTVs on CY 1975-2010 FARS, given similar physical insults, fatality risk increases by an estimated $3.11 \pm .08$ percent for each year that they get older.

GENDER

- The fatality risk of a female driver or RF passenger is, on the average, an estimated 17.0 ± 1.5 percent higher than for a male of the same age, given similar physical insults.

EFFECT BY SEAT POSITION

AGING

- Fatality risk increases by an estimated $2.83 \pm .08$ percent for each year that a driver of a car or LTV gets older, $3.39 \pm .07$ percent for RF passengers of cars and LTVs, and $3.46 \pm .14$ percent for back-seat outboard passengers of cars.

- The smaller effect for drivers may be, to some extent, because the healthier seniors continue to drive while less healthy seniors may ride only as passengers. FARS provides no information about the health history of the occupants that could shed light on this possibility.

GENDER

- Fatality risk is an average of 13.4 ± 2.0 percent higher for a female driver than for a male driver of the same age exposed to similar physical insults; the corresponding increase for RF passengers is 20.5 ± 2.2 percent and for back-seat passengers, 15.7 ± 6.1 percent.

EFFECT OF AGING BY GENDER

- Fatality risk increases by $3.28 \pm .08$ percent for each year that a male front-seat occupant ages, but by only $2.93 \pm .08$ percent for each year that a female gets older. The overall average rate of increase is lower for females because, at later ages, women's frailty and/or fragility generally increase at a slower pace than men's.

HOW THE EFFECTS OF AGING AND GENDER VARY WITH AGE

AGING

- The steady and irreversible trend of declining crash survivability characteristic of adulthood has set in by 21 or 22.

- For drivers, the added risk for aging one year remains fairly constant from 21 onward, intensifying just slightly at later ages. For male drivers 21 to 30, fatality risk increases by an average of 2.90 percent per year that they get older; for male drivers 65 to 74, 3.39 percent.

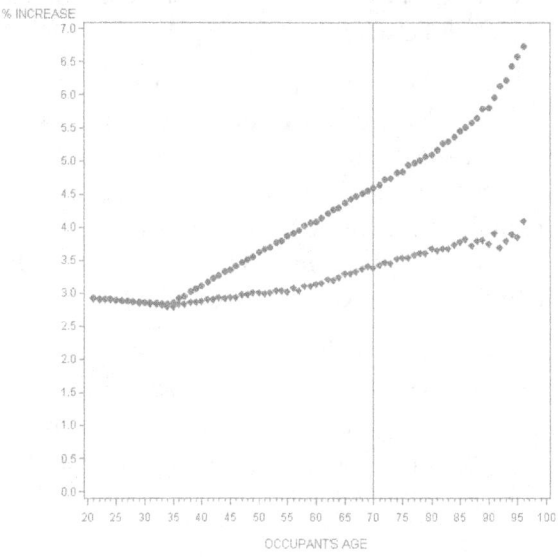

FIGURE 2: FATALITY INCREASE (%) FOR EACH YEAR
THAT A MALE DRIVER (blue diamonds) OR RF PASSENGER (green circles) GETS OLDER

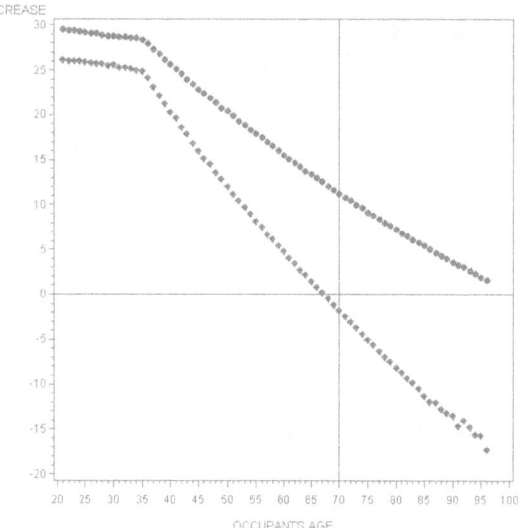

FIGURE 3: FATALITY INCREASE (%) FOR A FEMALE DRIVER
RELATIVE TO A MALE DRIVER (blue diamonds)
OR A FEMALE RF PASSENGER RELATIVE TO A MALE RF PASSENGER (green circles)

- For RF passengers, the effect intensifies more strongly at later ages. For male RF passengers 21 to 30, risk increases by an average of 2.90 percent per year that they get older; for male RF passengers 65 to 74, 4.58 percent. This difference between drivers and RF passengers may also reflect in part that older drivers may be, on the average, healthier than passengers the same age. Figure 2 shows how the relative effect of aging one year intensifies with age for drivers (blue diamonds) and RF passengers (green circles).

GENDER

- The substantial increase in fatality risk for young adult females relative to males the same age is already present by age 18 and perhaps even earlier.

- For drivers, females' added risk diminishes sharply after age 35. After their late 60s or early 70s, females are at lower risk than males the same age. For drivers 21 to 30, fatality risk averages 25.9 percent higher for females than for males the same age; for drivers 65 to 74, 1.4 percent lower.

- For RF passengers, females' added risk also diminishes after age 35, but not as sharply as for drivers. For RF passengers 21 to 30, fatality risk averages 29.2 percent higher for females than for males the same age; for RF passengers 65 to 74, it remains 11.4 percent higher. This difference between drivers and RF passengers may reflect that elderly female drivers may be exceptionally healthy for their age group. Figure 3 shows how the fatality risk for a female driver relative to a male driver changes with age (blue diamonds) and also for a female relative to a male RF passenger (green circles).

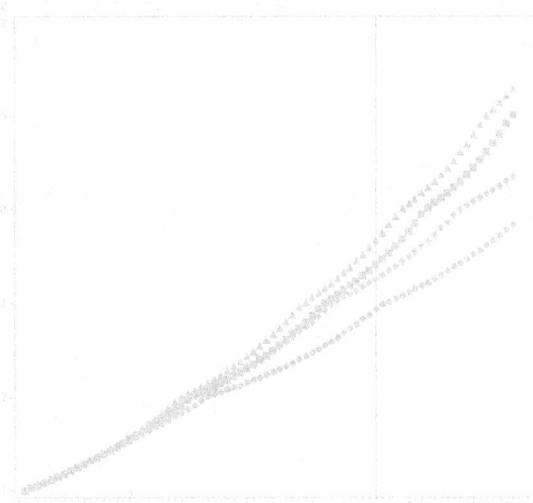

FIGURE 4: FATALITY RISK RELATIVE TO A 21-YEAR-OLD FOR A MALE DRIVER (blue diamonds), MALE RF PASSENGER (cyan triangles), FEMALE DRIVER (red circles), AND FEMALE RF PASSENGER (pink squares)

CUMULATIVE RISK INCREASE FROM AGE 21 TO 75

- Given similar physical insults, a 75-year-old male driver is, on the average, 5.04 times as likely to die as a 21-year-old male driver. For female drivers, the corresponding cumulative increase from age 21 to 75 is a factor of 3.87; for male RF passengers, 6.70; and for female RF passengers, 5.67.

- Figure 4 shows how the fatality risk relative to a 21-year-old increases with age for a male driver (blue diamonds), male RF passenger (cyan triangles), female driver (red circles), or a female RF passenger (pink squares).

EFFECTS OF AGING AND GENDER IN CARS VERSUS LTVs

AGING

- For drivers and RF passengers 21 to 96 of MY 1960-2011 vehicles on CY 1975-2010 FARS, given similar physical insults, the increase in fatality risk for aging one year is almost the same in cars (3.14 ± .08%) and LTVs (3.00 ± .17%).

GENDER

- The added risk for a female relative to a male of the same age is also almost the same in cars (16.8 ± 1.8%) as in LTVs (17.3 ± 3.2%).

HOW THE EFFECT OF AGING ONE YEAR HAS CHANGED OVER TIME (MY 1960-2011)

- The estimated increase in fatality risk for aging one year (average for drivers and RF passengers 21 to 96 years old, cars plus LTVs) has not changed much over the years. It may have increased slightly from vehicles of the 1960s to vehicles of the early 1990s and then perhaps diminished by a similar amount. These are not firm conclusions, given the small absolute size of the changes and the uncertainty in the individual estimates, as illustrated in Figure 5:

Model Year Cohort	Estimated Risk Increase (%)
1960-1966	2.89 ± .36
1967-1974	2.94 ± .17
1975-1979	3.17 ± .17
1980-1984	3.05 ± .20
1985-1989	3.24 ± .17
1990-1994	3.31 ± .25
1995-1999	3.00 ± .28
2000-2004	2.92 ± .33
2005-2011	3.15 ± .53

- The early belt systems and energy-absorbing devices (before air bags) were sometimes less effective for older occupants than for young people and may have contributed to intensifying the trend of risk increasing with age. Frontal and side air bags and belts with pretensioners and load limiters may be equally or even more effective for older than for younger occupants and have perhaps helped shrink the aging effect.

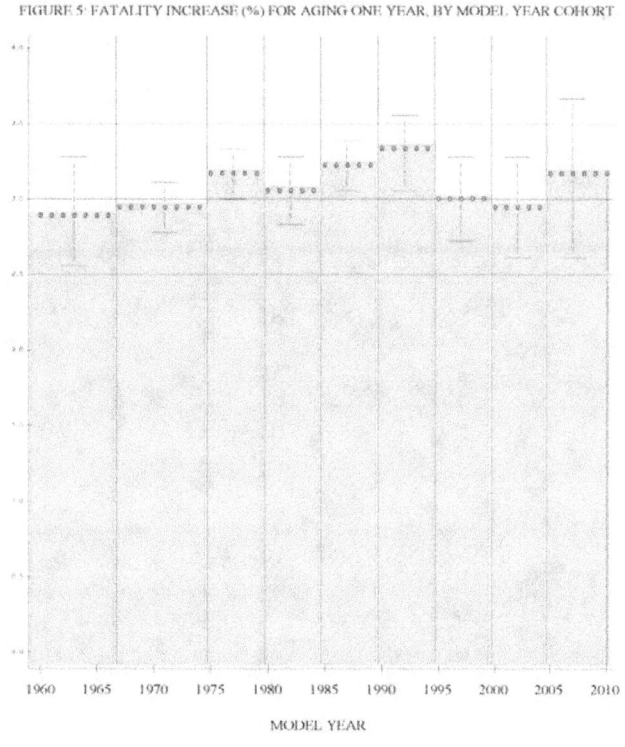

FIGURE 5: FATALITY INCREASE (%) FOR AGING ONE YEAR, BY MODEL YEAR COHORT

MODEL YEAR

HOW FEMALES' RISK RELATIVE TO MALES CHANGED OVER TIME (MY 1960-2011)

- The estimated increase in fatality risk for females relative to males of the same age perhaps grew a little bit from vehicles of the 1960s to vehicles of the late 1980s. This is not a firm conclusion, given the small absolute size of the early changes and the uncertainty in the individual estimates, as illustrated in Figure 6. Since the 1990s, females' risk relative to males has shrunk substantially, perhaps to half its original level or less; Figure 6 shows that the trend after 1990 is strong:

Model Year Cohort	Estimated Risk Increase (%)
1960-1966	16.7 ± 9.6
1967-1974	17.7 ± 4.5
1975-1979	17.8 ± 3.5
1980-1984	20.0 ± 2.9
1985-1989	20.5 ± 3.8
1990-1994	18.5 ± 5.2
1995-1999	12.0 ± 3.3
2000-2004	7.0 ± 4.2
2005-2011	8.8 ± 8.5

- The combination of air bags and belt use, especially belts with pretensioners and load limiters has greatly reduced the gender gap, at least for fatality risk.

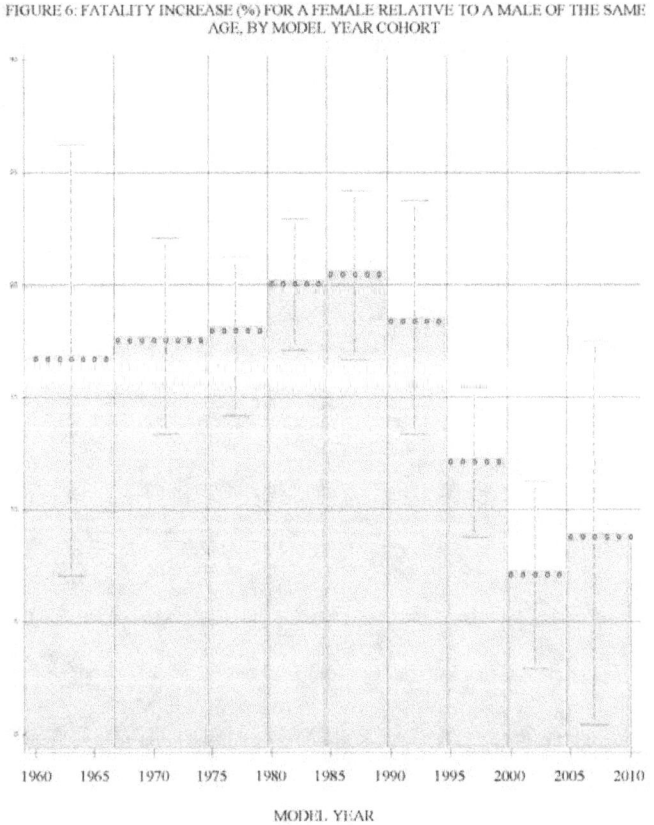

FIGURE 6: FATALITY INCREASE (%) FOR A FEMALE RELATIVE TO A MALE OF THE SAME AGE, BY MODEL YEAR COHORT

MODEL YEAR

HOW EFFECTS VARY BY THE TYPE OF IMPACT

AGING

- The estimated increase in fatality risk for aging one year varies little by the type of crash. It is slightly higher in frontals, nearside impacts and rollovers than in far-side and rear impacts:

Impact Type	Estimated Risk Increase (%)
Frontal	$3.20 \pm .09$
Nearside	$3.19 \pm .14$
Far-side	$2.84 \pm .14$
First-event rollover	$3.12 \pm .25$
Rear/other	$2.74 \pm .27$

GENDER

- The estimated increase in fatality risk for females relative to males of the same age is somewhat higher in nearside impacts and first-event rollovers than in frontals or far-side impacts:

Impact Type	Estimated Risk Increase (%)
Frontal	14.1 ± 2.2
Nearside	24.4 ± 4.7
Far-side	13.0 ± 4.1
First-event rollover	22.2 ± 4.9
Rear/other	17.7 ± 5.4

- However the increase in rollovers is primarily among <u>unbelted</u> females. They are at 33.3 ± 9.2 higher risk than unbelted males. Unbelted females are more likely to be ejected from the vehicles than unbelted males, presumably because females, being smaller on the average than males, can pass more easily through ejection portals such as the side-window area.

HOW EFFECTS VARY BY BELT USE

AGING

- The estimated **relative** increase in fatality risk for aging one year is perhaps slightly higher for 3-point belted front-seat occupants than for unrestrained; although, of course, the **absolute** fatality risk is much lower for belted occupants than unbelted occupants for any age group,[3] 3-point belts have in the past been slightly less effective for older occupants than for young adults; the relative increase is substantially higher with 2-point automatic belts, an obsolete system that is less effective for older occupants than for young people.

Belt Use	Estimated Risk Increase (%)
Unrestrained	2.99 ± .10
3-point belt	3.16 ± .15
2-point automatic belt	3.92 ± .35

GENDER

- The estimated increase in fatality risk for females relative to males of the same age is somewhat lower for 3-point belted occupants than for unrestrained; it appears to be higher with 2-point automatic belts.

Belt Use	Estimated Risk Increase (%)
Unrestrained	18.4 ± 2.1
3-point belt	13.2 ± 2.7
2-point automatic belt	27.7 ± 10.0

HOW EFFECTS FOR BELTED OCCUPANTS VARY BY AIR BAG AVAILABILITY

AGING

- The estimated relative increase in fatality risk in frontal impacts of passenger cars for 3-point belted occupants aging one year is more or less the same with and without frontal air bags (although, of course, the absolute fatality risk is much lower in frontal impacts with air bags than without air bags for adults of any age group[4]).

Frontal Air Bags	Estimated Risk Increase (%)
None	3.17 ± .10
Dual (w/o belt pretens & load lim)	3.19 ± .45
Dual (with belt pretens & load lim)	3.22 ± .61

---

[3] Kahane (2004), pp. 95-97, updated and confirmed by the analyses of Chapter 9 of this report.
[4] Ibid., p. 111, updated and confirmed by the analyses of Chapter 9 of this report.

- The estimated relative increase in fatality risk in side impacts for 3-point belted occupants aging one year may be lower with side air bags plus inflatable head protection (data still limited):

Side Air Bags	Estimated Risk Increase (%)
None (nearside impact)	$3.19 \pm .32$
None (far-side impact)	$2.94 \pm .30$
Curtain+torso or combo	2.28 ± 1.22

GENDER

- The estimated increase in fatality risk in frontal impacts of passenger cars for 3-point belted females relative to belted males of the same age may shrink with frontal air bags:

Frontal Air Bags	Estimated Risk Increase (%)
None	16.9 ± 6.3
Dual (w/o belt pretens & load lim)	7.5 ± 12.5
Dual (with belt pretens & load lim)	9.3 ± 10.6

- The estimated increase in fatality risk in side impacts for 3-point belted females relative to belted males of the same age may be lower with side air bags plus inflatable head protection (data still limited):

Side Air Bags	Estimated Risk Increase (%)
None (nearside impact)	15.5 ± 10.8
None (far-side impact)	18.5 ± 10.9
Curtain+torso or combo	2.6 ± 21.5

- In other words, side air bags with head protection might be especially helpful for older occupants and females; frontal air bags might be about equally effective across all ages and especially helpful for females.

EFFECTS FOR BACK-SEAT PASSENGERS OF CARS

AGING

- For back-seat outboard passengers 21 to 96, the estimated relative increase in fatality risk for aging one year is augmented by belt use – lap-belt-only and 3-point belts (although the analyses of this report will demonstrate that absolute fatality risk is lower for belted occupants than unbelted men or women for any age group):

Belt Use	Estimated Risk Increase (%)
Unrestrained	$3.18 \pm .18$
Lap belt only	$3.98 \pm .82$
3-point belt	$3.94 \pm .47$

GENDER

- The estimated increase in fatality risk for females relative to males of the same age likewise appears to be augmented by belt use:

Belt Use	Estimated Risk Increase (%)
Unrestrained	11.3 ± 7.2
Lap belt only	31.9 ± 28.3
3-point belt	28.0 ± 20.0

- In the back seat, current 3-point belt systems are not helping older occupants and possibly females as much as they help young males.

EFFECTS ON INJURIES THAT "CONTRIBUTE TO DEATH," BY BODY REGION

AGING

- The FARS-MCOD file lists a fatally injured occupant's injuries that "contributed to death." For drivers and RF passengers 21 to 96 of MY 1960-2008 cars and LTVs on CY 1987-2007 FARS-MCOD, given similar physical insults, the estimated increase in injury risk for aging one year is highest for injuries to the thorax, then the abdomen, then the neck; head injuries increase least with age:

Body Region Injured	Estimated Risk Increase (%)
Head	$2.49 \pm .17$
Chest	$4.06 \pm .21$
Abdomen	$3.43 \pm .32$
Neck	$3.19 \pm .37$

- In other words, older occupants are exceptionally vulnerable to torso injuries, especially chest injuries.

- The effects of aging are fairly similar across the various types of crashes: frontals, side impacts, and rollovers.

GENDER

- The estimated increase in injury risk for females relative to males of the same age is much larger for neck and abdominal injuries than for the head or chest:

Body Region Injured	Estimated Risk Increase (%)
Head	14.6 ± 3.1
Chest	8.8 ± 4.6
Abdomen	31.9 ± 8.3
Neck	39.4 ± 9.4

- In other words, women are exceptionally more vulnerable to neck and abdominal injuries. The high risk of neck injury appears to be related to the anatomy of a typical female: a male's neck has greater spinal-column strength than a female's, yet a female's neck is called upon to support and control the motion of a head that is almost as large and heavy as a male's. The high risk of abdominal injury is harder to explain.

- The effects of gender are fairly similar across the various types of crashes: frontals, side impacts, and rollovers.

- For belted occupants in vehicles without air bags, the estimated increase in neck-injury risk for females relative to males is exceptionally high: 90.9 ± 53.4 percent. So is the increase for aging one year: 4.39 ± 1.24 percent.

Figure 7 is a bar graph comparing the principal estimates, with confidence bounds, of the average relative effect of aging one year on fatality risk (FARS analyses) and on the risk of injuries that contribute to death (FARS-MCOD analyses) – overall and by seat position, crash type, belt use, and body region of the injury. In general, the effect of aging does not vary that much, but it is relatively low for head injuries, relatively high for chest injuries, with automatic 2-point belts in the front seat, and with 3-point belts in the back seat. (Figure 5 showed how the effect has changed over time in model years 1960-2011.)

Figure 8 is the corresponding bar graph comparing estimated risk increases for females relative to males of the same age (effect averaged across ages 21 to 96). The increases are relatively low for chest and head injuries, 3-point belted women in the front seat, and in frontal and far-side impacts. They are relatively high for neck and abdominal injuries, with automatic 2-point belts in the front seat, with 3-point belts in the back seat (not certain, due to the wide confidence bounds), and in nearside impacts and rollovers. Figure 6 showed how the effect has changed over time.

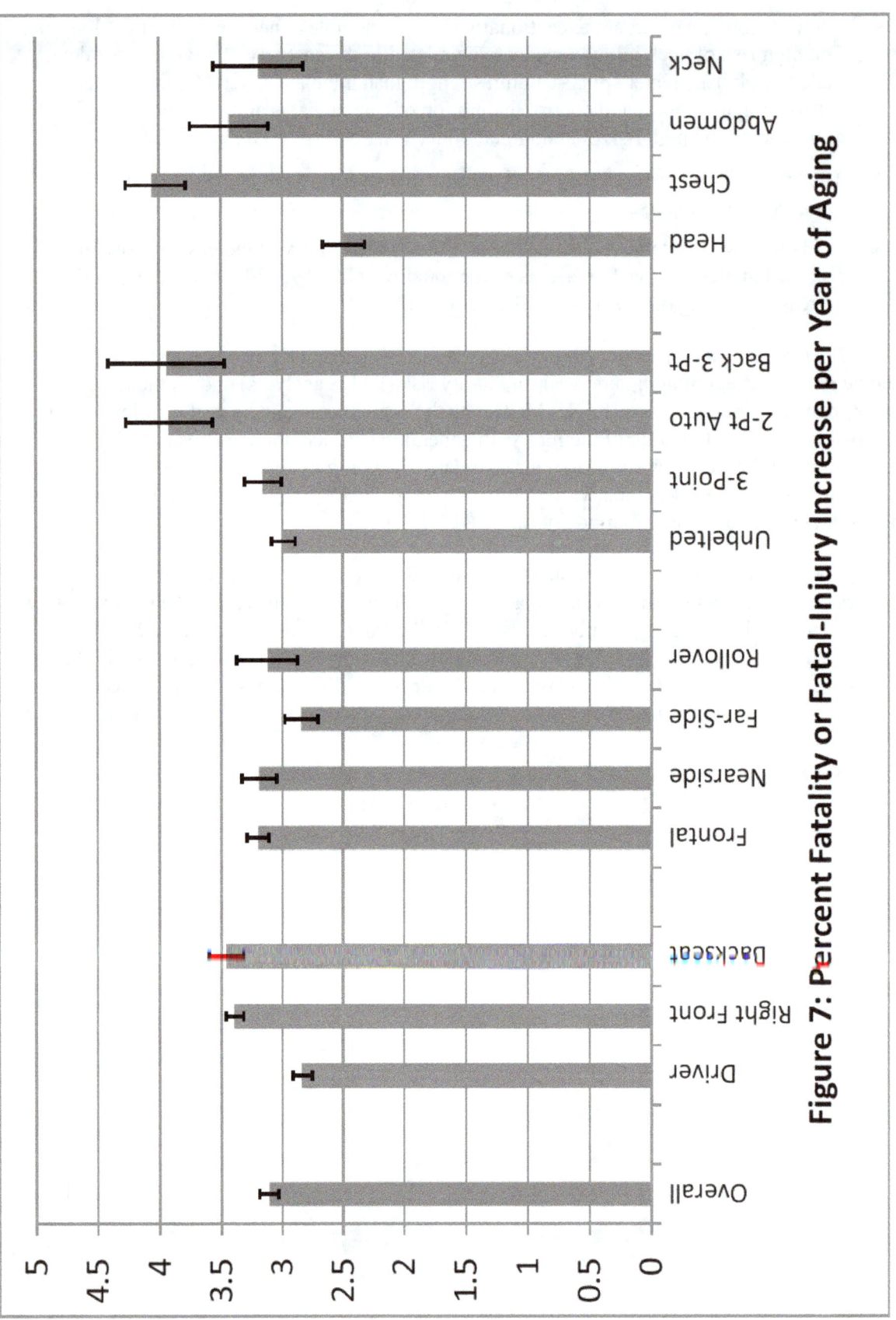

Figure 7: Percent Fatality or Fatal-Injury Increase per Year of Aging

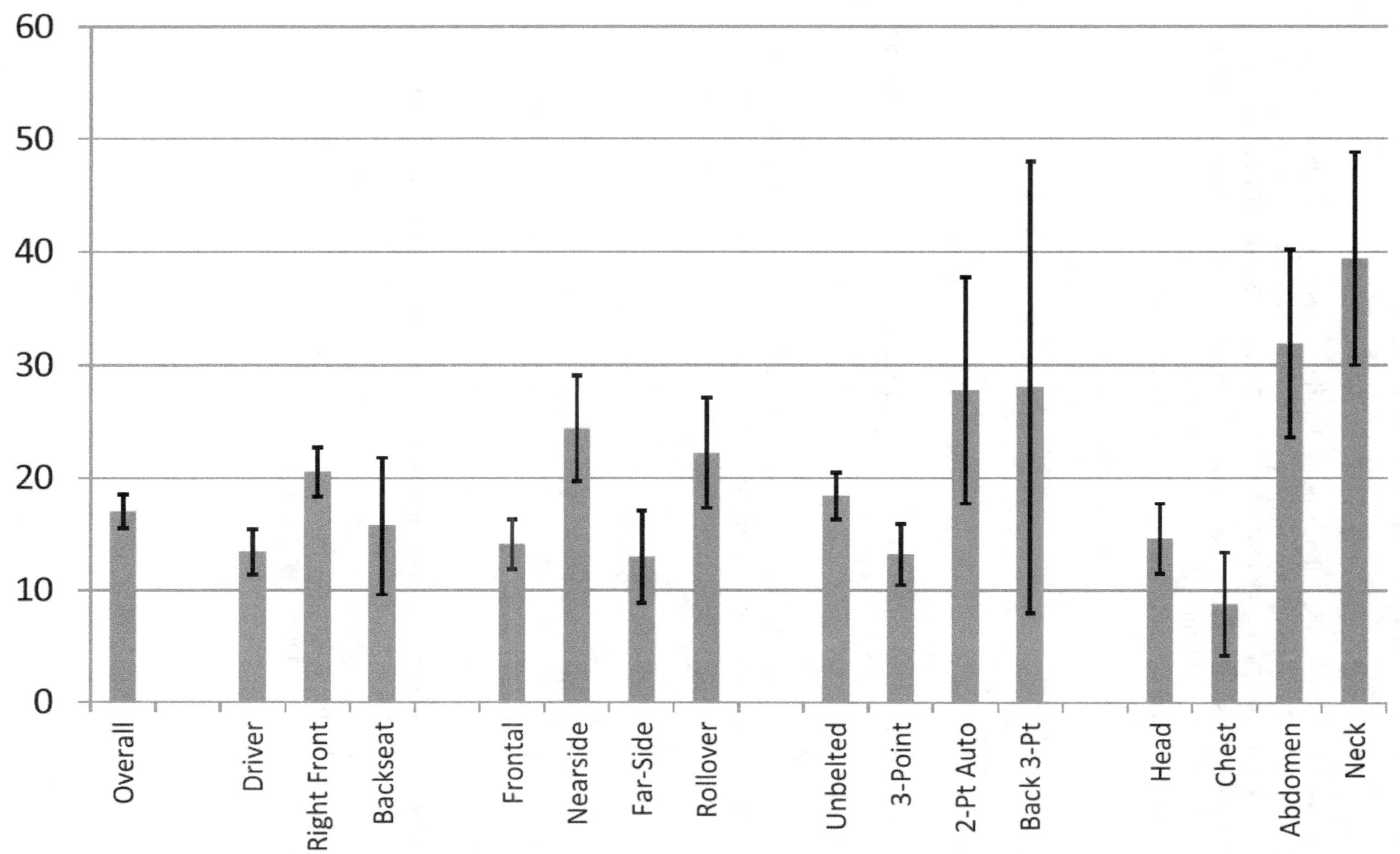

Figure 8: Fatality Increase (%) for a Female Relative to a Male of the Same Age
(Average Increase for Occupants Age 21 to 96)

EFFECTS ON FATAL AND NONFATAL INJURIES, BY AIS LEVEL

AGING

- The NASS-CDS file describes all occupants' injuries, including their AIS severity level and body region, for a probability sample of the nation's crashes. For drivers and RF passengers 21 to 96 of MY 1960-2011 cars and LTVs on CY 1988-2010 CDS, given similar physical insults, the estimated increase in injury risk for aging one year grows substantially as the maximum AIS of the occupant's various injuries increases in severity (the estimate for fatalities is based on FARS data, not CDS):

Injury Severity	Estimated Risk Increase (%)
MAIS ≥ 2 (moderate)	1.58 ± .35
MAIS ≥ 3 (serious)	2.29 ± .44
MAIS ≥ 4 (severe)	2.65 ± .61
Fatality	3.11 ± .08

- At the lower injury levels, older occupants' increasing frailty is not an issue, only their fragility.

GENDER

- The estimated increase in injury risk for females relative to males of the same age does not share the aging effects' pattern of growth as the injuries become more severe; in fact, the added risk for females is lowest at the fatal level (the estimate for fatalities is based on FARS data, not CDS):

Injury Severity	Estimated Risk Increase (%)
MAIS ≥ 2 (moderate)	28.8 ± 6.0
MAIS ≥ 3 (serious)	37.3 ± 10.5
MAIS ≥ 4 (severe)	28.9 ± 15.0
Fatality	17.0 ± 1.5

- The effect depends on the mix of injuries that constitute each AIS level. For example, there are many AIS 2-3 arm and leg injuries, but few AIS 4-6. Furthermore, the lower frailty (greater longevity) of older females compared to males of the same age is of no advantage at the lower injury levels. In fact, unlike the fatalities, females' added risk of some types of nonfatal injuries grows rather than shrinks with age.

EFFECTS ON AIS ≥ 2 INJURIES, BY BODY REGION

AGING

- In the CDS data on AIS ≥ 2 injuries (as in the FARS-MCOD on injuries that "contributed to death"), the estimated increase in injury risk for drivers and RF passengers aging one year is high for injuries to the thorax and low for head injuries; it is also low for leg injuries:

Body Region Injured	Estimated Risk Increase (%)
Head	1.35 ± .68
Chest	3.34 ± .68
Abdomen	2.57 ± .88
Neck	3.15 ± 1.24
Arm	2.44 ± .65
Leg	1.46 ± .53

GENDER

- The estimated increase in AIS ≥ 2 injury risk for female drivers and RF passengers relative to males of the same age is high for abdominal and neck injuries, but it is even higher for arm and leg injuries:

Body Region Injured	Estimated Risk Increase (%)
Head	22.1 ± 16.0
Chest	26.4 ± 13.6
Abdomen	38.5 ± 28.4
Neck	44.7 ± 34.0
Arm	58.2 ± 20.6
Leg	79.7 ± 16.3

- When the preceding analysis is limited to drivers, the increase in leg injuries is especially high for females relative to males: 98.5 ± 30.8 percent.

- The much higher overall injury risk for females than males at the AIS 2 and 3 levels may be largely due to their vulnerability to arm and leg injuries, which occur frequently but are rarely life-threatening.

Figure 9 is a bar graph comparing the principal estimates, with confidence bounds, of the average relative effect of aging one year on injury risk (NASS-CDS analyses) – by the occupant's overall injury severity (MAIS) and, for individual AIS ≥ 2 injuries, by the body region of the injury. The effect of aging increases as injury severity increases; it is relatively low for head and leg injuries, high for chest injuries. Figure 10 summarizes the corresponding risk increases for females relative to males of the same age. The increase is lower for fatalities than for nonfatal injuries; it is lower for head and chest injuries than for leg, arm, neck, and abdominal injuries. Confidence bounds are fairly wide for the CDS analyses, especially for the effects of aging or gender on injury risk to specific body regions.

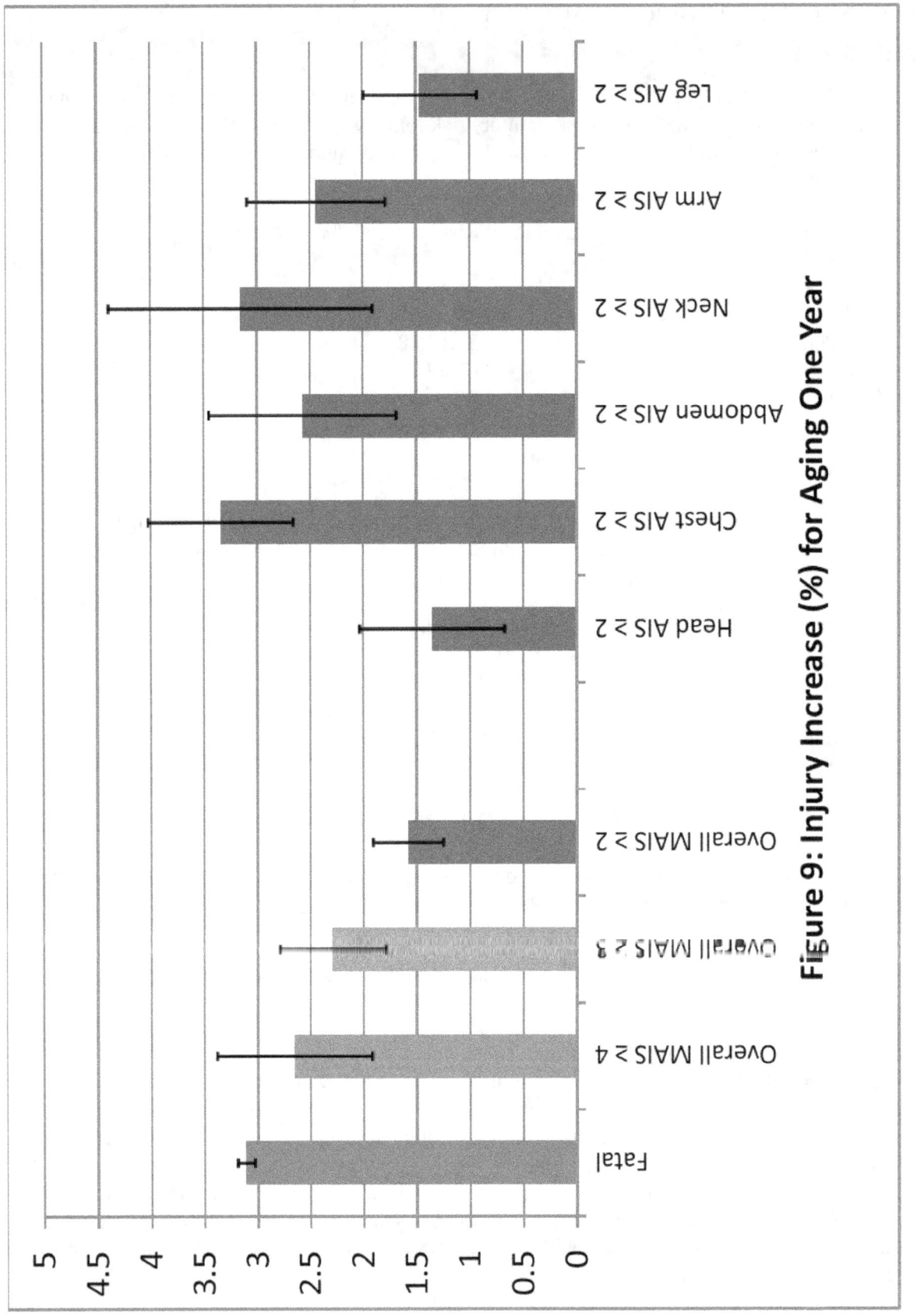

Figure 9: Injury Increase (%) for Aging One Year

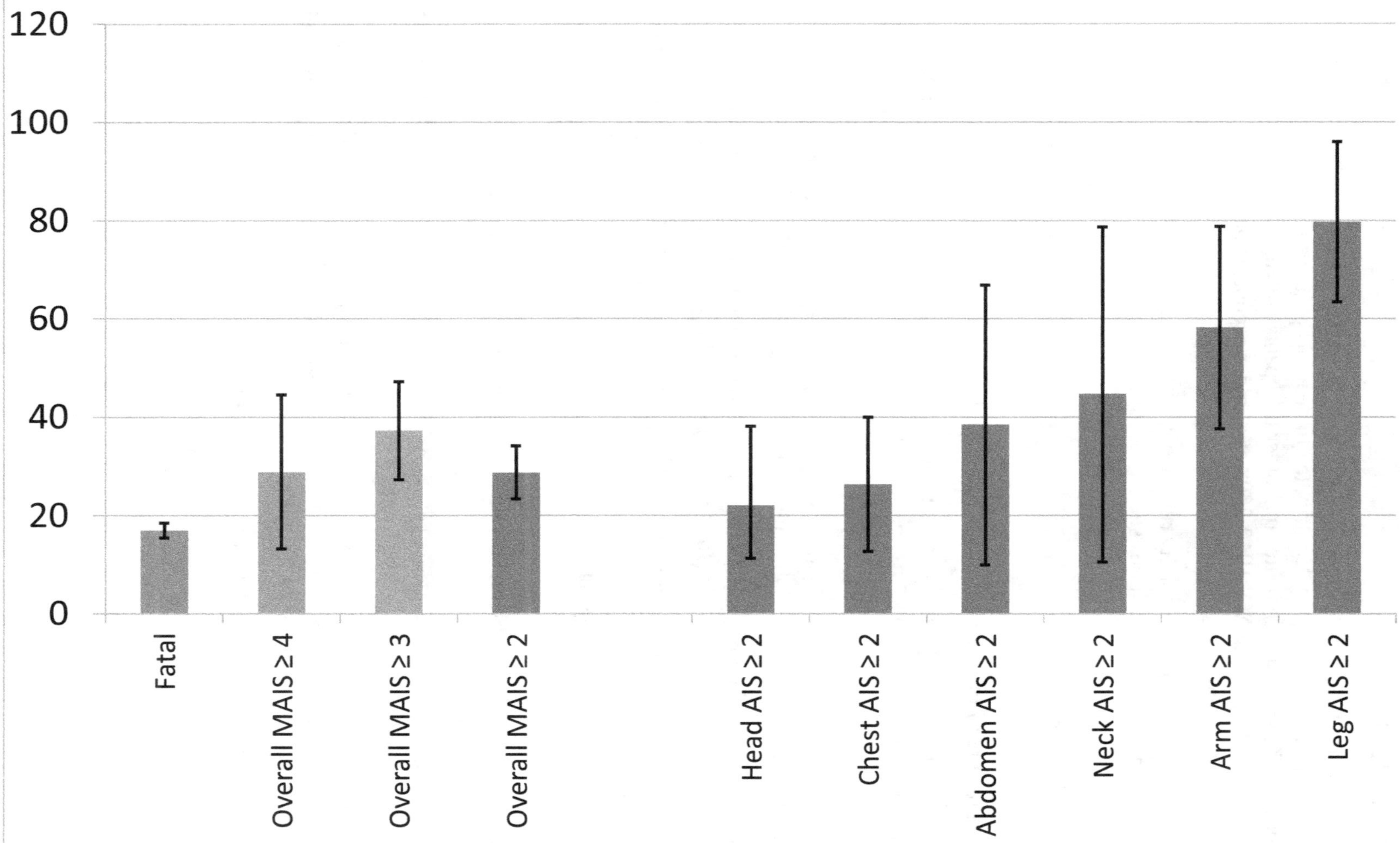

Figure 10: Injury Increase (%) for a Female Relative to a Male of the Same Age

(Average Increase for Occupants Ages 21 to 96)

INJURIES PREVALENT AMONG OLDER FATALITIES AND WOMEN

OLDER FATALITIES

- The CY 1999-2007 FARS-MCOD files all use ICDA-10 codes to describe a fatally injured occupant's injuries that "contributed to death." For any specific injury, a high average age of the victims indicates that older occupants are vulnerable. For drivers and passengers age 18 to 96 of MY 1960-2008 cars, older occupants are especially vulnerable to the following injuries:

  - Thoracic injuries, especially rib fractures, sternum fractures, and flail chest, plus resultant soft-tissue injuries such as pneumothorax, hemothorax, and heart injuries;

  - Hip fractures;

  - Subdural hemorrhage; and

  - Fracture of the second cervical vertebra (axis).

- Older occupants are not so vulnerable, relative to younger people, to most injuries to the head or the abdominal organs.

WOMEN

- For any specific injury, a high percentage of female victims indicates that females are vulnerable. Females are especially vulnerable to the following injuries:

  - Fracture of the first (atlas) or second (axis) cervical vertebra, even more so (relative to men, not in absolute terms) if the women are belted;

  - Injuries to abdominal organs, especially the spleen, liver, and gallbladder, more so if belted;

  - Arm and leg fractures; and

  - Traumatic shock.

- Women are not so vulnerable, relative to men, to most head injuries; and, in the front seat, to heart injuries.

- Women riding in the back seat are exceptionally vulnerable to flail chest and multiple rib fractures, but somewhat less to abdominal injuries.

  - However, when belted, they are more vulnerable than men to both thoracic and abdominal injury (possibly indicating that back-seat belts protect men better than women from these types of injury).

- Unbelted women are relatively more prone to some thoracic injuries such as hemopneumothorax.

AIS ≥ 2 INJURIES PREVALENT AMONG OLDER OCCUPANTS AND WOMEN

OLDER OCCUPANTS

- The CY 1988-2010 CDS files list each injury of crash-involved occupants (both fatalities and survivors). For any specific injury, a high average age of the victims indicates that older occupants are vulnerable. For drivers and passengers 18 to 96 of MY 1960-2011 cars, older occupants are especially vulnerable to the following AIS ≥ 2 injuries:
  - Thoracic injuries, especially rib and sternum fractures plus resultant soft-tissue injuries, especially heart injuries; and
  - Neck injuries.

- Older occupants are not so vulnerable, relative to younger people, to head injuries.

- Belted older occupants are also vulnerable, relative to young adults, to some abdominal or lower-torso injuries: diaphragm laceration, spleen rupture, diaphragm rupture, thoracolumbar fracture, and kidney contusion.

WOMEN

- Women riding in the front seat are highly vulnerable to arm and leg injuries, especially fractures, but are relatively less vulnerable to head injuries.

- Women riding in the back seat are relatively more susceptible to thoracic, dorsal, or shoulder fractures, but they are not quite so vulnerable to arm and leg injuries as in the front seat.

- Belted women are relatively more prone to neck fractures and certain types of abdominal injury than belted men of the same age.

- The FARS-MCOD and CDS findings are generally consistent on the injuries prevalent among older occupants and females (but CDS, which includes survivors, has far more cases of arm and leg injury).

INJURY SOURCES PREVALENT AMONG OLDER OCCUPANTS AND WOMEN

OLDER OCCUPANTS

- The CY 1988-2010 CDS files try to identify the injury source (vehicle component contacted) for each individual fatal or nonfatal injury. A high average age of the victims with AIS ≥ 2 injuries attributed to a specific source may indicate that older occupants are vulnerable to injury from that component. For drivers and passengers 18 to 96 of MY 1960-2011 cars, older occupants are especially vulnerable to AIS ≥ 2 injuries from the air bag in frontal impacts, the belt system in frontal and far-side impacts, and the side interior surface in nearside impacts.
  - This does not imply that air bags or belts are harmful for older people; rather, when they save an older occupant's life there may still be injuries ranging from moderate to severe (whereas a young person might walk away with little or no injury).

- Older occupants are also somewhat vulnerable to:
  - The steering assembly for torso, neck, and arm injuries (drivers only),
  - The side interior surface for torso injuries in nearside impacts,
  - Floor components for leg injuries in far-side impacts,
  - The instrument panel (drivers and RF passengers), and
  - The back of the front seat for leg injuries in frontal impacts (back-seat occupants only).

WOMEN

- One source of leg injuries has an exceptionally high proportion of female-driver victims: the floor and its components, including the toe pan and pedals. Female drivers are especially vulnerable to leg fractures from toe-pan intrusion.
- Women are also vulnerable to:
  - Arm injuries involving contact with the steering assembly, the air bag, or the A-pillar (drivers and unbelted RF passengers);
  - Torso injuries from the armrest or side hardware in nearside impacts; and
  - The back of the front seat for leg injuries in frontal impacts (back-seat occupants only).

CRASHWORTHINESS TECHNOLOGIES: FATALITY REDUCTION FOR OLDER OCCUPANTS AND WOMEN

- All of the major occupant protection technologies in current-model vehicles (seat belts, air bags, and energy-absorbing materials) have at least some benefit for adults of all age groups and of either gender; none of them are harmful for a particular age group or gender.
- But there are significant variations in effectiveness by age and gender for some technologies, especially those in earlier-model vehicles.

OLDER OCCUPANTS

- Seat belts, although effective for all age groups, have historically been somewhat less effective for older drivers and passengers than for young drivers and passengers in both cars and LTVs. In other words, the fatality reduction for a belted versus an unbelted older occupant can be significantly lower than the fatality reduction for a belted versus an unbelted young occupant.
  - A likely exception to the historical pattern is the latest generation of belts equipped with pretensioners and load limiters in vehicles with dual air bags. They may be about equally effective across the various adult age groups (but more data is needed for firm conclusions).
  - Back-seat lap belts and 3-point belts, on the other hand, are not as effective for older as for young occupants; so are 2-point automatic belts in the front seat.

- Frontal air bags are about equally effective across all age groups: that includes barrier-certified and sled-certified air bags, drivers as well as RF passengers, and car as well as LTV occupants.

- Side air bags with head protection (such as combination bags or separate head curtains and torso bags) are significantly more effective for older occupants than for young adults.

- By contrast, some of the earlier, non-inflatable occupant protection such as energy-absorbing steering assemblies and the energy-absorbing materials used for side-impact and head-impact protection may have been less effective for older occupants than for young adults.

- Belts without load limiters and the earlier non-inflatable occupant protection may exert force on the ribs or elsewhere that is beyond the tolerance limits of some older occupants, but not young adults – thus, lower effectiveness.

WOMEN

- For drivers, seat belts have historically been equally effective or even more effective for females than for males – from the early lap-belt systems to current 3-point belts with pretensioners and load limiters. In other words, the fatality reduction for a belted versus an unbelted female is often significantly higher than the fatality reduction for a belted versus an unbelted male.

- However, for RF passengers and back-seat passengers, seat belts have historically been less effective for females than for males.
  - A possible exception to the historical pattern is the latest generation of front-seat belts equipped with pretensioners and load limiters in vehicles with dual air bags; they may be about equally effective for females and males.

- It is unclear why belts are especially effective for female drivers and male passengers.

- Frontal air bags are about equally effective for females and males: that includes barrier-certified and sled-certified air bags, drivers as well as RF passengers, and car as well as LTV occupants. In fact, for LTV drivers, air bags may be even more effective for females.

- Side air bags with head protection are about equally effective for females and males.

- The earlier, non-inflatable occupant protection technologies are at least as effective for females as for males, possibly more effective.

CRASHWORTHINESS TECHNOLOGIES: CONCLUSIONS

- All the major safety technologies, except for the earlier belt systems for passengers, contribute to shrinking the added fatality risk for females relative to males.

- By contrast, many of the earlier technologies, although usually at least somewhat beneficial for all adult age groups were often less effective for the older occupants and tended to augment the original relative increase in fatality risk associated with aging. The more recent technologies developed with a deeper understanding of human tolerance limits, especially air bags, pretensioners, and load limiters have helped arrest this tendency and perhaps even begun to reverse it.

CHAPTER 1

CRASH SURVIVAL, OCCUPANT AGE, GENDER, AND VEHICLE DESIGN

1.1 Why study effects of age and gender on fatality risk?

The starting point and inspiration for this study is a 1988 paper by Leonard Evans, *Risk of Fatality from Physical Trauma Versus Sex and Age*, summarized and updated in his 1991 book, *Traffic Safety and the Driver*.[5] He quantified the effects of a person's age and gender on the risk of death from similar physical impacts or insults. One key to his analysis is a technique he called "double pair comparison," which he had originally used to estimate seat-belt effectiveness.[6] The other key is the Fatality Analysis Reporting System (FARS), started by NHTSA in 1975, which by the mid-1980s furnished detailed information on hundreds of thousands of fatal crashes.

Double-pair comparison allows unbiased measurement of the effect of, say, age or gender on fatality risk. It allowed him to separate how different people respond to similar physical insults from other issues – e.g., how safely they drive or what sort of crashes they get into. For example, in a group of cars in which the drivers are all 21-year-old males and there are also 21-year-old male right-front passengers, the ratio of passenger to driver fatalities might be R_1. But in another group of cars with 21-year-old male drivers and 21-year-old female right-front passengers, the ratio of passenger to driver fatalities might be R_2. In that case, at the right-front seat, at 21, a female has R_2/R_1 times as much risk of death from similar physical impacts or insults as a male. The 21-year-old male drivers (same age and gender in both groups of cars) act as a reference point or control group, thereby permitting the comparison of the male and female passengers.

Evans' basic empirical findings were that fatality risk from a given physical insult rises by a fairly constant percentage, like compound interest, for each year that a person gets older, starting at about age 20. Risk for females in their 20s or early 30s is substantially higher than for males of the same age, given similar physical insults, but at some point, perhaps in the 30s, the differential begins to shrink and eventually at another point, perhaps in the 60s, reverses in favor of females.

The empirical findings are consistent with the concepts of **fragility** and **frailty**. The more "fragile" a person, the more severe the injury they will sustain, given similar physical impacts. The "frailer" they are, the higher the likelihood of death given the same injury.[7] In essence, fragility and frailty both increase with age – a double minus. Females are more fragile than males, at least in the younger age groups. But females are less frail than males, at least in the older age groups, compensating for their fragility. This report will often mention "fragility" and "frailty," but as abstract concepts; the injury data is just not detailed enough to separately quantify a person's fragility and their frailty.

---

[5] Evans, L. (1988). Risk of Fatality from Physical Trauma Versus Sex and Age. *Journal of Trauma, 28*, pp. 368-378; Evans, (1991), pp. 22-28.

[6] Evans, L. (1986a). Double Pair Comparison – A New Method to Determine How Occupant Characteristics Affect Fatality Risk in Traffic Crashes. *Accident Analysis and Prevention, 18*, pp. 217-227.

[7] Kent, R., Trowbridge, M., Lopez-Valdes, F. J., Heredero Ordoyo, R., & Segui-Gomez, M. (2009). How Many People Are Injured and Killed as a Result of Aging? Frailty, Fragility and the Elderly Risk-Exposure Tradeoff Assessed via a Risk Saturation Model. *Annals of Advances in Automotive Medicine, 53*, pp. 41-50.

Evans noted that the age and gender trends were fairly similar at different seat positions or types of crashes. He surmised they showed the "age [or gender] dependence of basic physiological response to physical impact, with the specific details of the physical insult being of less central importance. There is every reason to expect that these same relationships apply to physical insults unrelated to occupant injuries; for example to [pedestrians] struck by vehicles or to injuries unrelated to traffic."[8]

He is, of course, correct in the big picture, but this report will focus on the details: the relatively small changes over time in the age and gender trends, or their variation by type of vehicle or crash. Conceptually, trends could change. Safety technologies are not always equally effective for occupants of all ages or for males and females. When technologies are introduced that are more effective for younger people, the trend of age to increasing fatality risk can become steeper. But technologies that are especially effective for older occupants and/or for females can shrink the trends. Evans, working with nine years of FARS, only had enough data for one snapshot. By now, there are 36 years of FARS data comprising large numbers of vehicles for about 50 model years, plenty of data to track the trends over time.

This report investigates the adequacy of current vehicle interiors for older occupants and female occupants. Have technologies been introduced that do not adequately protect these occupants and subject them to risk above and beyond their intrinsic fragility?

Well-known crashworthiness technologies, such as frontal air bags, share the following features:

- It is known what vehicles are equipped with them ("after") and which ones are not ("before").
- Effectiveness is large enough that a statistical before-versus-after analysis shows a significant fatality reduction.

In fact, NHTSA has already published evaluations of these technologies. Each evaluation defines a statistical procedure to estimate fatality reduction for that device. Here, the investigation consists of repeating those statistical procedures, with the latest FARS data, but now separately for males and females and for various age groups of occupants. Statistical tests will show if effectiveness is significantly different for older and younger occupants or for males and females. Chapter 9 of this report presents the effectiveness analyses for these individual crashworthiness devices.

But other technologies might not have a measurable, statistically significant effect on fatalities; or were gradually introduced in small increments; or it is unknown when they first appeared on various makes and models; or it is unknown in what types of crashes there would be an effect. Typically, they are not required by any specific FMVSS. Examples of these include the gradual shift to higher-strength steels, or incremental revisions in design to improve performance on crash ratings. Their effects cannot be investigated individually. Instead, the approach in Chapters 2 and 3 of this report is to monitor the trends in fatality risk by age and gender over the past 50 model years to see if trends have become more severe in any group of vehicles, possibly

---

[8] Evans (1991), p. 26.

indicating technologies that are not adequately benefiting or even harming older occupants or females. Chapters 4 and 5 monitor the trends of injury risk to individual body regions.

Chapters 6 to 8 identify specific injuries and their sources in vehicles (components contacted) that are especially prevalent among older occupants and/or females. Those injury mechanisms might be candidates for research on how to mitigate risk.

1.2   Age, gender, and injury mechanisms

Older occupants are especially vulnerable to thoracic injury. In NASS CDS data, Hanna and Hershman found that the proportion of crash-involved occupants with AIS 2+ thoracic injury increased five-fold from 1.5 percent of the occupants age 25 to 44 to 6.5 percent of the occupants 75 or older. By contrast, head-injury rates increased only from 1.9 percent to 2.6 percent. Abdomen, arm, and leg injury rates doubled and spinal-injury rates tripled. The risk of thoracic injuries appears to be intrinsic to aging, as it occurs for frontal as well as side impacts, in cars as well as LTVs, at all seat positions, for belted as well as unbelted occupants. Ribcage and sternum fractures are especially prevalent.[9] Wang and Rupp reported that bone density in the ribcage decreases strongly with age.[10]

According to Kent and Patrie, "Chest deflection injury threshold is strongly dependent on the age of the subject. This is true regardless of whether injury onset or severe injury is considered. A 30-year-old has a 50 [percent] risk of sustaining one rib fracture at a chest deflection level of 35 [percent]. This threshold drops to 13 [percent] deflection for a 70-year-old. A 30-year-old has a 50 [percent] risk of sustaining more than six rib fractures at a deflection level of 43 [percent], while a 70-year-old can tolerate only 33 [percent] deflection before reaching this threshold. These findings are…presumably due to multiple characteristics of aging. First, the failure strain of both cortical and trabecular bone decreases with age. Second, geometric changes associated with aging may predispose ribs to fracturing for older subjects under conditions where they might deflect non-injuriously in a younger subject. These geometric changes include a decrease in the proportion of the rib cross-section that is cortical bone and a general decrease in rib slope. Finally, material changes such as calcification of the costal cartilage and decreasing bone mineral density also are likely contributors to the decreased chest deflection tolerance."[11] The Center for Injury Biomechanics and Wake Forest University is currently quantifying the variation of rib cortical thickness, bone density, and ribcage geometry by age and gender, based on data collected through CT scans.[12]

Zhou, Rouhana, and Melvin statistically analyzed numerous existing sets of test results with cadavers of varying ages to estimate the effect of aging on thoracic-injury tolerance. The

---

[9] Hanna, R., & Hershman, L. (2009). *Evaluation of Thoracic Injuries Among Older Motor Vehicle Occupants.* (Report No. DOT HS 811 101). Washington, DC: National Highway Traffic Safety Administration.

[10] Wang, S., & Rupp, J. (2006). *Alterations in Injury Patterns and Body Composition With Aging.* (PowerPoint presentation. Ann Arbor, MI: University of Michigan Transportation Research Institute http://www.nhtsa.gov/DOT/NHTSA/NVS/CIREN/2006%20Presentations/MI_0306b.pdf

[11] Kent, R., & Patrie, J. (2005). Chest Deflection Tolerance to Blunt Anterior Loading Is Sensitive to Age but Not Load Distribution. *Forensic Science International, 149,* pp. 121-128.

[12] Morphometric Analysis of Age and Gender-Related Changes in the Rib Cage. Winston-Salem, NC: Wake Forest Baptist medical Center. Retrieved from www.wakehealth.edu/CIB/Rib-Morphometrics htm .

underlying hypothesis is that bones change with age: Their modulus of elasticity and bending strength decrease after age 30 due to increased porosity and demineralization. The rate of deterioration speeds up after age 40 and then even more after age 60. Muscles and arterial tissues also decrease in strength with age. However, the extent of deterioration in strength or tolerance is not uniform, but varies with the type of tissue and the manner of loading. Here are their estimates of the percent loss of strength or injury tolerance from age 25 to 75.[13]

Percent Reduction from Age 25 to 75

72	Belt force needed to produce AIS 3 injuries
55	Compact bone fracture toughness
50	Vertebral bone ultimate tensile strength (UTS)
40	Abdominal muscle wall UTS
35	Cardiac muscle UTS
29	Arterial tissue UTS
27	Side-impact velocity needed to produce AIS 3 injuries
21	Blunt-impact force needed to produce AIS 3 injuries

In other words, the effect of aging is more severe in belt loading than in blunt impact force. They believe that belt loading is a more static, less dynamic load than blunt impact, and thus has a less linear response. Also, belt force is concentrated on bone, rather than soft tissue. Bone deteriorates more rapidly with age than soft tissue. (This does not imply that belt use is harmful for older occupants, merely that belts can be relatively less effective for older than for young occupants.)

In FARS and CDS data, Austin and Faigin identified that side impacts account for an increasing share of the fatalities as occupants get older. For any given delta V, injury risk is significantly higher, at all AIS levels, for older occupants and females. Furthermore, the higher the delta V, the more risk increases with age.[14]

Ridella, Rupp, and Poland developed logistic regression analyses for 2000-2010 NASS-CDS data on MY 2000-2010 vehicles to estimate an occupant's odds of AIS ≥ 3 injury as a function of that occupant's age, gender, belt use, BMI, height, and seat position; the vehicle type (car or LTV); the impact type; and the delta v (or other measure of severity such as number of quarter turns in a rollover).[15] The database has one record per occupant, unlike the double-pair comparison analyses of Evans that consider pairs of occupants exposed to the same crash; on the other hand, the ability to control for delta v or other measures of severity in CDS data is useful for controlling for differences in the distribution of crash severities for occupants of different

---

[13] Zhou, Q., Rouhana, S. W., and Melvin, J. W. (1996). "Age Effects on Thoracic Injury Tolerance," *40th Stapp Car Crash Conference Proceedings*, Paper No. 962421. (Publication No. P-305). Warrendale, PA: Society of Automotive Engineers.

[14] Austin, R. A., & Faigin, B. M. (2003). Effect of Vehicle and Crash Factors on Older Occupants. *Journal of Safety Research, 34*, pp. 441-452.

[15] Ridella, S. A., Rupp, J. D., & Poland, K. (2012). Age-Related Differences in AIS 3+ Crash Injury Risk, Types, Causation and Mechanisms. *International IRCOBI Conference on the Biomechanics of Impact.*

ages. The backwards-stepwise regression method starts by considering all of the variables and then dropping those that are non-significant.

Table 1-1 shows the estimated increase in the odds of AIS ≥ 3 injury for aging 10 years. Aging significantly increases risk to almost every region of the body in all types of crashes. However, the largest age effect was observed for the thorax, in frontal crashes, for female occupants. Separate odds ratios were calculated for men and women for body region and crash mode combinations for which an age*gender interaction was significant.

Table 1-1: Adjusted Odds Ratios (and 95% CIs) for AIS ≥ 3 Injury to Different Body Regions Associated With a Decade Increase in Age[16]

	Head	Thorax	Abdomen	Spine	Upper Extremity	Lower Extremity
Frontal	1.29*** (1.21,1.37)	M: 1.37 (1.14,1.64) F:1.77 (1.31,2.39)	1.45* (1.10,1.90)	1.55*** (1.34,1.80)	M:1.46 (1.23,1.74) F: 1.09 (0.81,1.46)	1.22* (1.03,1.44)
Nearside	1.68* (1.14,2.48)	1.28*** (1.15,1.42)	NS	-	-	NS
Far-side	1.41** (1.15,1.73)	1.60** (1.23,2.07)	-	-	-	-
Rollover	1.17* (1.00,1.38)	1.23*** (1.12,1.34)	-	M:1.12 (0.88,1.42) F:1.51 (1.04, 2.18)	1.36*** (1.17,1.57)	NS

NS=Not Significant, M=Male, F=Female, *p<0.05, **p<0.001, ***p<0.0001

A supplementary analysis of injury causation in 1,289 CIREN cases found thoracic and spinal injuries increase with age and lower extremity and head injuries, as a percent of the total injuries experienced in each age group, decrease as age increases. There were no significant gender differences in these trends. For frontal crashes, thoracic injury type changed from soft-tissue injuries such as lung contusions to bony-tissue injuries such as rib and sternum fractures as age increased. The contribution of the air bag and steering wheel decreases and the seat belt begins to play a larger role in injury causation for the oldest age groups. Comorbid factors such as osteoporosis, osteopenia and obesity were more common for the older occupants. Crash severity was consistently lower for the older age groups and outcomes were worse in the older occupants even for similar crash severity and injury.

NHTSA's review of 122 front-seat occupant fatalities in CDS frontal impacts – belted, protected by air bags, and in MY 2000+ cars or LTVs – identified 16 cases where the occupant's age was a primary factor (i.e., the crash would otherwise not likely have been fatal) and 15 where it was a secondary factor. Here are two individual examples of older occupants who did not survive. Both

---

[16] Ibid.

occupants were quite fragile; the second, in addition, may have been exceptionally frail due to a pre-existing medical condition:

"As an example of a case in which the occupant's age was deemed critical, see Case No. 2004-50-147, in which the 80-year-old driver sustained a number of thoracic injuries leading to her demise. CDS did not report any pre-existing medical conditions. The crash was not very severe with a coded [delta v] of 22 mph…, and it did not result in any intrusions into the occupant's space. The team did not see signs of contact with the steering wheel and concluded that her numerous thoracic injuries were caused by loading from the shoulder belt. She suffered a number of fractured ribs and a heart laceration; the thoracic cage typically shows a greater tendency to sustain fracture with elevated age. Given that the chest injuries were responsible for her demise, and that this crash was not severe, the team felt that her age was a critical element leading to her death.

 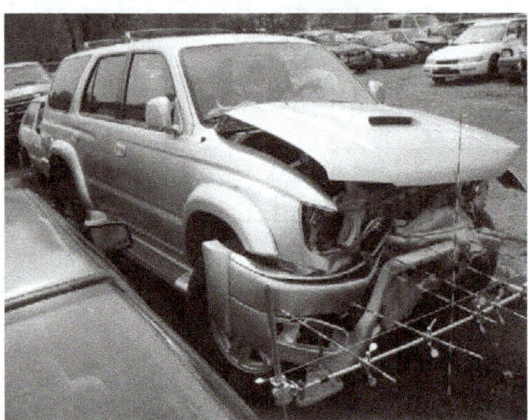

Case No. 2004-50-147   Case No. 2002-75-53

"Case No. 2002-75-53 is an example of a fatality that was likely due to [the occupant's fragility and/or frailty due to old age and] a pre-existing medical condition. The minor frontal crash, with a [delta v] of 14 mph, of the Toyota 4Runner led to the death of the right-front passenger, who suffered from advanced lung cancer that had metastasized to the liver.

"Her injuries included a heart laceration, a subdural hematoma and a cervical spine fracture. Although an older occupant (71 years old) who was of smaller stature (4'11"), she was in the right front passenger seat, making it unlikely that she was sitting close to the [instrument panel] (as may be the case for a driver of her stature). In fact, the seat was noted to be in the mid-to-rear track position, so [thoracic] interaction with a deploying bag was unlikely and there would be little reason for her type of injuries to occur in a typical adult [unless head interaction with the air bag was also a factor]. Due to her advanced cancer, the team concluded that her body's condition was weakened, making her exceptionally fragile and more susceptible to injury in this minor crash [and perhaps also frail, increasing the likelihood of a fatal outcome for these injuries]. The two other occupants in this vehicle were either not injured or only suffered minor contusions – suggesting that the pulse of this crash should not be injurious to a normal belted occupant."[17]

Bose, Segui-Gomez, and Crandall used logistic regression to analyze injury risk of belted drivers in 1998-2008 CDS data as a function of the driver's <u>gender</u>, controlling for a number of other variables. "Results from the multivariate regression analysis indicated that the odds of a belt-restrained female driver sustaining an MAIS 3+ and MAIS 2+ injury were 47 percent (95% CI=27%, 70%) and 71 percent (95% CI=44%, 102%) higher, respectively, than those of a belt-restrained male driver when we controlled for the effects of age, mass, BMI category, crash [delta v], vehicle body type, number of events, and crash direction…For chest and spine AIS 2+ injuries, the odds of an effectively belted female driver to sustain the injury was 38 percent (95% CI=1%, 89%) and 67 percent (95% CI=34%, 109%) higher, respectively, than those of a belted male driver in comparable crash conditions.

"To account for the correlation between sex and anthropometric size, the regression methodology used in the study specifically controlled for the effects of BMI and overall mass as measures of size. Tolerance to traumatic injury may also be predicted as a function of sex-specific properties. Specifically, female occupants are at a higher risk for sustaining whiplash injuries because of differences in neck anthropometry, strength, and musculature, and the relative positioning of the head restraint. Similarly, a higher risk of lower extremity injuries has been reported for female drivers as a result of their relatively short stature, preferred seating posture, and a combination of these factors yielding lower safety protection from the standard restraint devices." They did not find a statistically significant difference between belted females and males in the CDS data for fatality risk and AIS 2+ head injuries.[18]

NHTSA's review of 122 belted, frontal fatalities of late-model vehicles equipped with air bags in CDS cases did not attribute any fatalities to anatomical or physiological vulnerability specifically linked to gender – e.g., a female's neck injury that a male with a thicker neck might have avoided. But it rather often cited anatomical features that are more common in females than males. Short stature was a secondary factor in eight cases, six of them females. Four occupants had been displaced out of position, two of them lightweight females. Eleven occupants were obese and 5'6" or shorter, a combination that taxes existing restraint systems (because these people usually sit close to the steering wheel): ten of them were females. However, eight of these 11 cases also involved exacerbating circumstances such as exceedingly high delta v, corner/oblique impact, or underride. In two of the three remaining cases (nos. 2004-79-49 and 2007-12-180) it appeared that the driver's short stature and sitting close to the wheel contributed to poor driver-air bag interaction; in case no. 2006-41-64, the obese driver bottomed out the air bag (whose deployment was perhaps delayed because of the centered impact with a narrow object).[19]

[17] Bean, J. D., Kahane, C. J., Mynatt, M., Rudd, R. W., Rush, C. J., & Wiacek, C. (2009, September). *Fatalities in Frontal Crashes Despite Seat Belts and Air Bags.* (Report No. DOT HS 811 202, pp. 40-41). Washington, DC: National Highway Traffic Safety Administration Available at www-nrd.nhtsa.dot.gov/pubs/811102.pdf

[18] Bose, D., Segui-Gomez, M., & Crandall, J. F. (2011). "Vulnerability of Female Drivers Involved in Motor Vehicle Crashes: An Analysis of US Population at Risk," *American Journal of Public Health, 101,* pp. 2368-2373.

[19] Bean et al. (2009), Table 3-7 and Appendices A and C.

1.3   Evans' quantitative results and method

As stated above, Evans found that fatality risk from a given physical insult rises by a fairly constant percentage for each year that a person gets older, starting at about age 20. The increase is about 2.31 percent per year of age for males and about 1.97 percent for females. The rate of increase is lower for females for the same reason that the additional risk for a female relative to a male decreases with age. This additional risk for females, relative to males of the same age, reaches a peak of approximately 31 percent at age 30 and averages 25 percent from age 15 through 45. However, by age 45 the differential is clearly shrinking and may have already started shrinking at 35. Because the empirical data points are not on a straight line, it is difficult to say exactly at what age the trend crosses zero and females start to be at less risk than males, but that age appears to be somewhere between 55 and 65. Beyond age 70, Evans' observations are based on limited numbers of cases, but it looks like the trend may be leveling off.

Evans' estimates are based directly on individual double-pair comparison analyses: separate, discrete estimates for non-overlapping class intervals of age, without a technique such as logistic regression to smooth the results or to force them to follow a straight line or a specific curve. For example, to find the effect of gender for drivers, he classifies all male drivers into intervals of age centered on 20, 25, 30… and likewise all female drivers. He classifies all right-front (RF) passengers into just 8 groups: males age 16-24, 25-34, 35-54, 55+, and females age 16-24, 25-34, 35-54, and 55+. To find the effect of gender at age 30, he considers only vehicles where there is a driver 28 to 32 and an RF passenger 16 or older – and at least one or possibly both are fatalities. His first double-pair comparison is the driver-to-RF fatality ratio in vehicles with female drivers age 28-32 and male RF passengers age 16-24 divided by the corresponding ratio for vehicles with male drivers age 28-32 and male RF passengers age 16-24. He performs seven other double-pair comparisons, substituting the other groups of RF passengers for the males age 16-24. Then he takes the weighted average of the eight comparisons.[20]

To find the increase in fatality risk for a 30-year-old male driver relative to a 25-year-old male driver, his first double-pair comparison is the driver-to-RF fatality ratio in vehicles with male drivers age 28-32 and male RF passengers age 16-24 divided by the corresponding ratio for vehicles with male drivers age 23-27 and male RF age passengers 16-24. Again, he performs seven other double-pair comparisons and takes the weighted average.

The advantage of double-pair comparison is that it implicitly controls for crash proneness and crash severity. For example, 30-year-old male drivers experience a larger number of high-severity crashes than 30-year-old female drivers. But these male drivers drag their RF passengers into the very same high-severity crashes, whereas the female drivers share their lower-severity crash environment with their passengers. The ratio of driver to RF passenger fatalities depends only upon drivers' fragility and frailty relative to their passengers, because both the driver and the passenger are experiencing the same crashes.

---

[20] Evans (1988).

CHAPTER 2

TOOLS FOR CONSISTENTLY ESTIMATING AGE AND GENDER EFFECTS

2.1   Why double-pair comparison and logistic regression?

In 1988, Evans introduced "double-pair comparison analysis" as a technique to measure the effect of age and gender on fatality risk: to identify how different people respond to similar physical insults.[21] For his analyses of the fatality risk of drivers and of RF passengers, his database consists of all light vehicles on FARS in which the RF seat as well as the driver's seat was occupied and at least one or possibly both of these occupants were fatalities. For example, in a group of cars in which the drivers are all 21-year-old males and there are also 21-year-old male right-front passengers, the ratio of passenger to driver fatalities might be R_1. But in another group of cars with 21-year-old male drivers and 21-year-old female right-front passengers, the ratio of passenger to driver fatalities might be R_2. In that case, at the right-front seat, at 21, a female has R_2/R_1 times as much risk of death from similar physical impacts or insults as a male. The 21-year-old male drivers (same age and gender in both groups of cars) act as a reference point or control group, thereby permitting the comparison of the male and female passengers. He called the analysis "double-pair comparison" because one set of vehicles pairs a male driver with a male RF passenger while a second set of vehicles pairs a male driver with a female RF passenger. As discussed in Section 1.1, double-pair comparison allows unbiased measurement of the effect of, say, age or gender on fatality risk. It separates how different people respond to similar physical insults from other issues – e.g., how safely they drive or what sort of crashes they get into.

As Section 1.3 describes in detail, Evans based his estimates directly on individual double-pair comparison analyses: separate, discrete estimates for non-overlapping class intervals of age, without a technique such as logistic regression to smooth the results or to force them to follow a straight line or a specific curve.

He presents results for drivers as two graphs. The first, the additional risk for being female, as a function of occupant age, is a set of discrete points at five-year intervals centered on 20, 25, 30… The second, the cumulative additional risk of being a male in the age group centered on N years relative to a male in the group centered on 20 years, is likewise a set of discrete points at five-year intervals.[22]

Because he ran only a single set of analyses on all FARS data then available (namely 1975-1983), the discrete points are fairly precise estimates that tend to follow a curve or line to some extent. Thus, it is absolutely clear that the fatality increase for being female is not constant across all ages, but substantially higher at age 20 and for quite a few years thereafter, but then dropping off until it equals or even falls below the risk for a male. However, the sequence of points is not precise enough to judge whether the drop-off begins closer to 30 or 40, or whether the age of equal risk is closer to 50 or 70.

---

[21] Evans (1988); Evans (1991), pp. 22-28.
[22] Evans (1991), p. 25 and p. 27.

Similarly, the cumulative fatality increase for males as a function of age, graphed on log paper, is fairly close to a straight line, although the trend perhaps levels out a bit for drivers and perhaps accelerates for RF passengers. Evans does not speculate whether the trend is intrinsically log-linear – i.e., a constant percent risk increase for each additional year of age.

As discussed in Section 1.1, Evans' objective was to capture the big picture, the overall relationship of age and gender to fatality risk: thus, only a single analysis using all the data. This report, on the other hand, focuses on the details of how the relationships might change for different vehicles or crash types. As the database is split into subgroups of vehicles or crash types, there will not be that much data in any subgroup. Simply performing individual double-pair comparison analyses and graphing the results as discrete points at five-year intervals of occupant age might create indecipherable scatterplots rather than Evans' relatively smooth curves for the full database.

That suggests the use of a technique such as logistic regression that will fit the effects of gender or aging – as a function of age – along a curve generated by a few key parameters. The logistic regression is still implicitly a double-pair comparison. But whereas Evans estimates age and gender effects for class intervals of age and graphs the results as discrete points, logistic regression, so to speak, performs a double-pair comparison on each individual data point and fits the results to a continuous curve. Here is a description of the FARS database used for the logistic regressions, the formulation of the model, and how the regression results are used to estimate the overall average effect of gender or aging on fatality risk.

2.2   FARS database

As in Evans' analyses, the database initially consists of all FARS cases of light vehicles (cars or LTVs) involved in fatal crashes and in which the driver and RF passenger seats were occupied[23] (and possibly other seats, too). Either the driver, or the RF passenger, or both were fatalities. The range of calendar years is 1975 to 2010; the range of model years, 1960 to 2011. Both occupants' age and gender must be known, and age must initially be in the range of 12 to 96.

Furthermore, the driver and RF passenger should have similar belt use and air bag availability. Either both are belted or neither (cases where either or both occupants' belt use is unknown are excluded). The vehicle must have no air bags or dual air bags. Vehicles with air bags only for the driver or with an on-off switch for the passenger are excluded. Vehicles with 3-point belts for the driver and 2-point automatic belts for the passenger are also excluded. The vehicle type (car, pickup truck, SUV, or van) is decoded from the VIN using NHTSA's analysis programs or defined from BODY_TYP if no VIN is available. The type of occupant protection available, including belts, frontal air bags, side air bags, belt pretensioners, and load limiters is decoded from the VIN or inferred from the model year, if possible, when no VIN is available.

The initial screening created a file of 242,641 vehicle records, including 97,450 where the drivers died but the RF passengers survived, 102,650 where the RF passengers died but the drivers survived, and 42,541 where both were fatalities.

---

[23] When FARS lists two or more people occupying the same seat position (e.g. child sitting on someone's lap), only the first is included in the analysis.

2.3   Early logistic regression models

In 1987-1988, soon after PROC LOGISTIC became available in SAS, NHTSA modeled the probability of a RF passenger fatality in vehicles involved in crashes fatal to the RF passenger, the driver, or both.[24] The primary objective was not to estimate the effect of age or gender, but the safety of the RF seat relative to the driver seat as a function of the vehicle's model year. Age and gender were control variables (covariates). The dependent variable, FATAL3 is whether or not the RF passenger was a fatality. The occupant-age variable in the regressions is formulated as:

$$\Delta\_AGE = \log(120\text{-}AGE1) - \log(120\text{-}AGE3)$$

where AGE1 is the age of the driver and AGE3 is the age of the RF passenger. In diagrams of passenger cars, the driver's seat is traditionally labeled no. 1, [the center-front seat, no. 2,] and the RF passenger seat no. 3 – thus, the variables for driver and RF passenger age will be called AGE1 and AGE3 throughout the report. This formulation makes the effect of aging on fatality risk increase with age: the constant 120 in the formula makes the effect approximately twice as high at age 80 as at 40. The motivation was Evans' 1988 paper, which showed the risk increase for the RF passenger relative to the driver increased with age (although it is not clear that the effect in Evans' data was twice as high at 80 as at 40).[25] The gender variable is formulated as:

Δ_FEM = 0 if driver & RF same gender, 1 if driver male & RF female, -1 if driver female & RF male

In other words, the gender effect is assumed to be constant at all ages, even though Evans' data suggests otherwise.

In 1994, NHTSA again used logistic regression to estimate the difference in the relative fatality risk of the two belted drivers in a head-on collision between two cars with different NCAP ratings, controlling for the relative weights of the cars and the relative ages and genders of the drivers.[26] As in the 1987-1988 study, the primary objective was not to estimate the effect of age or gender precisely, but to control for age and gender, at least at a first-order level, to better estimate the effect of NCAP ratings. The database consisted of head-on collisions, both drivers belted, fatal to at least one and possibly both drivers. The dependent variable was whether the driver of the "case" vehicle was a fatality. Δ_FEM was formulated the same as in the previous analysis, but the formulation of Δ_AGE was simpler, namely, the arithmetic difference of the ages of the two drivers. Again, the motivation was Evans' 1988 paper, which showed, for drivers only (as opposed to RF passengers), an effect of aging that is relatively constant across all ages.[27]

---

[24] Kahane, C. J. (1988, January). *An Evaluation of Occupant Protection in Frontal Interior Impact for Unrestrained Front Seat Occupants of Cars and Light Trucks.* (Report No. DOT HS 807 203, pp. 140-153). Washington, DC: National Highway Traffic Safety Administration. Available at www-nrd nhtsa.dot.gov/Pubs/807203.pdf
[25] Ibid., p. 130.
[26] Kahane, C. J. (1994, January). *Correlation of NCAP Performance with Fatality Risk in Actual Head-On Collisions.* (Report No. DOT HS 808 061). Washington, DC: National Highway Traffic Safety Administration. Available at www-nrd.nhtsa.dot.gov/Pubs/808061.PDF, pp. 35-42.
[27] Ibid., p. 39.

11

2.4 Formulation of the "simple" models

The linear age and constant gender terms in the NCAP analysis comparing drivers of two different vehicles may also be applied to compare the driver and RF passenger of the same vehicle in the database for this report. There are two models, one for each of the two dependent variables. FATAL1 = 1 if the driver is a fatality, = 2 if a survivor. FATAL3 = 1 if the RF passenger is a fatality, = 2 if a survivor. In both cases, there are four independent variables: AGE1 and AGE3, the actual age in years of the driver and RF passenger; FEM1 = 1 if the driver is female, = 0 if male. FEM3 = 1 if the RF passenger is female, = 0 if male. The model is slightly more complex than the NCAP analysis, because having separate variables for AGE1 and AGE3 allows estimation of different age effects for the driver and passenger, whereas having a single variable Δ_AGE essentially constrains the effect to be the same for both (or takes the average of the two effects).

Here, for example, are the estimated regression coefficients for the two simple models applied to the subset of the database consisting of the 154,467 vehicles in which the driver's and RF passenger's age are both in the range of 21 to 96:

```
               Analysis of Maximum Likelihood Estimates


                                 Standard          Wald
Parameter      DF    Estimate      Error     Chi-Square    Pr > ChiSq


FATAL1 (Driver): 87,567 fatal, 66,900 survived

Intercept       1     0.5727      0.0136     1764.7073       <.0001
AGE1            1     0.0391     0.000511    5864.5994       <.0001
AGE3            1    -0.0430     0.000493    7613.5608       <.0001
FEM1            1     0.1268      0.0126      101.9116       <.0001
FEM3            1    -0.2040      0.0113      325.2364       <.0001


FATAL3 (RF passenger): 95,647 fatal, 58,820 survived

Intercept       1     0.0296      0.0138        4.5871      0.0322
AGE1            1    -0.0379     0.000533    5065.5310       <.0001
AGE3            1     0.0459     0.000525    7643.8821       <.0001
FEM1            1    -0.1364      0.0128      112.8986       <.0001
FEM3            1     0.2431      0.0115      444.9983       <.0001
```

The 154,467 vehicle involvements resulted in 87,567 driver and 95,647 RF passenger fatalities. In other words, the RF seat may present slightly higher risk than the driver's seat or, perhaps, the population of RF passengers includes a higher proportion of more-vulnerable older and/or female occupants.

In the model to estimate FATAL1, the log-odds of a driver fatality, AGE1 and FEM1 both have positive coefficients with high statistical significance, as evidenced by Wald chi-squares of 5,865 and 102 (where just 3.84 is enough for statistical significance). In other words, the older the driver, the higher the fatality risk; risk also increases if the driver is female. But AGE3 and FEM3 have effects of similar magnitude in the opposite direction. At first glance, the aging of the RF passenger should have no effect on the driver's risk. These FARS cases, however, are a sort of zero-sum game: they include only vehicles where the driver, the RF passenger, or both, were fatalities. Only a fairly small proportion of the involvements (28,747 of 154,467, in this instance) were fatal to both. Thus, in most cases, fatality to the RF passenger implies survival of the driver; anything that increases the RF passenger's risk, such as being older or female, implicitly reduces the probability of the driver's fatality.

In the model to estimate FATAL3, the log-odds of a RF passenger fatality, it is AGE3 and FEM3 that have the positive coefficients: the older the passenger, the higher the risk. AGE1 and FEM1 now have negative coefficients. Each of the four coefficients is approximately, but not exactly, the opposite of what it was in the FATAL1 model.

Max-rescaled R-square and overall Wald chi-square, two indicators of the models' overall fit, are .0801 and 8132 for drivers, .0850 and 8382 for RF passengers. These statistics can be compared to corresponding numbers for the more complex models that will be defined later in this chapter.

Logistic regression estimates a linear relationship between the log-odds that the dependent variable equals 1 and the independent variables. In this case, "the dependent variable equals 1" when the driver is a fatality. If p is the probability of a driver fatality, the "log-odds" of a driver fatality is $\log(p/[1-p])$. In other words, the log-odds of a driver fatality are:

$$.5727 + .0391 \text{ AGE1} - .0430 \text{ AGE3} + .1268 \text{ FEM1} - .2040 \text{ FEM3}$$

Whereas the log-odds of a RF passenger fatality are:

$$.0296 - .0379 \text{ AGE1} + .0459 \text{ AGE3} - .1364 \text{ FEM1} + .2431 \text{ FEM3}$$

The regression coefficients do not exactly correspond to the statistics that are the goal of the analysis, namely "the additional risk of aging one year" and "the additional risk of being female." That will require some more arithmetic, presented in Section 2.8. However, the coefficients turn out to be about the same order of magnitude as the desired statistics. Thus, roughly speaking, the regression equations of the simple model suggest the effect of aging a year increases fatality risk on the order of 3 to 4 percent, while being female increases it on the order of 10 to 25 percent.

2.5   Empirical study of how age and gender effects vary with age

The simple models may be run for any subset of the FARS driver-RF database. In particular, in a subset of cases where the driver's and passenger's age are both within the same narrow range, say 60-79, the simple models will estimate a constant effect of age and gender that ought to be fairly accurate for people at the midpoint of that range, close to 70. By repeating this procedure for heavily overlapping but gradually older subsets of occupants (61-80, 62-81, 63-82…), the sequence of estimated age and gender coefficients from the models generate a moving average that should indicate how the effects of age and gender actually vary with age. Knowing these actual relationships (which are complex and not based on a mathematical formula) will allow selection of a regression model (based on relatively few mathematical parameters) that tracks the actual trends fairly well.

Because few drivers are younger than 16, this will be the lower limit of the age ranges in any subset. The subsets included in the analysis are age 16-19 (average close to 18), 16-21, 16-23...16-35 (case-weighted average close to 23), 17-36 (average close to 24), 18-37, 19-38...77-96 (case-weighted average close to 83). In other words, these subsets will yield a moving average of the estimated age and gender effects for each year of age (more or less) from 18 to 83. There are estimates for a total of eight coefficients, namely:

- D_AGE1 is the effect on the driver's fatality risk (D_) of the driver's age (AGE1) – i.e., the coefficient that the regression with FATAL1 as the dependent variable estimates for the independent variable AGE1.
- D_AGE3 is the effect on the driver's fatality risk (D_) of the RF passenger's age (AGE3) – i.e., the coefficient that the regression with FATAL1 as the dependent variable estimates for the independent variable AGE3.
- D_FEM1 is the effect on the driver's fatality risk (D_) of the driver being a female (FEM1) – i.e., the coefficient that the regression with FATAL1 as the dependent variable estimates for the independent variable FEM1.
- D_FEM3 is the effect on the driver's fatality risk (D_) of the RF passenger being a female (FEM3).
- P_AGE1 is the effect on the RF passenger's fatality risk (P_) of the driver's age (AGE1) – i.e., the coefficient that the regression with FATAL3 as the dependent variable estimates for the independent variable AGE1.
- P_AGE3 is the effect on the RF passenger's fatality risk (P_) of the RF passenger's age (AGE3).
- P_FEM1 is the effect on the RF passenger's fatality risk (P_) of the driver being a female (FEM1).
- P_FEM3 is the effect on the RF passenger's fatality risk (P_) of the RF passenger being a female (FEM3).

The first question is at what minimum age the characteristic adult patterns of the effects of aging and gender begin to appear – i.e., a substantial increase in risk for each year of aging and for a female relative to a male. The minimum age can be empirically found by carefully examining the estimated coefficients for the subsets at the low end of the age ranges:

AGE RANGE	AVG DRV AGE	D_AGE1	D_FEM1	D_AGE3	D_FEM3	AVG RF AGE	P_AGE1	P_FEM1	P_AGE3	P_FEM3
16-19	17.6067	0.015050	0.26664	0.003473	-0.30126	17.4150	-0.006797	-0.31942	0.011469	0.34951
16-21	18.4822	0.031271	0.28755	-0.004079	-0.27307	18.2269	-0.021534	-0.34305	0.014989	0.32531
16-23	19.2483	0.036950	0.29856	-0.018244	-0.29709	18.9519	-0.026835	-0.34946	0.024394	0.33822
16-25	19.9225	0.036639	0.30141	-0.021122	-0.29549	19.5902	-0.026283	-0.34424	0.025670	0.34181
16-27	20.5492	0.037464	0.29742	-0.024770	-0.30817	20.1764	-0.030422	-0.33951	0.030501	0.34941
16-29	21.1390	0.037049	0.29767	-0.027311	-0.30629	20.7250	-0.032273	-0.34415	0.032420	0.34900
16-31	21.6810	0.036667	0.30731	-0.027331	-0.30921	21.2431	-0.033202	-0.34889	0.032630	0.35249
16-33	22.1975	0.037118	0.31326	-0.028475	-0.30864	21.7363	-0.034669	-0.34767	0.032617	0.34807
16-35	22.6937	0.038118	0.31412	-0.029666	-0.30922	22.2174	-0.036358	-0.35031	0.033996	0.34748
17-36	23.5138	0.037439	0.30442	-0.030639	-0.31035	23.0837	-0.035952	-0.34012	0.033349	0.34573

For example, in the first line of numbers, the subset of cases where the driver's and RF passenger's ages are both in the 16-to-19 range, the average age of the drivers is 17.6067 and the average age of the passenger is 17.4150. At that age (approximately 17½), D_AGE1 is .015, D_FEM1 is .267, etc.

The gender coefficients are quite stable throughout these low age ranges, running close to .30 for D_FEM1, -.30 for D_FEM3, -.34 for P_FEM1, and .34 for P_FEM3. In other words, the characteristic adult pattern of substantially higher risk for females than males is already well-established at age 17½ and it may have started even earlier (but this approach does not supply data to ascertain how much earlier).

By contrast, the age coefficients take several years to stabilize. D_AGE1 starts at .015 at age 17.6, reaches .031 at age 18.5, and already stabilizes at the apparent full adult level of .037 at age 19.2. But D_AGE3 needs more time: it is close to zero at ages 17.6 and 18.5, -.018 at 19.2, -.021 at 19.9, -.025 at 20.5, and only begins to approach the adult level at 21.1 or perhaps even a little later. P_AGE1 and P_AGE3 likewise stabilize at approximately 21. In other words, the peak of human survivability occurs at age 18 or perhaps even earlier, and the decline is already setting in at 18. But the full, steady declining trend characteristic of adulthood does not set in until about 21. That is where "full adulthood" may be said to start in the context of irreversible, steady trends toward ever greater fragility and frailty. That is why all the principal estimates of the effect of aging and gender in Chapters 3 to 5 are weighted averages for the age range 21 to 96 (the highest age reportable on FARS), based on regression analyses on the subset of vehicle-involvement cases where both the driver and the RF passenger's age is in the range 21 to 96.

The second question is what trends the coefficients exhibit from 21 onwards. Evans' data suggests the _FEM coefficients are likely to decline in magnitude after some age and perhaps even cross zero at some later age, but it is less clear about the _AGE coefficients. The trend can be studied by continuing the above table of coefficients all the way to the last subset, ages 77-96 (average close to 83). It is easier to see in a graph than in a table. Figure 2-1a is a graph of the D_FEM1 coefficients for all the subsets, with the average age for each subset on the horizontal axis. D_FEM1 stays fairly close to .30 in the subsets of younger drivers and passengers. When the average age reaches 35, give or take a few years, the estimated D_FEM1 coefficients begin a downward trend that is quite steady despite a few wiggles.

FIGURE 2-1a: TREND OF D_FEM1 COEFFICIENT BY CRASH SUBSET

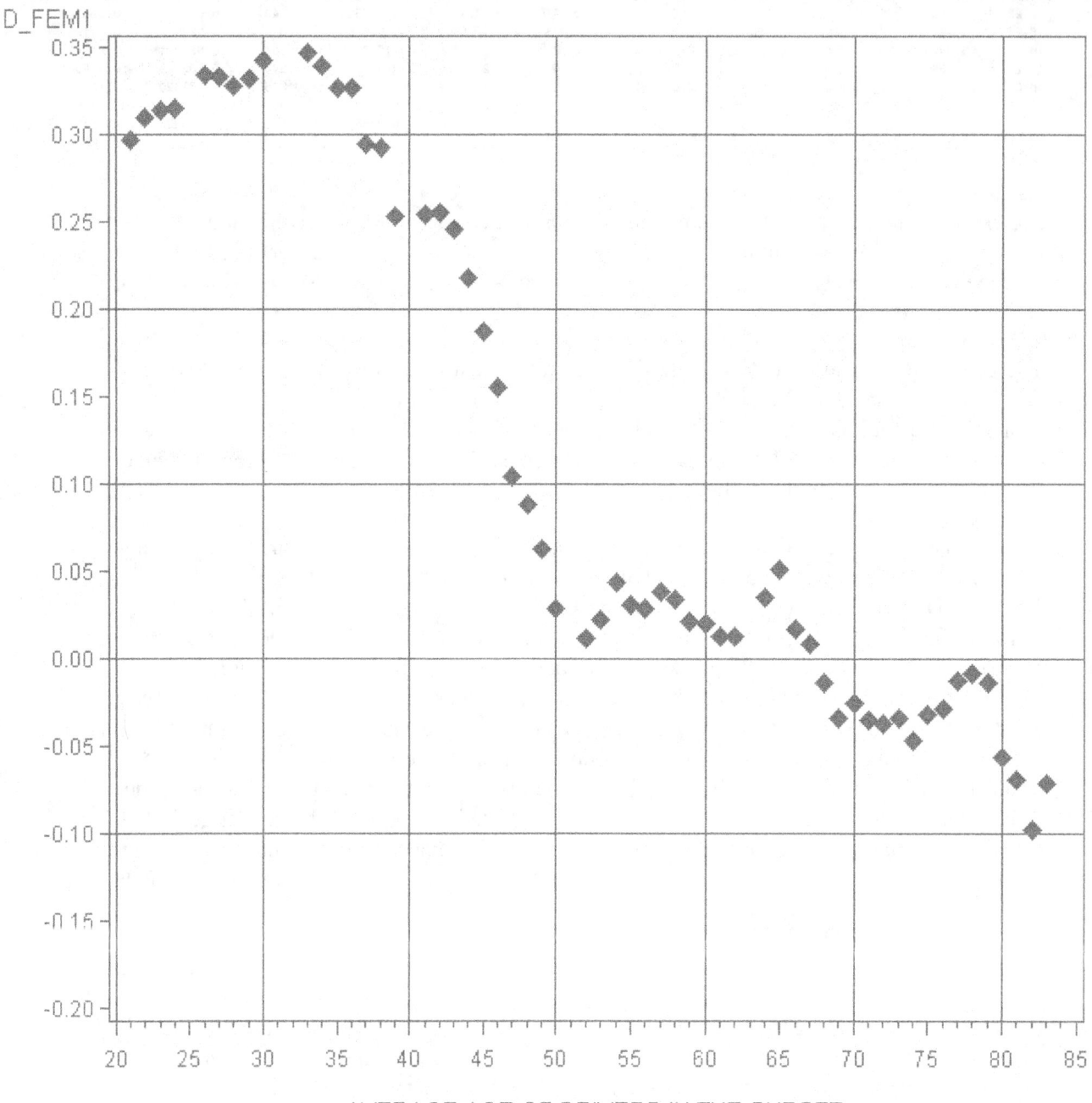

AVERAGE AGE OF DRIVERS IN THE SUBSET

Likewise, the other seven coefficients tend to be fairly constant up to a certain age, perhaps 35, although there is more ambiguity about that for some of them, and then they begin to trend in mostly one direction. The other three _FEM coefficients, like D_FEM1, become steadily weaker in magnitude at those higher ages, sometimes crossing zero. The four _AGE coefficients, on the other hand, show trends of increasing strength after age 35 or so. (Of course, D_AGE1, D_FEM1, P_AGE3, and P_FEM3 start out positive, whereas D_AGE3, D_FEM3, P_AGE1, and P_FEM1 start out negative, as shown in the short table above).

2.6   The principal models: piecewise linear with breakpoints at age 35

The trends of the coefficients obtained by many separate regressions on many subsets suggest that the actual effects of aging and gender are not constant over all ages, but can be well estimated by two single regressions, one for drivers and one for RF passengers, that treat age and gender as piecewise linear variables: constant effects up to age 35, and after that linear trends toward stronger effects for age and weaker effects for gender.

Such models can be formulated by adding just four more independent variables to the simple models of Section 2.4. A total of eight independent variables is still parsimonious when there are thousands of crash cases available. In addition to the first-order terms AGE1, AGE3, FEM1, and FEM3 in the simple models, these formulas define four second-order terms.

AGE1_35 = $(AGE1 - 35)^2$ for drivers older than 35; = 0 otherwise
AGE3_35 = $(AGE3 - 35)^2$ for RF passengers older than 35; = 0 otherwise
FEM1_35 = $(AGE1 - 35)$ for female drivers older than 35; = 0 for females < 35 and all males
FEM3_35 = $(AGE3 - 35)$ for female RF passengers older than 35; = 0 for females < 35 and all
        Males

Here are the estimated regression coefficients when these two models are applied to the same data as the two simple models in Section 2.4: the 154,467 vehicles in which the driver's and RF passenger's age are both in the range of 21 to 96:

```
                    Analysis of Maximum Likelihood Estimates

                                       Standard          Wald
       Parameter      DF     Estimate      Error    Chi-Square    Pr > ChiSq

       FATAL1 (Driver): 87,567 fatal, 66,900 survived

       Intercept       1       0.3590     0.0256     197.0662       <.0001
       AGE1            1       0.0366    0.000896    1669.3235      <.0001
       AGE3            1      -0.0335    0.000872    1475.3720      <.0001
       FEM1            1       0.2955     0.0168     310.5027       <.0001
       FEM3            1      -0.3010     0.0143     441.5819       <.0001
       AGE1_35         1      0.000144   0.000023     40.8749       <.0001
       AGE3_35         1      -0.00035   0.000022    252.2949       <.0001
       FEM1_35         1      -0.00961   0.000796    145.7701       <.0001
       FEM3_35         1      0.00461    0.000706     42.7351       <.0001


       FATAL3 (RF passenger): 95,647 fatal, 58,820 survived

       Intercept       1       0.2088     0.0259      64.8248       <.0001
       AGE1            1      -0.0345    0.000909    1438.4570      <.0001
       AGE3            1       0.0364    0.000906    1614.8532      <.0001
       FEM1            1      -0.2816     0.0168     279.5587       <.0001
       FEM3            1       0.3325     0.0143     541.4469       <.0001
       AGE1_35         1      -0.00016   0.000024     47.3909       <.0001
       AGE3_35         1      0.000381   0.000024    243.5127       <.0001
       FEM1_35         1       0.00820   0.000832     97.0562       <.0001
       FEM3_35         1      -0.00531   0.000782     46.0343       <.0001
```

All eight variables have statistically significant effects in both models, as evidenced by Wald chi-squares greater than 3.84. In both models, AGE1_35 has the same sign as AGE1 and AGE3_35 has the same sign as AGE3 – i.e., the risk-increasing effect of aging a year becomes ever stronger after age 35. In both models, FEM1_35 has the opposite sign from FEM1 and FEM3_35 has the opposite sign from FEM3 – i.e., the risk increase for being female becomes ever weaker after age 35.

For drivers, max-rescaled R-square and overall Wald chi-square are .0849 and 8520, both higher than the .0801 and 8132 in the simple model. For RF passengers, these statistics are .0891 and 8574, likewise higher than the .0850 and 8382 in the simple model.

The simple model estimated a sort of "average" coefficient across all ages. The coefficient for AGE1, for example, was .0391 for drivers. In this model, the AGE1 coefficient is weaker, just .0366, because at every age above 35, the effect of AGE1 is supplemented by multiples of the effect of AGE1_35. Conversely, the coefficient of FEM1 in the simple model, .1268, is now strengthened to .2955, because at every age above 55, this effect is diminished by multiples of the negative effect of FEM1_35.

The coefficients for the original variables AGE1, AGE3, FEM1, and FEM3 in the new models are fairly close to the corresponding coefficients, in the table in Section 2.5, for what the simple model calculated for the subsets of crashes centered around ages 21, 22, and 23. They are good estimates of the effects of aging and gender for occupants in the 21-35 age range. Above age 35, these initial effects are modified by the effects of AGE1_35, AGE3_35, FEM1_35, and FEM3_35, which become ever larger as age increases.

Figure 2-1b recapitulates Figure 2-1a's D_FEM1 coefficients for the separate regressions on the various subsets (red diamonds) and compares them to the composite effect of the driver's gender on the driver's risk, as a function of the driver's age, from the new regression (blue dots), namely .2955 - .0961 FEM1_35, where .2955 and -.0961 are the estimated coefficients for FEM1 and FEM1_35. The effects from the regression, constant up to age 35 and then linearly decreasing, track the red diamonds quite well. Here, a single regression with eight variables generates essentially the same information as the 63 separate regressions, each with four variables, on subsets of the data.

FIGURE 2-1b: TREND OF D_FEM1 COEFFICIENT BY CRASH SUBSET (red diamonds)
AND EFFECT OF DRIVER'S GENDER ON DRIVER'S RISK IN THE PRINCIPAL REGRESSION MODEL (blue dots)

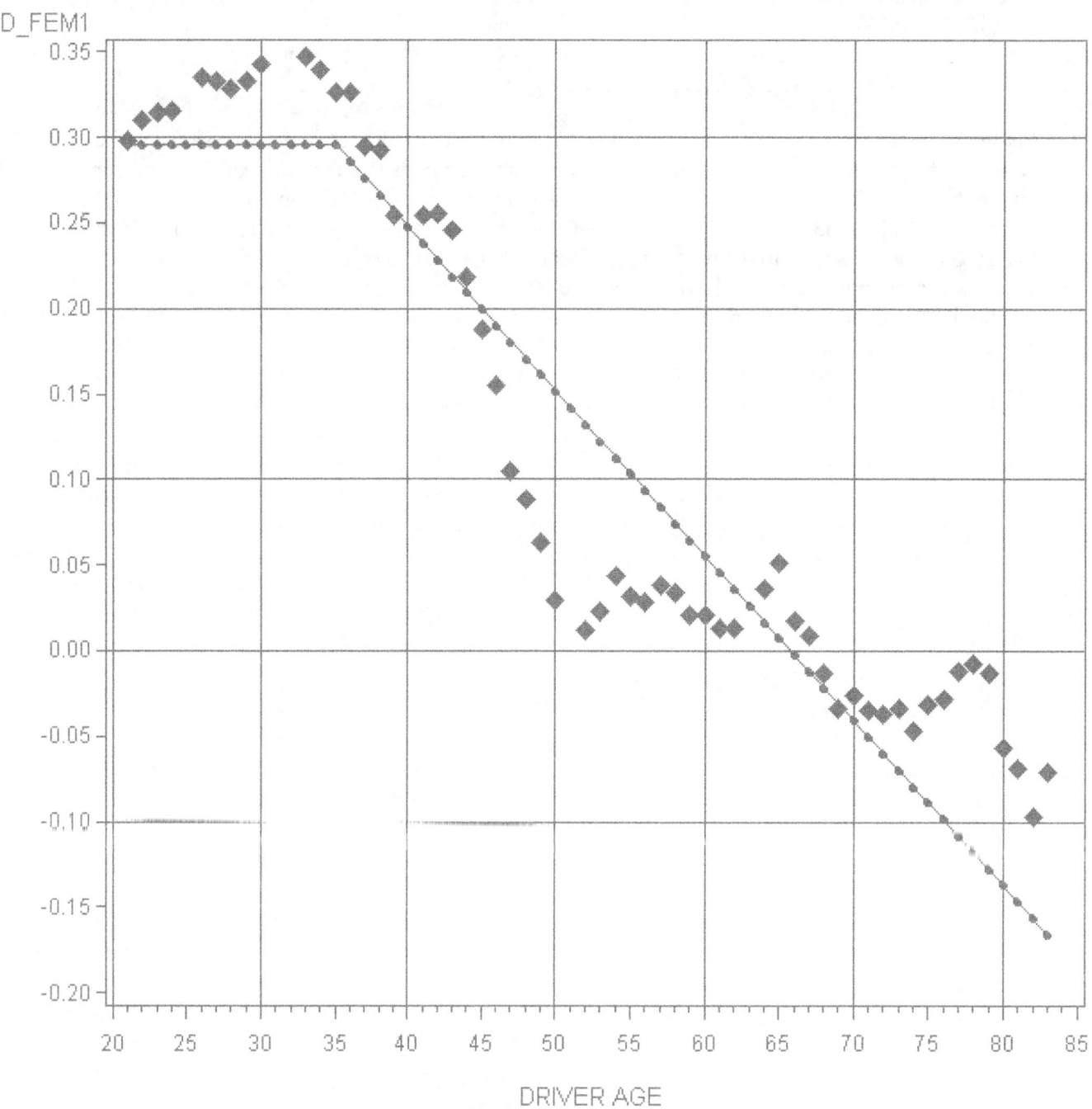

DRIVER AGE

Similarly, Figures 2-2 to 2-8 compare, for the other seven parameters, the coefficients for the separate regressions on the various subsets (red diamonds) to the corresponding composite effects, as a function of the subject's age, from the new regressions (blue dots). Here, too, the blue dots track the red diamonds, but not always as well as in Figure 2-1b.

Figure 2-2 tracks the effect of the driver's age on the driver's risk (D_AGE1). The blue dots and red diamonds start and end together, but diverge in between: probably the poorest match among the eight parameters. The effect of the driver's aging simply does not have a clear-cut relationship with age; it peaks at 50 but then drops way down by 60 and only increases steadily after 60. The relatively weak relationship is also reflected in the AGE1_35 coefficient in the regression, which is small in absolute terms and also, although significant, has the lowest chi-square (40.87) of the 16 coefficients in the two regressions.

In Figure 2-3, the "effect" of the RF passenger's age on the driver's risk (D_AGE3), the red diamonds conform reasonably well to a piecewise linear pattern, first constant and then dropping at a linear rate. The blue dots, by design, have the same pattern, but they are consistently lower than the diamonds rather than passing through the middle of the diamonds' scatterplot. In Figure 2-4, the "effect" of the RF passenger's gender on the driver's risk, the dots track the diamonds well except at the oldest ages (where there are fewer crash cases and, as a consequence, less influence on the single regression); nevertheless, the trend in the diamonds gets quite steep after age 50, unlike the upward trend in the dots, which, despite its head start, eventually falls behind the diamonds.

Figure 2-5 (P_AGE3, the effect of the RF passenger's age on the RF passenger's risk) is more or less the inverse of Figure 2-3, with the blue dots running parallel, but consistently above the red diamonds. Figure 2-6 (P_FEM3), on the other hand, shows excellent agreement between the dots and the diamonds, except at the oldest ages.

In Figure 2-7 (P_AGE1), the blue dots are consistently below the red diamonds between ages 50 and 80 but track fairly well at the other ages. In Figure 2-8 (P_FEM1) the dots track the diamonds extremely well from age 35 onwards, but not so well for the young drivers: here the diamonds are not level like the dots but appear to follow the same linear trend as they do at higher ages.

In other words, our single regressions with four first-order terms and four second-order terms all starting at age 35 plausibly model the actual patterns in the data (as evidenced by the regressions on multiple subsets) even though they do not track it exactly. But they are an improvement on the simple models, which assume the same effect of aging and gender at all ages.

A second, perhaps even more important advantage of the regressions with the second-order terms is that they are more "portable" across data sets. The simple models assume the effects of aging and gender are constant, when they are not. Thus, the older the occupants in the calibration database, the higher an effect for aging and the lower an effect for gender will be estimated by the simple model. This can be a problem when comparing two different sets of vehicles, say cars of the 1960s and late-model cars. Because the population has aged, the simple models would estimate higher age effects and lower gender effects in the late-model cars – merely because the calibration database for the late-model cars has, on the average, older occupants, not because the vehicles have changed. But the regressions with the second-order terms, which estimate effects as a function of age, will not be affected that way by a shift in the age distribution of the calibration databases.

FIGURE 2-2: TREND OF D_AGE1 COEFFICIENT BY CRASH SUBSET (red diamonds) AND EFFECT OF DRIVER'S AGE ON DRIVER'S RISK IN THE PRINCIPAL REGRESSION MODEL (blue dots)

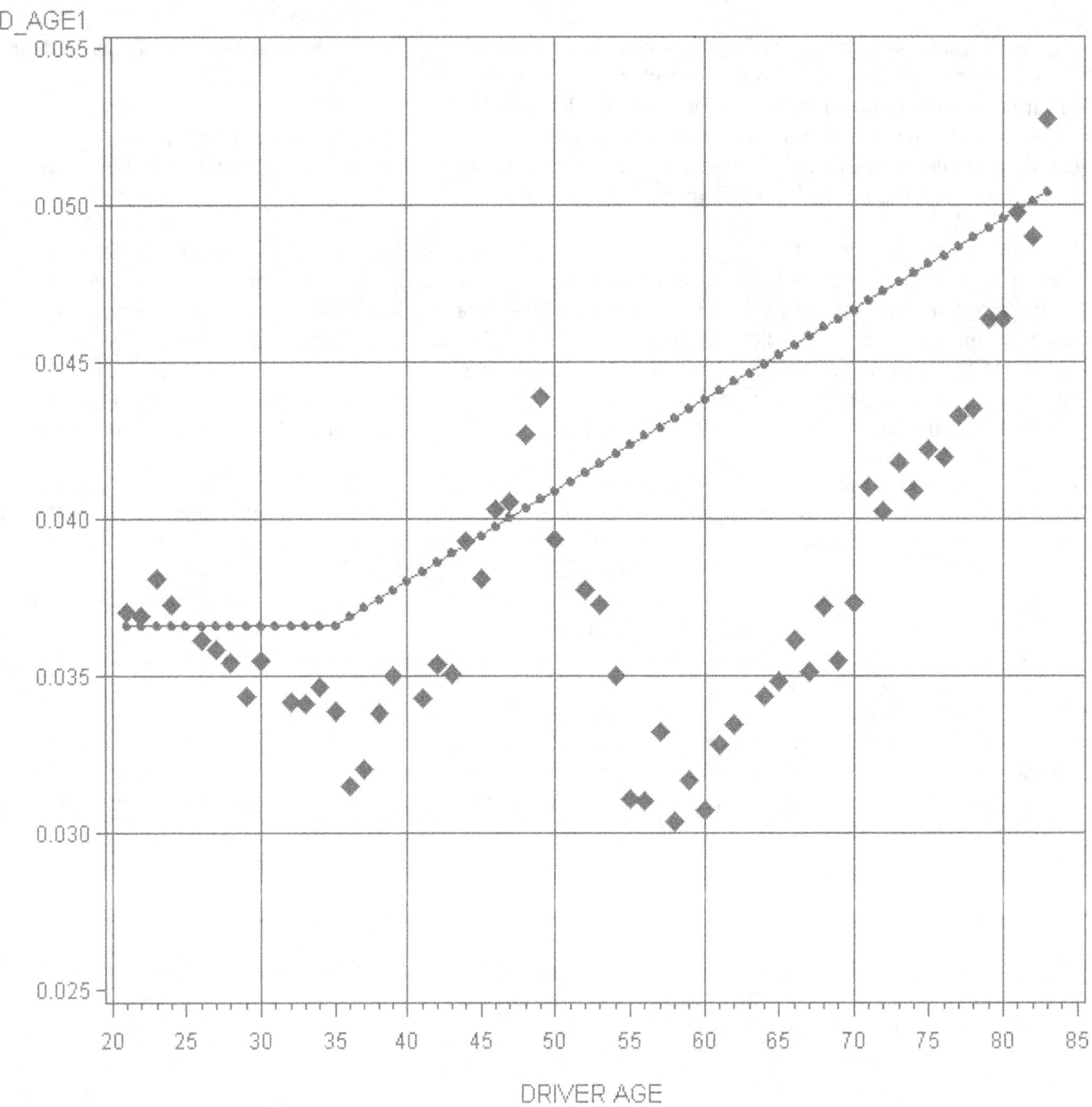

FIGURE 2-3: TREND OF D_AGE3 COEFFICIENT BY CRASH SUBSET (red diamonds)
AND EFFECT OF RF PASSENGER'S AGE ON DRIVER'S RISK IN THE PRINCIPAL REGRESSION MODEL (blue dots)

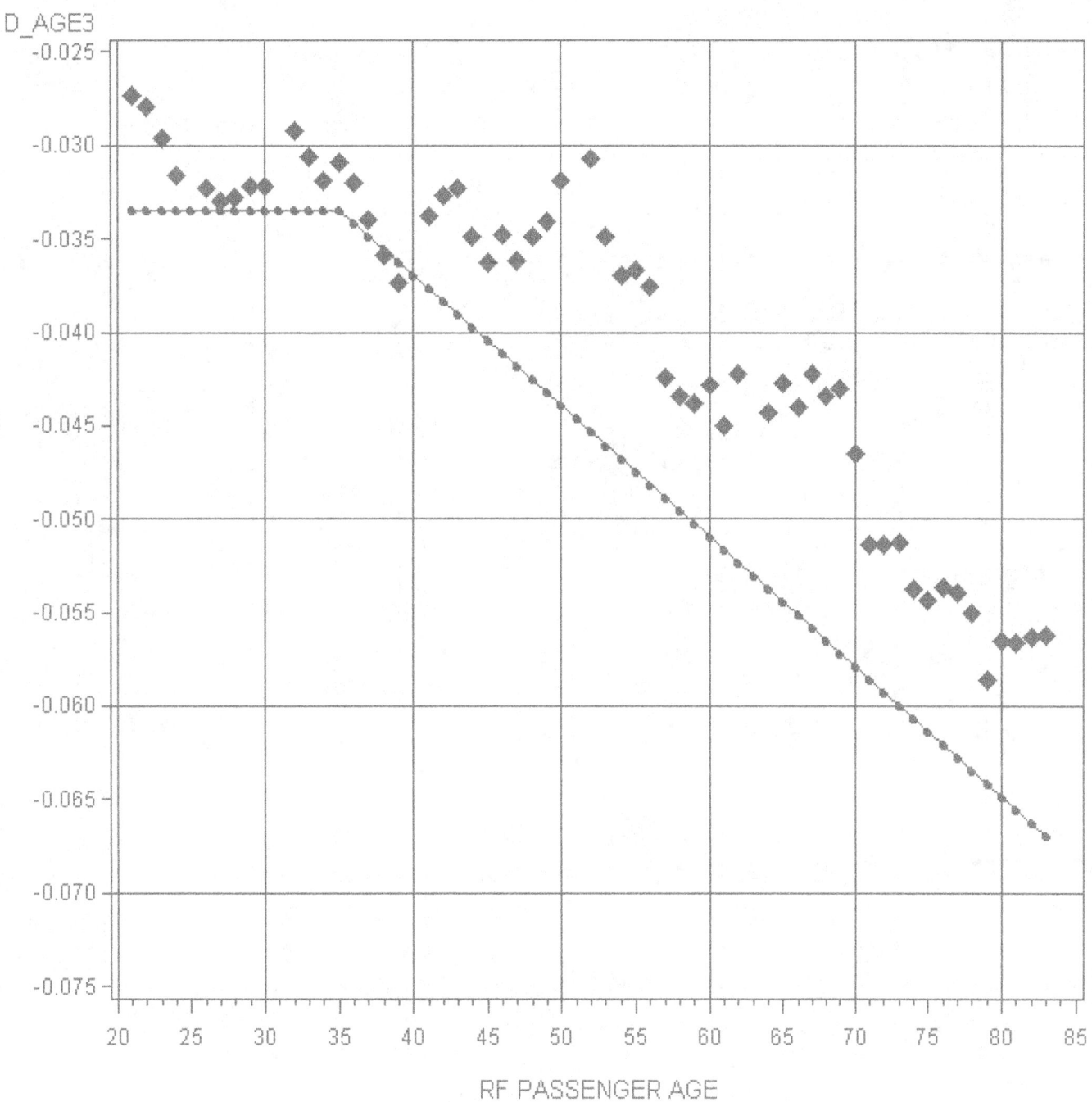

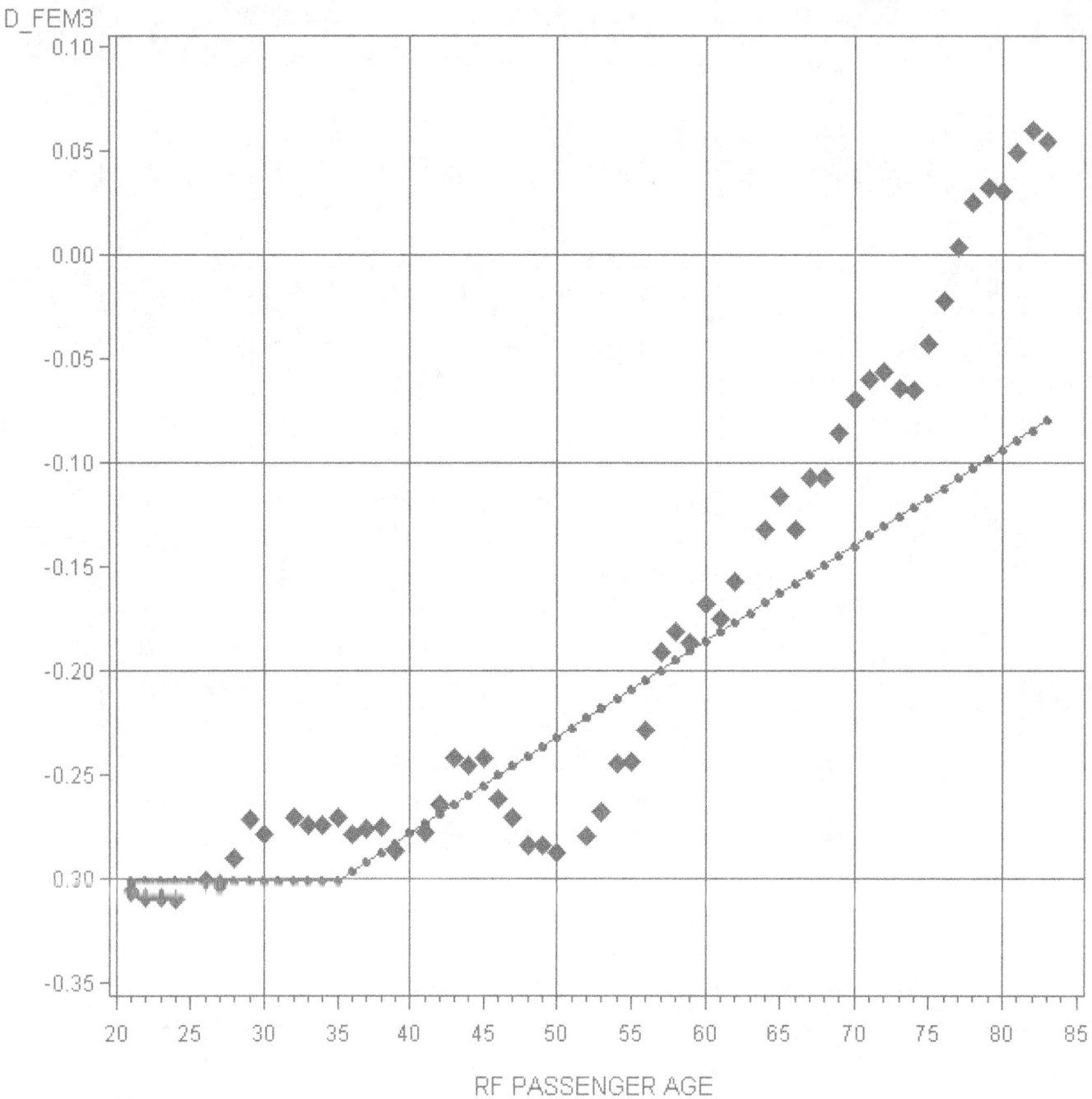

FIGURE 2-4: TREND OF D_FEM3 COEFFICIENT BY CRASH SUBSET (red diamonds)
AND EFFECT OF RF PASSENGER'S GENDER ON DRIVER'S RISK IN THE PRINCIPAL REGRESSION MODEL (blue dots)

D_FEM3

RF PASSENGER AGE

FIGURE 2-5: TREND OF P_AGE3 COEFFICIENT BY CRASH SUBSET (red diamonds)
AND EFFECT OF RF PASSENGER'S AGE ON RF PASSENGER'S RISK IN THE PRINCIPAL REGRESSION MODEL (blue dots)

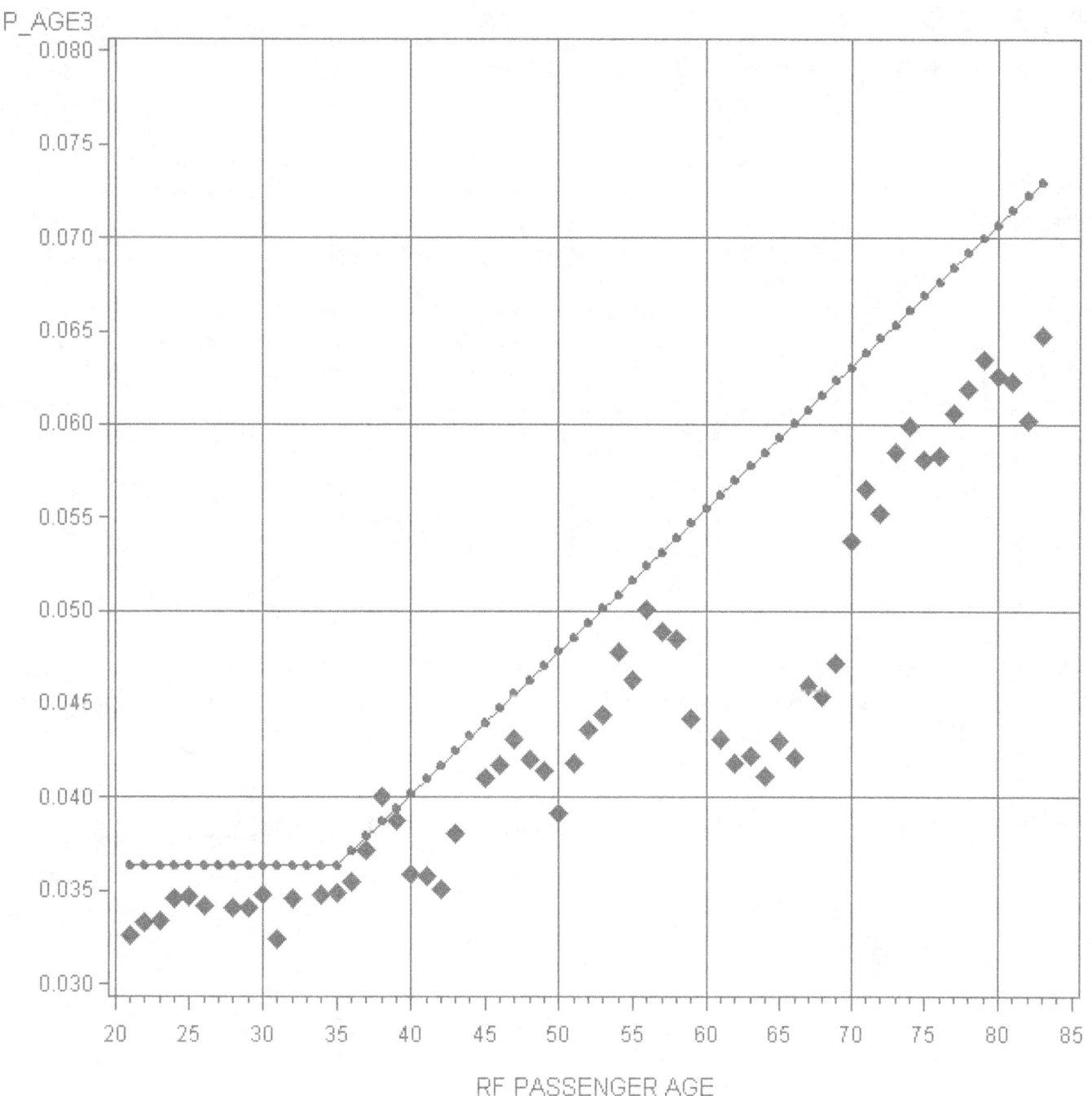

FIGURE 2-6: TREND OF P_FEM3 COEFFICIENT BY CRASH SUBSET (red diamonds)
AND EFFECT OF RF PASSENGER'S GENDER ON RF PASSENGER'S RISK IN THE PRINCIPAL REGRESSION MODEL (blue dots)

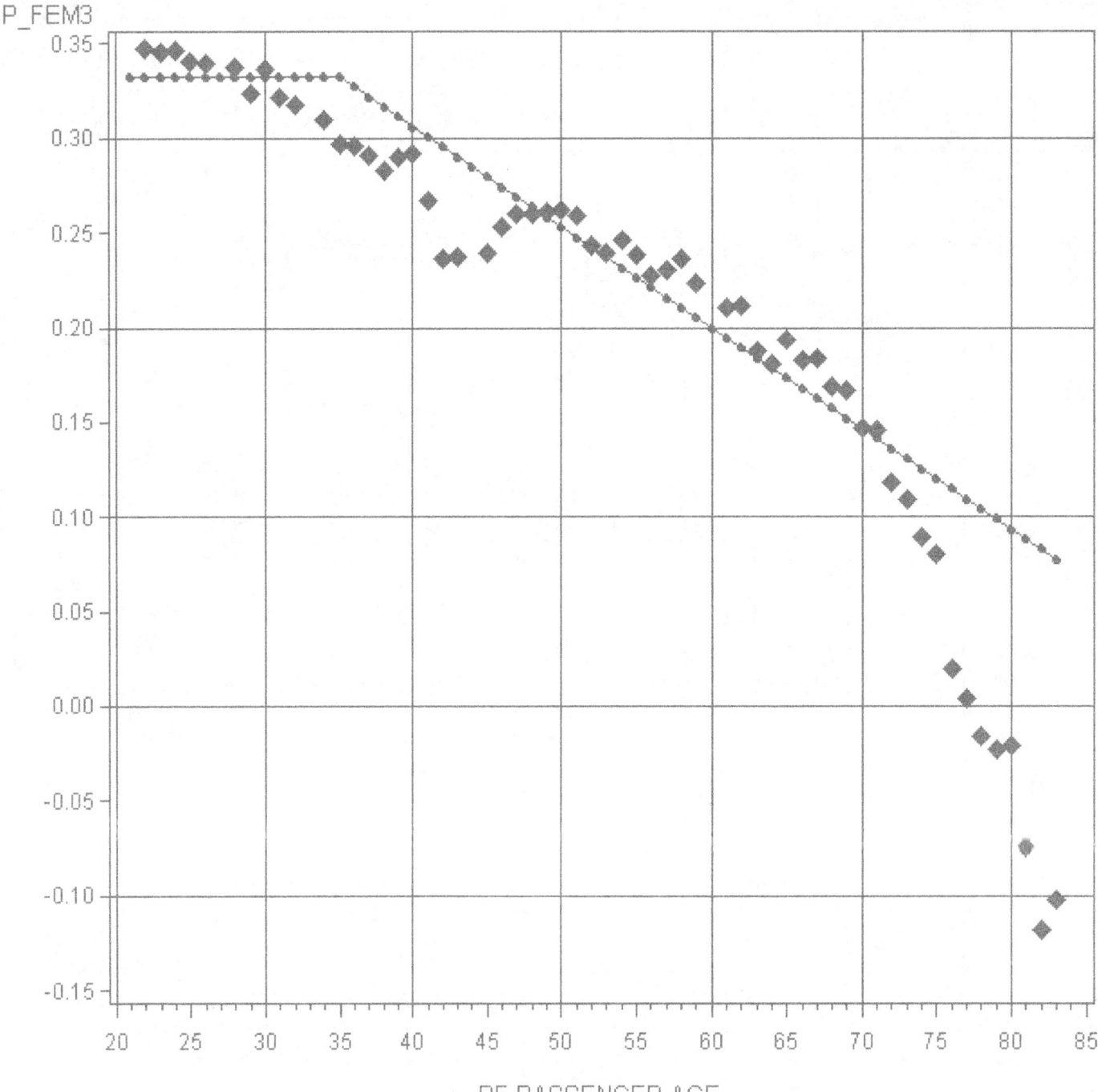

FIGURE 2-7: TREND OF P_AGE1 COEFFICIENT BY CRASH SUBSET (red diamonds)
AND EFFECT OF DRIVER'S AGE ON RF PASSENGER'S RISK IN THE PRINCIPAL REGRESSION MODEL (blue dots)

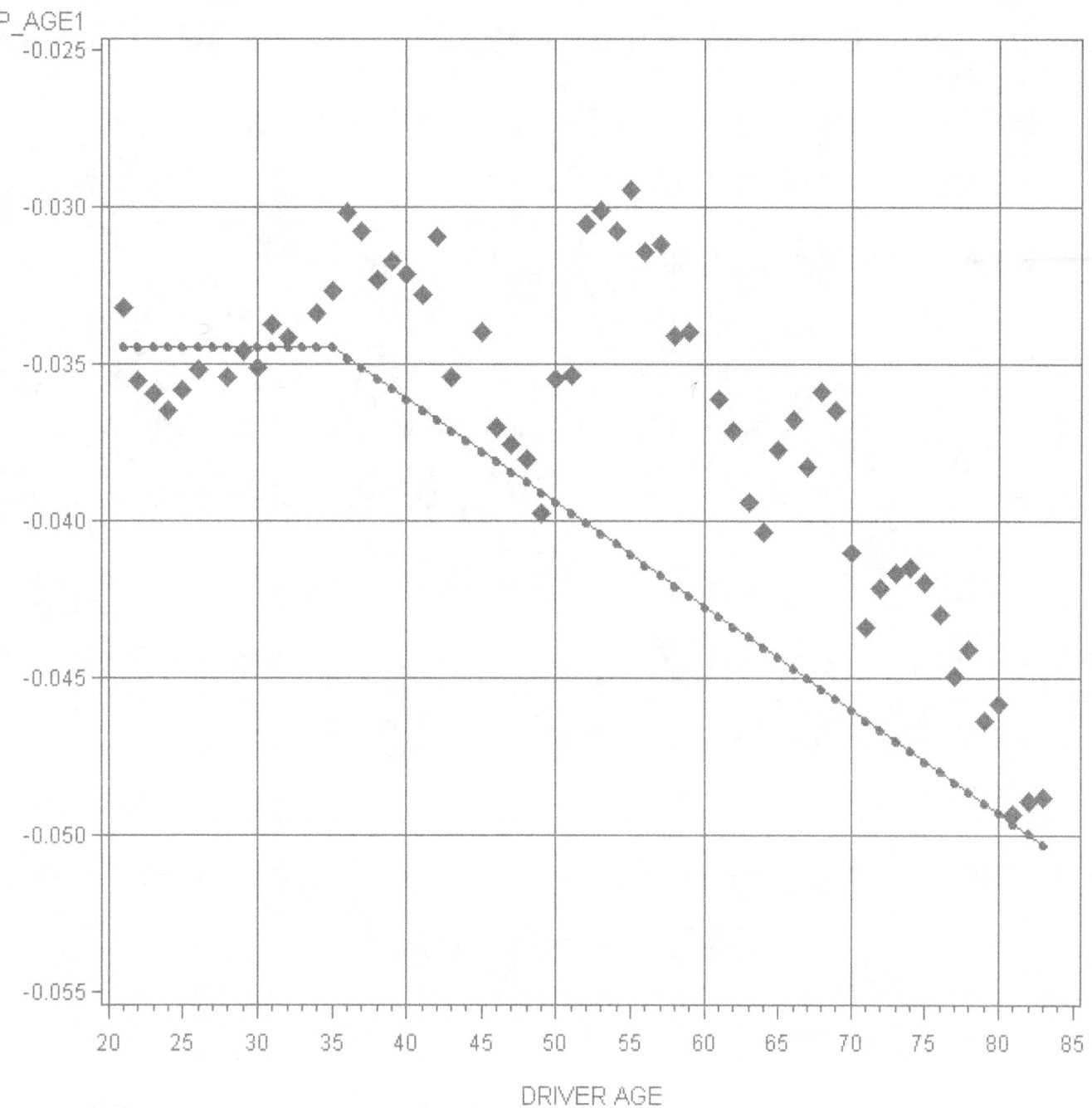

P_AGE1

DRIVER AGE

FIGURE 2-8: TREND OF P_FEM1 COEFFICIENT BY CRASH SUBSET (red diamonds)
AND EFFECT OF DRIVER'S GENDER ON RF PASSENGER'S RISK IN THE PRINCIPAL REGRESSION MODEL (blue dots)

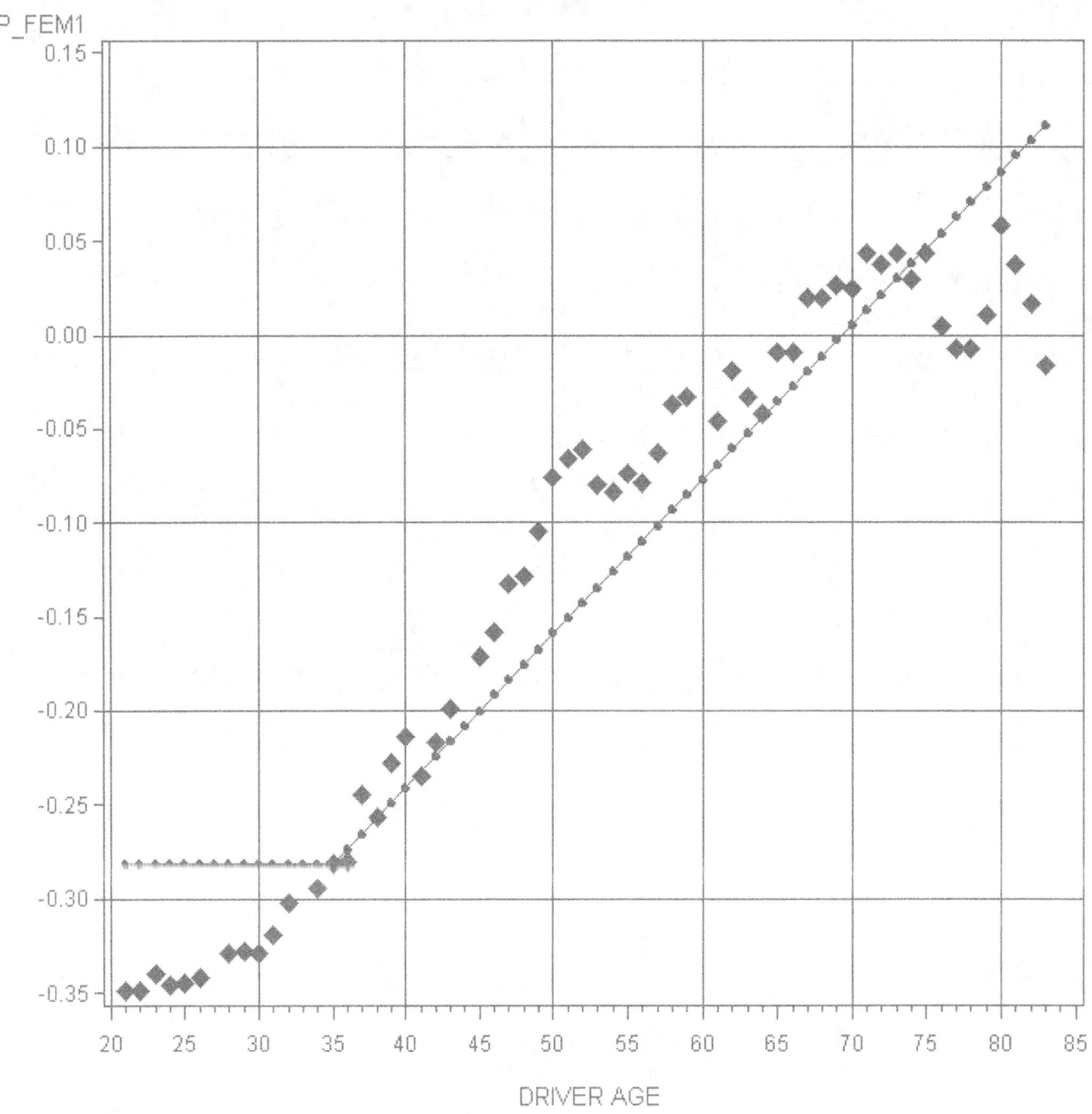

2.7   More complex models: piecewise linear with optimized breakpoint-ages

The preceding models assumed all the parameters had a constant effect up to age 35, and after that linear trends toward stronger effects for age and weaker effects for gender. That tracked quite well for the effect of the driver's gender on the driver's risk (Figure 2-1b), but not so well for some of the other parameters. For example, Figures 2-3 and 2-4 suggest the linear upward or downward trend may start somewhat later than 35 for their parameters, whereas Figure 2-8 hints at an even earlier start.

One way to potentially improve fit, while still retaining piecewise linear models, is a regression procedure with forward selection that allows the data to "pick" the breakpoint-ages where constant effects shift to linear trends. In addition to the first-order terms AGE1, AGE3, FEM1, and FEM3 and the second-order terms AGE1_35, AGE3_35, FEM1_35, and FEM3 35 defined in Section 2.6, the regression will also have initially available other potential second-order terms AGE1_30, AGE3_30, FEM1_30, FEM3_30, AGE1_40, AGE3_40, FEM1_40, FEM3_40, AGE1_45, AGE3_45, FEM1_45, FEM3_45, AGE1_50, AGE3_50, FEM1_50, FEM3_50, AGE1_55, AGE3_55, FEM1_55, and FEM3_55, all defined like the _35 terms, but substituting the age shown for 35 in the definition. While this does not allow complete freedom in selecting breakpoints, it does allow a wide range of possibilities.

The first step in the forward regression procedure is to require the model to include the four first-order terms AGE1, AGE3, FEM1, and FEM3 and then allow it to pick from the 24 potential second-order terms the single one that most increases the model's overall chi-square. The regression for drivers picks FEM1_35 first (confirming, in this case, that 35 is the best breakpoint and consistent with the excellent fit shown in Figure 2-1b). Having selected FEM1_35, we do not want the model to include any additional FEM1_ terms, and drop FEM1_30, FEM1_40, FEM1_45, FEM1_50, and FEM1_55 from further consideration.

The second step is to require the model to include AGE1, AGE3, FEM1, FEM3, and FEM1_35 and then allow it to pick the strongest of the 18 remaining second-order terms. The regression for drivers picks AGE3_45. The remaining AGE3_ variables are dropped. The third step picks FEM3_55 from the 12 remaining second-order terms. The fourth and final step picks AGE1_55.

Here are the estimated regression coefficients when the two models with breakpoints selected by forward regression are applied to the same data as the two models with breakpoints set at age 35 in Section 2.6 and the two simple models in Section 2.4: the 154,467 vehicles in which the driver's and RF passenger's age are both in the range of 21 to 96:

```
                Analysis of Maximum Likelihood Estimates

                                    Standard           Wald
      Parameter     DF    Estimate     Error    Chi-Square    Pr > ChiSq

      FATAL1 (Driver): 87,567 fatal, 66,900 survived


      Intercept      1     0.3403     0.0201     286.3516      <.0001
      AGE1           1     0.0387    0.000666   3371.2410      <.0001
      AGE3           1    -0.0351    0.000713   2427.2370      <.0001
      FEM1           1     0.2891     0.0168     297.8194      <.0001
      FEM3           1    -0.2815     0.0127     487.5373      <.0001
      AGE1_55        1    0.000289   0.000045     41.8347      <.0001
      AGE3_45        1    -0.00051   0.000032    259.2014      <.0001
      FEM1_35        1    -0.00921   0.000801    132.1185      <.0001
      FEM3_55        1     0.00843   0.00129      42.9578      <.0001



      FATAL3 (RF passenger): 95,647 fatal, 58,820 survived


      Intercept      1     0.3459     0.0231     223.3828      <.0001
      AGE1           1    -0.0371    0.000675   3018.2220      <.0001
      AGE3           1     0.0343    0.000892   1474.6037      <.0001
      FEM1           1    -0.2875     0.0178     260.5761      <.0001
      FEM3           1     0.3160     0.0128     613.5840      <.0001
      AGE1_55        1    -0.00031   0.000047     43.4732      <.0001
      AGE3_35        1    0.000454   0.000028    267.7137      <.0001
      FEM1_30        1     0.00705   0.000763     85.2981      <.0001
      FEM3_55        1    -0.0115    0.00149      59.4689      <.0001
```

All eight variables have statistically significant effects in the same directions as the corresponding variables in the models with breakpoints set at age 35. For drivers, max-rescaled R-square and overall Wald chi-square are still .0849 and 8520, the same as for the models with breakpoints set at 35 and no further improvement over the .0801 and 8132 in the simple model. For RF passengers, these statistics are .0893 and 8595, a slight improvement over the .0891 and 8574 for the model with breakpoints set at 35, not nearly as great as that model's improvement over the .0850 and 8382 in the simple model.

A more empirical way to assess the three groups of models is to compare their fit with the regression coefficients obtained at the individual ages by the regressions on the subsets of data centered on those ages, the difference between the x's and the A's in the preceding figures. The effects of aging and gender, as a function of age are tabulated for the regressions on subsets (D_AGE1, D_FEM1, etc.) at each year of age from 21 to 83 and computed by the three sets of regression formulas. The square of the difference is computed at each age. The weighted mean of these square differences is computed across 21 to 83; the weight factor for the driver terms is the proportion of all car and LTV driver fatalities in 2001-2010 FARS at that age. For the RF terms, the weight factor is the proportion of RF passenger fatalities in 2001-2010. Here are the weighted mean-square errors for the three sets of single-regression models relative to the coefficients observed in the regressions on subsets. The model with the least error is typed green and bold:

	SIMPLE	BREAKPOINT-35	OPTIMIZED-BREAKPOINT
D_AGE1	0.000019850	0.000027759	0.000027465
D_FEM1	0.0220400	0.0024651	0.0024852
D_AGE3	0.000092172	0.000044093	0.000052357
D_FEM3	0.0094751	0.0018612	0.0019967
P_AGE1	0.000022067	0.000018904	0.000019434
P_FEM1	0.0220921	0.0024083	0.0018678
P_AGE3	0.000101593	0.000043689	0.000061363
P_FEM3	0.0132635	0.0023886	0.0019732

The regression model with the breakpoints set at age 35 has the lowest error for five of the eight parameters, while the more complex model is best only twice. The simple model usually has much larger error, sometimes even 10 times as large, except for D_AGE1. Here, as seen in Figure 2-2, the empirical results do not follow a clear pattern and it is apparently more accurate to just draw a horizontal line through them than to attempt to fit a piecewise linear model. These statistics support choice of the model with breakpoints set at age 35, because the optimized model offers no improvement in accuracy in return for its additional complexity; it also increases the risk of over-fitting the data if applied to small subsets of vehicles.

2.8 From regression coefficients to estimated effects of aging and gender

Logistic regression estimates a linear relationship between the log-odds that the dependent variable equals 1 and the independent variables. When the models with breakpoints set at age 35 are applied to the full database of 154,467 MY 1960-2011 cars and LTVs on 1975-2010 FARS in which the driver's and RF passenger's age are both in the range of 21 to 96 and at least one of them was a fatality, this model predicts the log-odds of a driver fatality to be:

$$Z1 = .3590 + .0366\,AGE1 - .0335\,AGE3 + .2955\,FEM1 - .3010\,FEM3$$
$$+ .000144\,AGE1\_35 - .00035\,AGE3\_35 - .00961\,FEM1\_35 + .00461\,FEM3\_35$$

Whereas the log-odds of a RF passenger fatality are:

$$Z3 = .2088 - .0345\,AGE1 + .0364\,AGE3 - .2816\,FEM1 + .3325\,FEM3$$
$$- .00016\,AGE1\_35 + .000381\,AGE3\_35 + .0082\,FEM1\_35 - .00531\,FEM3\_35$$

Based on the regression formulas, given a crash in which the driver's characteristics are AGE1 and FEM1, the passenger's characteristics are AGE3 and FEM3, and at least one of them is a fatality, the probability of a driver fatality is E_FATAL1 = exp(Z1)/(1+exp(Z1)) and the probability of a RF passenger fatality is E_FATAL3 = exp(Z3)/(1+exp(Z3)). The expected ratio of the driver's risk to the passenger's risk is E_FATAL1/E_FATAL3. These expected probabilities of a fatality can be computed for any driver-RF pair of occupants that was involved in an actual fatal crash in FARS.

Probabilities can also be calculated for imaginary pairs of occupants that are similar to the FARS pairs, but an attribute has been changed. For example, what would have happened if the male driver in a FARS case had instead been a female of the same age? Or a male driver, but one year older? Being female will increase the driver's probability of fatality, except for the oldest drivers. Aging by a year will also increase the driver's probability of death. But the probabilities will change for the RF passenger, too. As explained in Section 2.4, FARS is a sort of zero-sum game, because at least one person – and usually only one – has to be a fatality: whatever increases the probability that the driver died, implicitly reduces that probability for the RF passenger. The risk ratio will change twofold, because the probability of death has changed for both occupants.

The regression formulas compute the probabilities for any conceivable imaginary pair. If, in the preceding formulas of Z1 and Z3, the driver had definitely been a male, but the age of the driver and the age and gender of the RF passenger had not changed, the log-odds for the driver and the RF passenger would now be:

$$MZ1 = .3590 + .0366\,AGE1 \quad -.0335\,AGE3 \qquad\qquad\qquad -.3010\,FEM3$$
$$\qquad\quad +.000144\,AGE1\_35 \; -.00035\,AGE3\_35 \qquad\qquad +.00461\,FEM3\_35$$

$$MZ3 = .2088 - .0345\,AGE1 \quad +.0364\,AGE3 \qquad\qquad\qquad +.3325\,FEM3$$
$$\qquad\quad -.00016\,AGE1\_35 \; +.000381\,AGE3\_35 \qquad\qquad -.00531\,FEM3\_35$$

because the FEM1 and FEM1_35 terms would drop out. The probability of a driver fatality would now be M1_FATAL1 = exp(MZ1)/(1+exp(MZ1)) and the probability of a RF passenger fatality would now be M1_FATAL3 = exp(MZ3)/(1+exp(MZ3)). The risk ratio is M1_FATAL1/ M1_FATAL3.

However, if the driver had definitely been a female, but the age of the driver and the age and gender of the RF passenger had not changed, the log-odds would now be:

$$FZ1 = .3590 + .0366\,AGE1 \quad -.0335\,AGE3 \quad +.2955 \qquad -.3010\,FEM3$$
$$\qquad\quad +.000144\,AGE1\_35 \; -.00035\,AGE3\_35 \; -.00961\,X \qquad +.00461\,FEM3\_35$$

$$FZ3 = .2088 - .0345\,AGE1 \quad +.0364\,AGE3 \quad -.2816 \qquad +.3325\,FEM3$$
$$\qquad\quad -.00016\,AGE1\_35 \; +.000381\,AGE3\_35 \; +.0082\,X \qquad -.00531\,FEM3\_35$$

where X = AGE1 – 35 if AGE1 > 35, 0 otherwise. The probability of a driver fatality would now be F1_FATAL1 = exp(FZ1)/(1+exp(FZ1)) and the probability of a RF passenger fatality would now be F1_FATAL3 = exp(FZ3)/(1+exp(FZ3)). The risk ratio is F1_FATAL1/F1_FATAL3.

The two risk ratios (or four probabilities) may be used to compute the risk for a female driver relative to a male driver of the same age by double-pair comparison, a technique discussed in Chapter 1 and Section 2.1. The RF passenger fatality counts (in this case, fractional probabilities of a fatality: M1_FATAL3 and F1_FATAL3, respectively) are the control group. The risk for a female age AGE1 relative to a male of the same age, given that the RF passenger has age AGE3 and gender FEM3, is:

$$(F1\_FATAL1/F1\_FATAL3) / (M1\_FATAL1/M1\_FATAL3)$$

The added risk for a female driver relative to the male driver, expressed as a percentage, is:

$$100 \times \{[(F1\_FATAL1/F1\_FATAL3) / (M1\_FATAL1/M1\_FATAL3)] - 1\}$$

If the male driver had not been replaced by a female of the same age, but instead by a male driver one year older, and the age and gender of the RF passenger had not changed, the log-odds would now be:

$$OZ1 = .3590 + .0366\,(AGE1+1) \quad -.0335\,AGE3 \quad\quad -.3010\,FEM3$$
$$+ .000144\,Y \quad -.00035\,AGE3\_35 \quad\quad +.00461\,FEM3\_35$$

$$OZ3 = .2088 - .0345\,(AGE1+1) \quad +.0364\,AGE3 \quad\quad +.3325\,FEM3$$
$$-.00016\,Y \quad +.000381\,AGE3\_35 \quad\quad -.00531\,FEM3\_35$$

where $Y = (AGE1 - 34)^2$ if $AGE1 > 34$, 0 otherwise. The probability of a driver fatality would now be $OM1\_FATAL1 = \exp(OZ1)/(1+\exp(OZ1))$ and the probability of a RF passenger fatality would now be $OM1\_FATAL3 = \exp(OZ3)/(1+\exp(OZ3))$. The risk ratio is $OM1\_FATAL1/OM1\_FATAL3$. By double-pair comparison, the added risk for a male age AGE1+1 relative to a male of the age AGE1, given that the RF passenger has age AGE3 and gender FEM3, is:

$$100 \times \{[(OM1\_FATAL1/OM1\_FATAL3) / (M1\_FATAL1/M1\_FATAL3)] - 1\}$$

With similar formulas, it is possible to compute the added risk for a female driver becoming one year older; also, for a female RF relative to a male RF passenger of the same age and for either a male RF or a female RF passenger becoming one year older while the driver stays the same – for one specific crash and the imaginary pairs derived from that crash.

The risk increases for aging one year or for a female relative to a male will vary from crash to crash, depending on the ages and genders of both occupants, since all of these terms will appear in the regression formulas. The next step is to estimate the _average_ risk increase for aging or for females over a large, representative set of crash involvements. This can be accomplished by summing up the fatality probabilities in each crash and then performing the double-pair comparison on the sums of the fatality probabilities – e.g., for a female relative to a male driver:

$$100 \times \{[(\Sigma\,F1\_FATAL1/\,\Sigma\,F1\_FATAL3) / (\Sigma\,M1\_FATAL1/\,\Sigma\,M1\_FATAL3)] - 1\}$$

The "large, representative set of crash involvements" that will be used in _every_ analysis of Chapters 3 to 5, even those not based on FARS data, is the 58,438 records of cars and LTVs on 2001-2010 FARS that had a driver and a RF passenger, at least one of them a fatality. Unlike the databases in the regression analyses, the vehicles may be any model year and the occupants need not match in terms of belt use or air bags. Always using the same database for summing up the effects helps make the results more "portable": the effects of aging a year can be directly compared for current vehicles and for cars of the past, or for fatalities and for nonfatal injuries.

This data set provides the actual distribution by age and gender of RF passengers in fatal crashes for the last 10 years of FARS, plus the actual joint distribution of driver and RF passenger age and gender. It does not, however, provide the age and gender distributions of all drivers in fatal crashes, because many drivers are not accompanied by an RF passenger. A weight factor, based on the ratio of all driver fatalities to accompanied-driver fatalities, is computed by age and gender (but with single computations for males age 86-96 and for females age 86-96, because there are not many cases at each individual age). In computing statistics for drivers, this weight factor is applied to each of the 58,438 pairs; it inflates the 32,624 accompanied driver fatalities to 175,027, the total count of driver fatalities and makes the drivers' age and distributions representative of all drivers, not just the accompanied ones. No such inflation is necessary for the RF passenger, because every RF passenger is accompanied by a driver and may be found in one of the 58,438 pairs.

Here are some examples of the computations. If the regression formulas are applied to the 58,438 pairs, but with every RF passenger replaced by a male of the same age, the fatality probabilities sum up to 34,654 driver fatalities and 34,546 RF passenger fatalities. If the same formulas are applied, but now with every RF passenger replaced by a _female_ of the same age, the probabilities sum up to 31,554 drivers and 37,918 RF passengers. In other words, RF passenger fatalities are expected to increase; also, FARS being a zero-sum game where there is most often only one fatality per vehicle, the increase in RF passenger fatalities is compensated by a decrease in driver fatalities. Based on a double-pair comparison that uses the driver fatalities as a control group, the average additional risk for a female RF passenger relative to a male of the same age, expressed as a percentage, is:

$$100 \times \{[(37{,}918/31{,}554) / (34{,}546/34{,}654)] - 1\} = 20.5\% \text{ increase}$$

On the other hand, if the same formulas are applied, but now with every RF passenger replaced by a _male one year older_, the probabilities sum up to 34,070 drivers and 35,154 RF passengers. The average additional risk for a male RF passenger aging one year is:

$$100 \times \{[(35{,}154/34{,}070) / (34{,}546/34{,}654)] - 1\} = 3.51\% \text{ increase}$$

If the formulas are applied to the 58,438 pairs, but with every _driver_ replaced by a male of the same age, the _weighted_ fatality probabilities sum up to 175,797 driver fatalities and 197,654 RF passenger fatalities. If the same formulas are applied, but now with every driver replaced by a female of the same age, the weighted sums of the probabilities are 187,340 drivers and 185,673 RF passengers. The average additional risk for a female driver relative to a male of the same age is:

$$100 \times \{[(187{,}340/185{,}673) / (175{,}797/197{,}654)] - 1\} = 13.4\% \text{ increase}$$

If, instead, every driver is replaced by a _male one year older_, the weighted probabilities sum up to 178,723 drivers and 194,975 RF passengers. The average additional risk for a male driver aging one year is:

$$100 \times \{[(178{,}723/194{,}975) / (175{,}797/197{,}654)] - 1\} = 3.06\% \text{ increase}$$

Similar formulas may be applied to estimate the average additional risks of a female driver or RF passenger aging one year.

2.9   Estimation of confidence bounds

The average increase in risk for aging or for females is estimated by a complex algorithm that consists of: (1) running regression analyses on a database of FARS driver-RF-pair cases; (2) using the estimated regression coefficients to build formulas that estimate an occupant's probability of fatality; (3) applying the formulas to another FARS database (all CY 2001-2010 driver-RF pairs) to estimate the number of driver and RF passenger fatalities that would likely have occurred in various imaginary situations, such as all the drivers being male, or all the drivers being a year older than they actually were; and (4) using double-pair comparison to estimate the increase in fatalities if, for example, all drivers had been female, relative to if all drivers had been male.

These estimates can be construed to have sampling errors and confidence bounds. Even though all statistics are based on FARS, which is a census, not a sample, NHTSA analyses often treat FARS data as if it were a sample and apply customary statistical tests. For example, a chi-square could be applied to see if the actual joint distribution of two categorical variables is close to what would be expected based on the marginal proportions. Sometimes, the crashes that actually occur in the course of some epoch (in this case, 1975 to 2010) are construed as a sample of the crashes that would have occurred if more-or-less the same national crash environment of that epoch had been repeated over and over – in parallel universes, so to speak – each time resulting in somewhat different numbers and distributions of fatal crashes. Specifically, it is the first step of the algorithm – the regression analyses – that produces coefficients that may have considerable uncertainty, as evidenced by their standard errors and Wald chi-squares generated by the SAS® LOGISTIC procedure. The remaining steps are purely mechanistic or, at most, might contribute negligible uncertainty (the distribution of CY 2001 to 2010 fatalities by age and gender).

Thus, sampling error can be estimated by running the regression analyses for various alternative sets of FARS data instead of the master calibration database always used up this point, the one consisting of 154,467 vehicles in which the driver's and RF passenger's age are both in the range of 21 to 96. Each alternative database will produce slightly different regression coefficients. The remaining steps of the estimation procedure are executed for each of the alternative sets of regression coefficients, and the variation in the final estimates of average increased risk is assessed to estimate confidence bounds.

Specifically, the various alternative sets of FARS data are generated by a jackknife technique, because of the complexity of the estimator (a logistic regression coefficient) and the need for ample data to drive the regression. The FARS cases are subdivided into 10 systematic random subsamples of equal size, based on the last two digits of the case number, ST_CASE – e.g., one of these subsamples might consist of all FARS cases with ST_CASE ending in 09, 20, 39, 52, 53, 71, 78, 79, 84, or 95. Ten regressions are performed, each using the 9/10 of the FARS data that remain after one of the subsamples is removed. The subsample is then replaced before the next subsample is removed. The 10 regressions yield 10 estimates of the regression coefficients – the intercept and the coefficients for AGE1, AGE3, FEM1, FEM3, AGE1_35, AGE3_35, FEM1_35, and FEM3_35 – each of which is slightly different from the original coefficient based on the full FARS data. The 10 sets of alternative coefficients are each used in the remainder of the estimation algorithm to obtain 10 estimates of the key statistics, such as the risk increase for female drivers relative to males, or the risk increase for a male aging one year. If, for example, the original key statistic is x and that statistic changes to x + h when all FARS cases are used except those with ST_CASE ending in 09, 20, 39, 52, 53, 71, 78, 79, 84, or 95, a "pseudo-estimate" x – 9h is generated for the subsample including only the FARS cases with ST_CASE ending in 09, 20, 39, 52, 53, 71, 78, 79, 84, or 95 (because if a regression could have been run using only these cases, its coefficients would have generated the key statistic x – 9h in order for it and the x + h generated from the other 9/10 of the data to average out to x). The standard error of these 10 pseudo-estimates serves as standard deviation of the original key statistic and it can be treated as a t-distribution with 9 degrees of freedom (df).

However, the variance estimate obtained by running through the procedure just once could be too high or too low by chance, depending on what cases happened to get into the 10 subsamples. A second iteration of the same procedure, but with FARS split up into subsamples in a different way, might generate a lower or higher estimate. Numerous iterations, each with a different splitting of FARS into subsamples, will generate a range of estimates of the standard error, and the median of these estimates will be used. Specifically, the last two digits of ST_CASE were used to subdivide FARS into 100 groups (numbered 0 to 99). The numbers 0 to 99 were randomly re-ordered by a SAS random-number generator and listed in the new order. The FARS cases whose last two ST_CASE digits were among the first 10 on the new list became subsample 1, the next 10 became subsample 2, and so on. After these 10 subsamples were created, the numbers 0 to 99 were randomly reordered anew and another set of 10 subsamples was created. In all, the procedure was repeated 11 times and it created 11 sets of 10 subsamples each.

For example, in Section 2.8, the point estimate (based on regressions on the full 154,467 FARS cases) of the additional risk for a female driver relative to a male of the same age is 13.442%. The first of the 10 subsamples in the first of the 11 iterations consists of cases with ST_CASE ending in 09, 20, 39, 52, 53, 71, 78, 79, 84, or 95. Based on regressions on the remaining 9/10 of the data, the estimated additional risk for females changes to 13.158%. Thus the pseudo-estimate of females' added risk for the subsample of cases ending in 09, 20, 39, 52, 53, 71, 78, 79, 84, or 95 is 15.998%. The remaining 9 pseudo-estimates for the first of the 11 iterations range from 8.650% to 17.770%. The standard error of these 10 pseudo-estimates is .8854 percentage points. The remaining 10 iterations of the same procedure yield standard errors ranging from .6247 to 1.0438 percentage points. The median of these 11 estimates of the standard error is .8962 percentage points.[28] The confidence bounds for the average additional risk of a female driver relative to a male of the same age are:

$$13.442 \pm 2.262 \times .8962 = 13.4 \pm 2.0 \text{ percent}$$

where 2.262 is the 97.5th percentile of a t-distribution with 9 df.

---

[28] Kahane, C. J. (2012, August). *Relationships Between Fatality Risk, Mass, and Footprint in Model Year 2000-2007 Passenger Cars and LTVs – Final Report*. (Report No. DOT HS 811 665). Washington, DC: National Highway Traffic Safety Administration. Available at www-nrd.nhtsa.dot.gov/Pubs/811665.PDF P. 82 uses a jackknife technique to estimate confidence bounds for statistics derived from logistic regressions on FARS; Kahane, C. J. (2009). *The Long-Term Effect of ABS in Passenger Cars and LTVs*. (Report No. DOT HS 811 182). Washington, DC: National Highway Traffic Safety Administration. Available at www-nrd.nhtsa.dot.gov/Pubs/811182.PDF. Pp. 26-27 iterates the error estimation 11 times and takes the median of the 11 results.

2.10 Additional statistics

The average risk increase for a male aging one year, for a female aging one year, and for a female relative to a male of the same age – each computed for a driver and for an RF passenger – constitute the six basic key statistics. These will be supplemented by six "averages of the averages": (1) the overall average effect for a male aging one year, which is the average of the effects for the driver and the RF passenger; (2) the overall average effect for a female aging one year; (3) the overall average effect for a driver aging one year, which is the average of the effects for a male driver and a female driver; (4) the overall average effect of a RF passenger aging one year; (5) the grand overall average effect of aging one year, which is the average of the effects for the male driver, male RF passenger, female driver, and female RF passenger; and (6) the overall average increase for a female relative to a male of the same age, which is the average of these effects for the driver and the RF passenger. In all cases, these averages are simple arithmetic means of the point estimates and do not take into account that there are more drivers than RF passengers or more males than females in fatal crashes. (Arithmetic averages have the advantage that they would not change over time even if the gender or seat-position distribution of the population changes.) In all cases, the confidence bounds for the averages are computed by the same jackknife procedure as for the six basic statistics – i.e., by computing the point estimates for each average based on regression coefficients estimated from 11 iterations of 10 subsets.

Another area of interest is how the effect of aging or being female changes with age. The average effect of, say, male RF passengers aging by one year is estimated in Section 2.8 by summing probabilities of fatality over the file of 58,438 driver-RF pairs in 2001-2010 FARS and then performing a double-pair comparison analysis on the sums. But this computation could also be performed on any subset of the 58,438 pairs. For example, a computation limited to the pairs where the RF passenger is 21 years old would yield an estimate of the average effect of a 21-year-old male RF passenger aging one year. Separate computations for each year of RF passenger age will trace how the effect of a male aging one year changes with age; likewise for female RF passenger aging one year; likewise for male or female drivers; or for the added risk of a female relative to a male of the same age. Table 2-1 tracks the percent risk increase year by year from 21 to 96: (1) the increase for male drivers and RF passengers aging one year; (2) the increase for a female driver or RF passenger relative to a male of the same age in the same seat position; and (3) the increase for female drivers and RF passengers aging one year. Figure 2-9 graphs the effects of a male driver (blue diamonds) or passenger (green circles) aging one year; Figure 2-10 graphs the added risk for a female driver relative to a male driver of the same age (blue diamonds) or a female passenger relative to a male passenger of the same age (green circles).

By design of our regression models, the effects in the first four columns of Table 2-1 are nearly constant up to age 35 and then follow a nearly linear trend: the effect of aging intensifies with age, whereas the added risk for a female steadily decreases and, in the case of drivers, eventually becomes negative.

TABLE 2-1: PERCENT RISK INCREASE FOR A MALE AGING ONE YEAR,
RISK INCREASE FOR A FEMALE RELATIVE TO A MALE OF THE SAME AGE, AND
PERCENT RISK INCREASE FOR A FEMALE AGING ONE YEAR
DRIVERS AND RF PASSENGERS, ALL CARS AND LTVs, BY OCCUPANT AGE

	MALE AGING 1 YEAR		FEMALE RELATIVE TO MALE		FEMALE AGING 1 YEAR	
AGE	DRIVER	RF	DRIVER	RF	DRIVER	RF
21	2.9264	2.9226	26.2113	29.4832	2.8900	2.8555
22	2.9152	2.9160	26.0951	29.4123	2.8778	2.8502
23	2.9135	2.9146	26.0923	29.4052	2.8787	2.8513
24	2.9145	2.9076	26.1097	29.3254	2.8815	2.8445
25	2.8968	2.8950	25.9205	29.1773	2.8610	2.8305
26	2.8922	2.8871	25.8770	29.0903	2.8569	2.8237
27	2.8828	2.8866	25.7877	29.0900	2.8481	2.8246
28	2.8761	2.8713	25.7231	28.9118	2.8418	2.8084
29	2.8575	2.8642	25.5307	28.8320	2.8215	2.8015
30	2.8617	2.8593	25.5906	28.7790	2.8293	2.7971
31	2.8356	2.8524	25.3045	28.7068	2.7976	2.7925
32	2.8353	2.8492	25.3093	28.6740	2.7989	2.7899
33	2.8278	2.8446	25.2372	28.6201	2.7920	2.7846
34	2.8038	2.8380	24.9870	28.5511	2.7650	2.7798
35	2.8034	2.8557	24.8657	28.4115	2.0714	2.4000
36	2.8400	2.9195	24.1535	27.9731	2.1093	2.4665
37	2.8450	2.9603	23.1205	27.2749	2.1174	2.5080
38	2.8615	3.0195	22.2224	26.7905	2.1370	2.5692
39	2.8689	3.0678	21.2587	26.1880	2.1489	2.6161
40	2.8783	3.1152	20.3236	25.5947	2.1613	2.6648
41	2.9165	3.1727	19.6460	25.1134	2.2022	2.7265
42	2.9138	3.2205	18.6513	24.5447	2.2059	2.7773
43	2.9392	3.2638	17.8810	23.9458	2.2341	2.8232
44	2.9291	3.3179	16.8703	23.4396	2.2299	2.8779
45	2.9380	3.3567	16.0058	22.8267	2.2421	2.9205
46	2.9450	3.4174	15.1573	22.3868	2.2549	2.9832
47	2.9834	3.4682	14.5043	21.8797	2.2957	3.0379
48	2.9842	3.5123	13.6186	21.3228	2.2989	3.0825
49	3.0076	3.5531	12.8974	20.7575	2.3270	3.1254
50	3.0053	3.6310	12.0341	20.4477	2.3285	3.2095
51	2.9981	3.6721	11.1976	19.8910	2.3305	3.2513
52	3.0067	3.6935	10.4172	19.2474	2.3415	3.2814
53	3.0453	3.7624	9.7592	18.8583	2.3783	3.3461
54	3.0461	3.7950	8.9815	18.2957	2.3851	3.3859
55	3.0196	3.8654	8.1455	17.9236	2.3692	3.4560
56	3.0792	3.9104	7.5731	17.4255	2.4252	3.5026
57	3.0364	3.9500	6.7108	16.9212	2.3941	3.5500
58	3.1054	4.0244	6.1609	16.5552	2.4581	3.6190
59	3.1083	4.0651	5.4229	16.0642	2.4640	3.6671
60	3.1415	4.0757	4.7836	15.4477	2.4981	3.6821
61	3.1586	4.1400	4.1060	15.0566	2.5174	3.7472
62	3.2123	4.2031	3.4510	14.6616	2.5651	3.8118
63	3.1975	4.2664	2.7384	14.2605	2.5578	3.8739
64	3.2388	4.2941	2.1666	13.7499	2.6014	3.9084
65	3.2944	4.3678	1.5073	13.3888	2.6514	3.9812
66	3.2900	4.4269	0.8098	12.9829	2.6519	4.0423
67	3.3257	4.4717	0.1580	12.5335	2.6866	4.0893
68	3.3636	4.5032	-0.4387	12.0574	2.7253	4.1264
69	3.4143	4.5457	-1.1280	11.6166	2.7712	4.1726
70	3.3781	4.6004	-1.7502	11.2119	2.7469	4.2305
71	3.4266	4.6343	-2.3884	10.7542	2.7924	4.2664
72	3.4701	4.7236	-3.0479	10.4278	2.8336	4.3533
73	3.4543	4.7407	-3.6547	9.9508	2.8257	4.3782
74	3.5252	4.8230	-4.3620	9.6109	2.8899	4.4601
75	3.5382	4.8346	-5.0044	9.1296	2.9058	4.4770
76	3.5446	4.9312	-5.5644	8.8115	2.9165	4.5702
77	3.5834	4.9628	-6.2357	8.3824	2.9548	4.6074
78	3.6128	5.0017	-6.8660	7.9713	2.9853	4.6504
79	3.6135	5.0682	-7.4739	7.6008	2.9905	4.7162
80	3.6793	5.0972	-8.1869	7.1791	3.0532	4.7492
81	3.6530	5.1627	-8.6998	6.8174	3.0346	4.8160
82	3.6842	5.2671	-9.2949	6.4969	3.0667	4.9170
83	3.6762	5.2919	-9.8101	6.0883	3.0636	4.9469
84	3.7354	5.3693	-10.5034	5.7314	3.1217	5.0229
85	3.7886	5.4529	-11.2763	5.3965	3.1769	5.1066
86	3.8231	5.5044	-11.9047	5.0191	3.2120	5.1592
87	3.7306	5.5747	-11.9895	4.6560	3.1302	5.2281
88	3.7987	5.6524	-12.7385	4.3000	3.1960	5.3041
89	3.8050	5.7974	-13.2325	4.0064	3.2061	5.4434
90	3.7543	5.8134	-13.5361	3.5986	3.1610	5.4605
91	3.9031	5.9577	-14.6468	3.2872	3.3095	5.5975
92	3.6929	6.1367	-14.0615	2.9875	3.1079	5.7671
93	3.7895	6.2197	-14.7233	2.6179	3.1939	5.8467
94	3.8900	6.4410	-15.5474	2.3013	3.2926	6.0550
95	3.8573	6.5760	-15.7383	1.9355	3.2604	6.1819
96	4.1015	6.7323	-17.2278	1.5598	3.4879	6.3297

FIGURE 2-9: FATALITY INCREASE (%) FOR EACH YEAR
THAT A MALE DRIVER (blue diamonds) OR RF PASSENGER (green circles) GETS OLDER

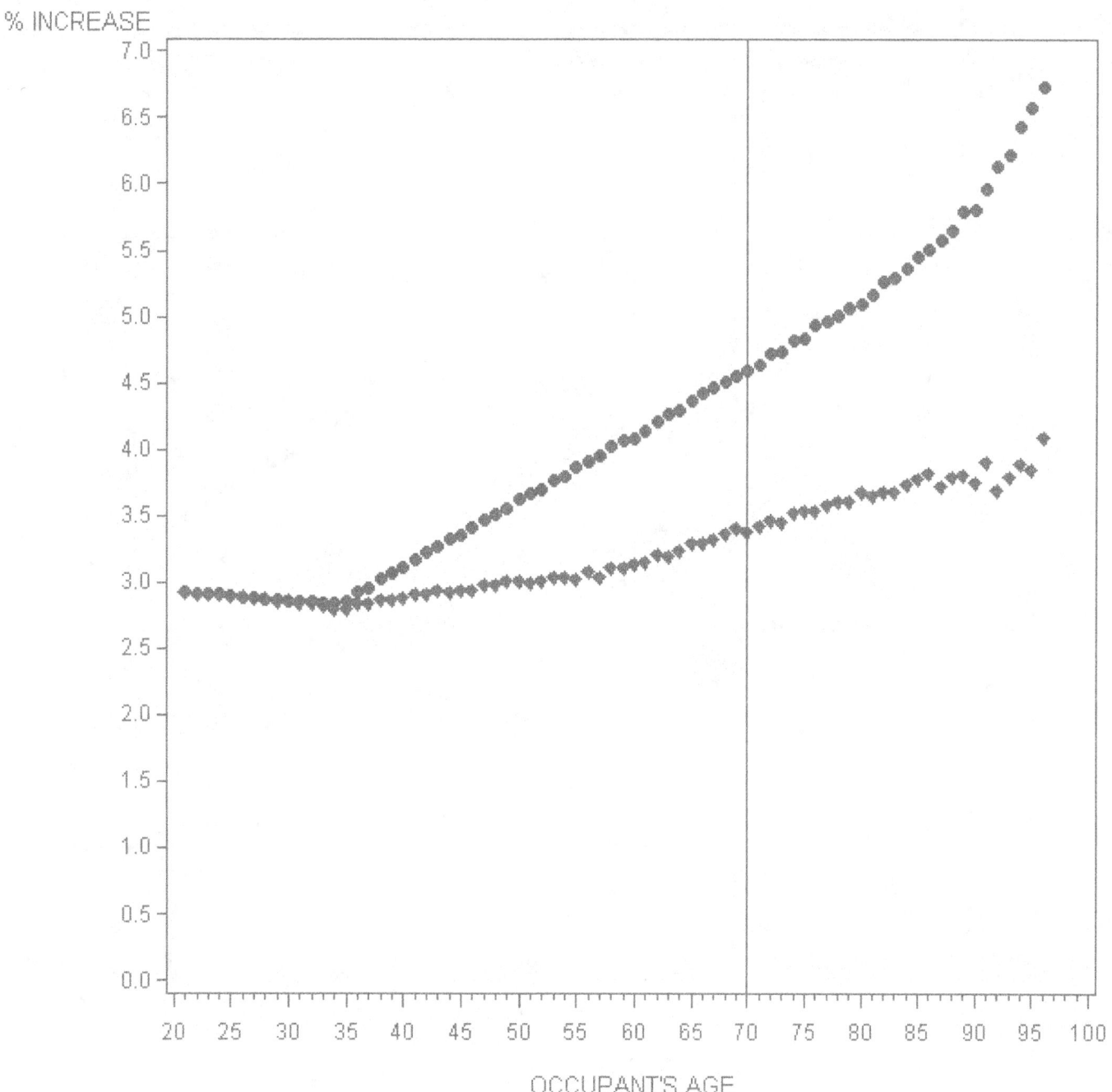

FIGURE 2-10: FATALITY INCREASE (%) FOR A FEMALE DRIVER
RELATIVE TO A MALE DRIVER (blue diamonds)
OR A FEMALE RF PASSENGER RELATIVE TO A MALE RF PASSENGER (green circles)

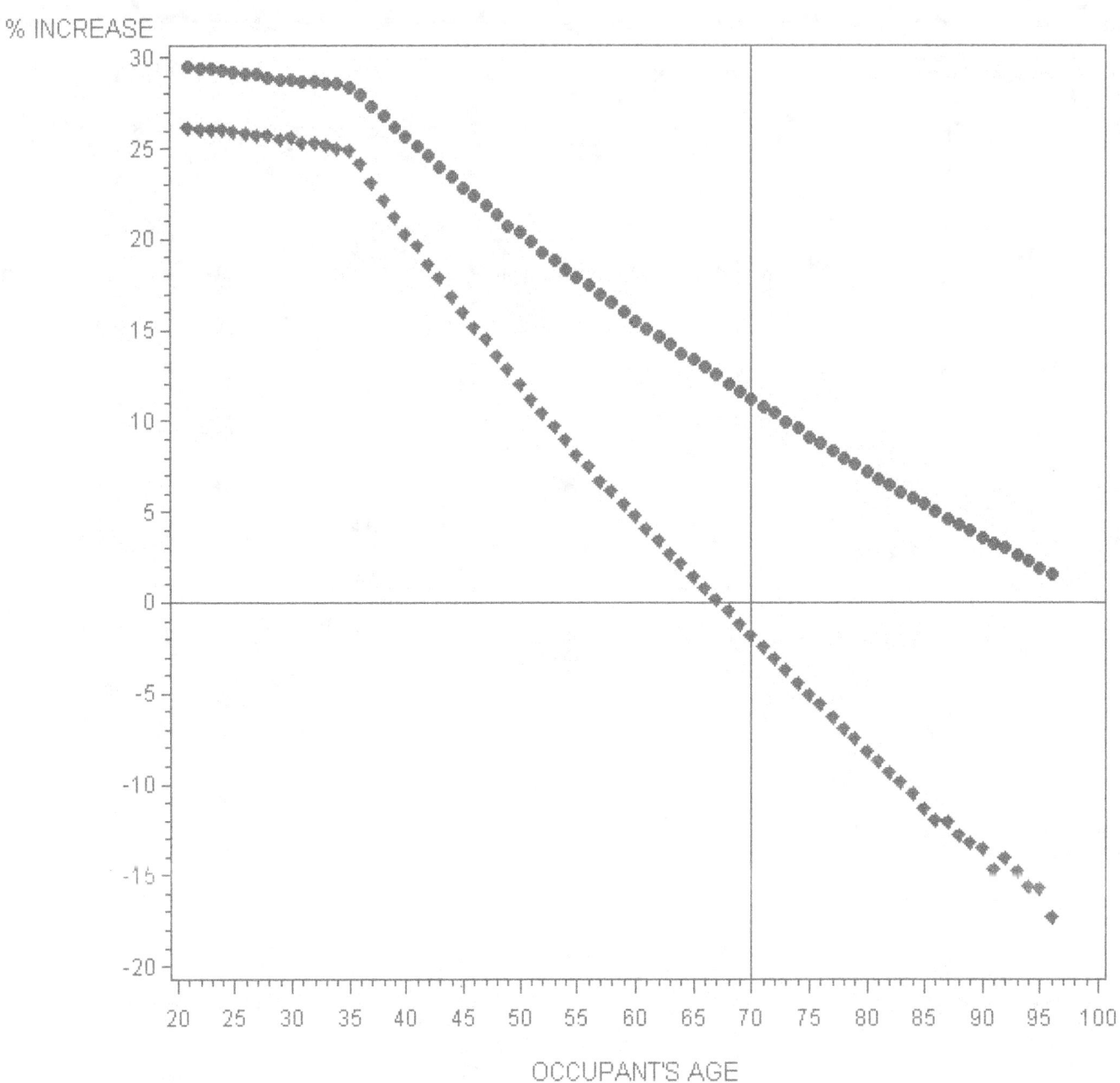

The slight year-to-year variations in the first four columns of Table 2-1 and in Figures 2-9 and 2-10 before age 35 as well as the slight departures from linear trends after 35 are because: (1) the effects are not simply the regression coefficients, but double-pair comparisons derived from complex formulas using these coefficients; (2) the results depend to a slight extent on the actual pairs of occupants in the 2001-2010 FARS data, and at the higher ages, where there are relatively few cases, the age distribution of passengers riding with, say, 91-year-old drivers may differ from the corresponding distribution riding with 92-year-old drivers. In reality, of course, the trend would not abruptly change from constant to linear at exactly age 35 but would transition over a period of several years; however, our regression models have made the simplifying assumption of a linear term beginning at age 35 to create a model with relatively few parameters that comes close to tracking the real trend, as discussed in Section 2.7.

The last two columns of Table 2-1 track the percent risk increase for a female aging one year, from 21 to 96, year by year. By design of our regression models, these effects are essentially derived from the first four columns (and are shown in italics for that reason): the effect for a female aging a year is the effect for a male aging one year minus the rate of change in the added risk of a female relative to a male. By design, that rate of change is close to zero up to age 35 and then fairly constant: about .7 percentage points per year for drivers and about .35 percentage points per year for RF passengers. Thus, the model's estimate of the year-to-year increase for females is almost identical to males' up to age 35 and then drops abruptly by .7 percentage points below the males' rate for drivers, .35 for RF passengers. Again, in reality, the rate of increase for females would not drop abruptly at exactly age 35 but would transition to the lower level over a span of several years; the estimates in Table 2-1 reflect the simplifying assumptions of our regression models.

Intuitively, these findings are consistent with the concepts of **fragility** (likelihood of injury given a specific physical insult) and **frailty** (likelihood of death given a specific injury), but they do not quantify what role each one plays. It is reasonable to presume that risk increases with age because people become more fragile; however, the intensification of the aging effect at the higher ages could indicate an accelerating trend to greater fragility and/or an added effect of frailty. It is likely that the increased risk for females is highest at ages 21 to 35 because, in the prime of life, males are substantially less fragile; but the steady and rapid closing of this gap at older ages could indicate that males' fragility increases more rapidly with aging and/or that females are substantially slower to become frail.

Figures 2-9 and 2-10 reveal another pervasive trend in the data: diverging results for drivers and RF passengers at the higher ages. Figure 2-9 shows the risk increase of aging one year is nearly identical for drivers and RF passengers, approximately 2.9 percent, up to age 35. But after 35, the added risk increases for passengers at a much steeper rate than for drivers. By age 70, for example, the annual increase is 4.6 percent for passengers but still only 3.4 percent for drivers. Figure 2-10 shows that the added risk for females relative to males is fairly similar for drivers and RF passengers up to age 35, approximately 26 and 29 percent, respectively – a difference that is probably within the sampling error of these statistics. But after age 35, the added risk drops off much faster for drivers – crossing zero and indicating lower risk for females than males from age 68 onward – than for RF passengers, where it never quite reaches zero.

These effects may to some extent be due to genuine differences in the crash environments faced by the driver and the RF passenger. But they may also be an artifact of differences between the occupants of these seats. Just because two people are the same age and gender does not necessarily make them equally fragile and frail. When two elderly people of similar age but substantially different health ride together in a vehicle it seems reasonable that the healthier person would more often be the driver. For example, a husband used to do most of the driving, but now he is quite ill and his wife does the driving. Thus, the effect of aging could accelerate more rapidly for RF passengers because they are the more fragile and frail subset of their age cohort; the gender gap could close more quickly for drivers if it is the healthier females who drive more.

Because FARS provides no information about the health history of the occupants, we cannot quantify to what extent the diverging results for drivers and RF passengers are due to the vehicle and to what extent they are due to differences among the occupants.

Just as risk may be computed at each individual year of age, as in Table 2-1, it may also be computed by age group. Table 2-2 shows the average effects of a male aging one year and the added risk for a female relative to a male of the same age for age groups 21-30, 31-44, 45-54, 55-64, 65-74, and 75-96. The age groups 21-30 and 65-74 are of particular interest, as the results may be considered benchmarks of the effects for "young adults" and the "young old."

TABLE 2-2: PERCENT RISK INCREASE FOR A MALE AGING ONE YEAR AND
RISK INCREASE FOR A FEMALE RELATIVE TO A MALE OF THE SAME AGE
DRIVERS AND RF PASSENGERS, ALL CARS AND LTVs, BY OCCUPANT AGE GROUP

	MALE AGING 1 YEAR		FEMALE RELATIVE TO MALE	
AGE GROUP	DRIVER	RF	DRIVER	RF
21-30	2.8983	2.8975	25.9392	29.2074
31-44	2.8639	3.0117	22.0921	26.7781
45-54	2.9929	3.5682	12.6183	20.7257
55-64	3.1209	4.0722	5.3297	15.8467
65-74	3.3933	4.5820	-1.4265	11.4084
75-90	3.0704	5.1683	-9.0323	6.5472

Another valuable statistic is the <u>cumulative</u> increase in fatality risk for aging from 21 up to a specific age. It is computed by consecutively applying the risk increases for each of the individual years – as shown in Table 2-1 for males, for example – up to that age: like calculating the accumulation of compound interest at variable rates. Table 2-3 shows the risk for an N-year-old relative to a 21-year-old, for male and female drivers and RF passengers. Figure 2-11 graphs the risk ratios for male drivers (blue diamonds), female drivers (red circles), male RF passengers (cyan triangles) and female RF passengers (pink squares), with a logarithmic scale for the y-axis (which depicts a rate that grows by a constant percentage each year – like compound interest – as a straight line). Age 75 is of particular interest, as the results may be considered a benchmark of the vulnerability of an elderly occupant relative to a young adult.

TABLE 2-3: FATALITY RISK, GIVEN SIMILAR PHYSICAL INSULTS, RELATIVE TO A
21-YEAR-OLD OCCUPANT OF THE SAME GENDER IN THE SAME SEAT POSITION
DRIVERS AND RF PASSENGERS, ALL CARS AND LTVs, BY OCCUPANT AGE

	MALES		FEMALES	
AGE	DRIVER	RF	DRIVER	RF
21	1.000	1.000	1.000	1.000
22	1.029	1.029	1.029	1.029
23	1.059	1.059	1.059	1.058
24	1.090	1.090	1.089	1.088
25	1.122	1.122	1.120	1.119
26	1.154	1.154	1.152	1.151
27	1.188	1.188	1.185	1.183
28	1.222	1.222	1.219	1.217
29	1.257	1.257	1.254	1.251
30	1.293	1.293	1.289	1.286
31	1.330	1.330	1.326	1.322
32	1.368	1.368	1.363	1.359
33	1.407	1.407	1.401	1.397
34	1.446	1.447	1.440	1.435
35	1.487	1.488	1.480	1.475
36	1.529	1.530	1.510	1.511
37	1.572	1.575	1.542	1.548
38	1.617	1.622	1.575	1.587
39	1.663	1.671	1.609	1.628
40	1.711	1.722	1.643	1.670
41	1.760	1.776	1.679	1.715
42	1.811	1.832	1.716	1.761
43	1.864	1.891	1.753	1.810
44	1.919	1.953	1.793	1.861
45	1.975	2.017	1.833	1.915
46	2.033	2.085	1.874	1.971
47	2.093	2.156	1.916	2.030
48	2.155	2.231	1.960	2.091
49	2.220	2.310	2.005	2.156
50	2.287	2.392	2.052	2.223
51	2.355	2.478	2.099	2.295
52	2.426	2.569	2.148	2.369
53	2.499	2.664	2.199	2.447
54	2.575	2.765	2.251	2.529
55	2.653	2.870	2.305	2.614
56	2.733	2.980	2.359	2.705
57	2.818	3.097	2.416	2.800
58	2.903	3.219	2.474	2.899
59	2.993	3.349	2.535	3.004
60	3.086	3.485	2.598	3.114
61	3.183	3.627	2.662	3.229
62	3.284	3.777	2.729	3.350
63	3.389	3.936	2.799	3.477
64	3.498	4.104	2.871	3.612
65	3.611	4.280	2.946	3.753
66	3.730	4.467	3.024	3.903
67	3.853	4.665	3.104	4.060
68	3.981	4.873	3.187	4.226
69	4.115	5.093	3.274	4.401
70	4.255	5.324	3.365	4.585
71	4.399	5.569	3.458	4.778
72	4.550	5.827	3.554	4.982
73	4.708	6.103	3.655	5.199
74	4.870	6.392	3.758	5.427
75	5.042	6.700	3.867	5.669
76	5.220	7.024	3.979	5.923
77	5.405	7.371	4.095	6.193
78	5.599	7.736	4.216	6.479
79	5.801	8.123	4.342	6.780
80	6.011	8.535	4.472	7.100
81	6.232	8.970	4.608	7.437
82	6.460	9.433	4.748	7.795
83	6.698	9.930	4.894	8.178
84	6.944	10.456	5.044	8.583
85	7.203	11.017	5.201	9.014
86	7.476	11.618	5.366	9.474
87	7.762	12.257	5.539	9.963
88	8.051	12.941	5.712	10.484
89	8.357	13.672	5.895	11.040
90	8.675	14.465	6.084	11.641
91	9.001	15.306	6.276	12.277
92	9.352	16.217	6.484	12.964
93	9.698	17.213	6.685	13.712
94	10.065	18.283	6.899	14.513
95	10.457	19.461	7.126	15.392
96	10.860	20.741	7.358	16.344

43

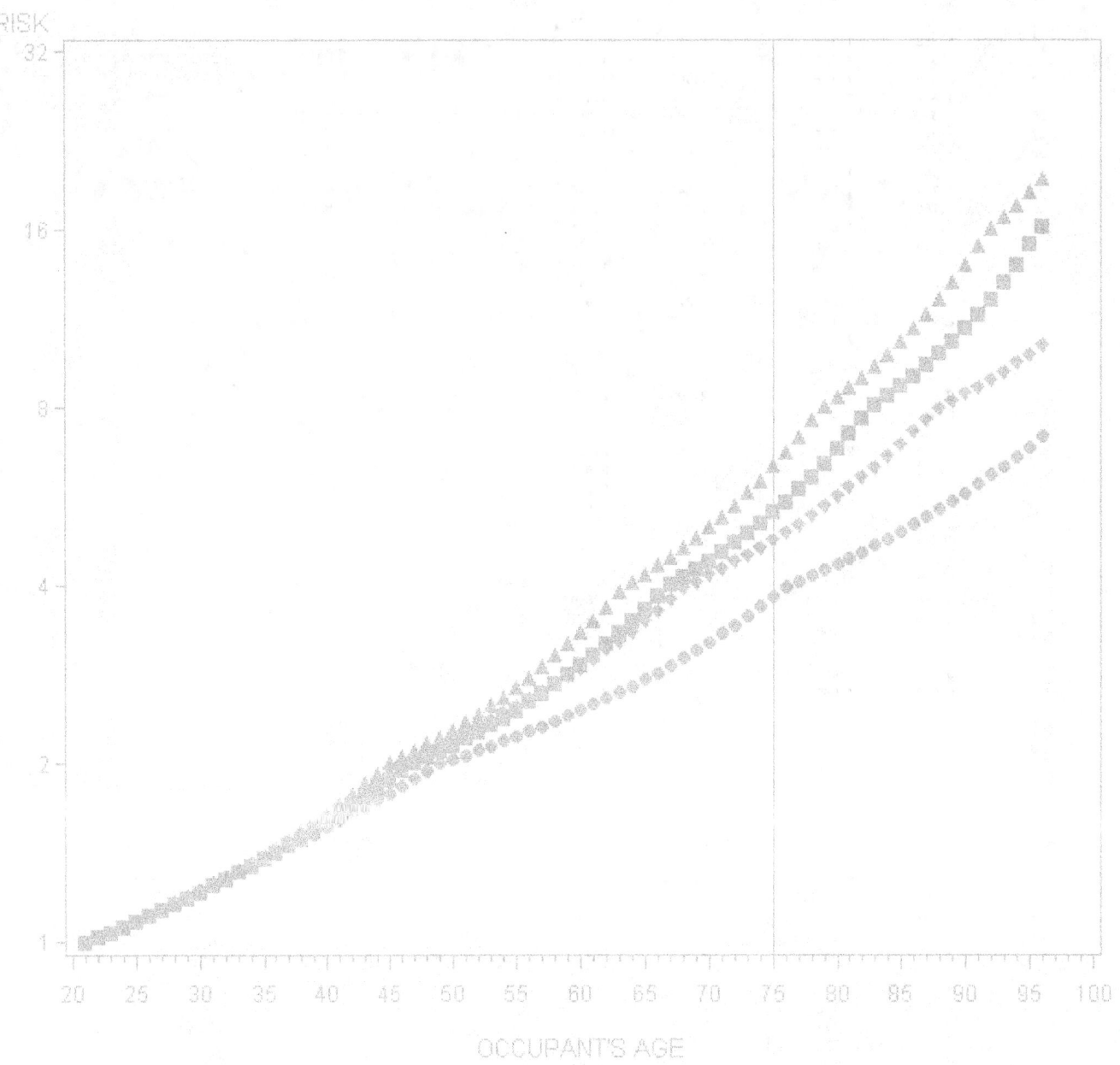

FIGURE 2-11: FATALITY RISK RELATIVE TO A 21-YEAR-OLD
FOR A MALE DRIVER (blue diamonds), MALE RF PASSENGER (cyan triangles),
FEMALE DRIVER (red circles), AND FEMALE RF PASSENGER (pink squares)
WITH A LOGARITHMIC SCALE FOR THE Y-AXIS

Table 2-4 summarizes the statistics that will be used to demonstrate the added risk of aging one year and the added risk for a female relative to a male of the same age: (1) the average effects across ages 21 to 96, and their confidence bounds; (2) the average effects for age groups 21 to 30 and 65 to 74, for a male aging one year and for a female relative to a male, as point estimates (no confidence bounds); and (3) fatality risk for a 75-year-old relative to a 21-year-old, as point estimates. The statistics are computed here for the complete database of MY 1960-2011 cars and LTVs in CY 1975-2010 FARS. Chapter 3 will present and compare the corresponding statistics for numerous subgroups of vehicles, crashes, and occupant protection systems.

A characteristic of Table 2-4 and the majority of the analyses in Chapters 3 to 5 is that the effects of aging one year are lower for females than for males at the same seat position. For example, the average risk increase is 3.06 percent a year for male drivers, 2.60 percent for female drivers. Also, the cumulative increase up to 75 is lower for females than for males. Intuitively, females age more slowly than males, at least in terms of vulnerability to physical insult. Computationally, the increased risk for a female relative to a male begins to shrink at age 35 by approximately .7 percentage points per year for drivers and .35 percentage points for passengers; essentially, from age 35 onwards, these .7 and .35 percentage points are subtracted from males' risk increase with age (which runs 3 to 6% per year) to derive females' risk increase with age.

2.11   Modifications to analyze other data files

Sections 3.1 to 3.6 of Chapter 3 will apply exactly the methods described above to produce statistics on the effects of aging and gender for drivers and RF passengers in various subgroups of vehicles, crashes, and occupant protection systems. In other words, the regression analyses (piecewise linear with breakpoints set at age 35) are applied each time to some subset of the database of 154,467 MY 1960-2011 cars and LTVs on 1975-2010 FARS in which the driver's and RF passenger's age are both in the range of 21 to 96 and at least one of them was a fatality, generating a list of regression coefficients. The coefficients, however, are always applied to the full database of 58,438 records of cars and LTVs on 2001-2010 FARS that had a driver and a RF passenger, at least one of them a fatality, to obtain, by double-pair comparison, estimates of the risk increase for aging one year or for females relative to males of the same age.

Section 3.7 will provide comparable statistics for back-seat, outboard occupants of passenger cars, also based on FARS. Chapter 4 examines the specific injuries of fatally injured front-seat occupants, based on CY 1987-2007 FARS-MCOD (Multiple Cause of Death) data and Chapter 5 analyzes injuries, most of them nonfatal, by AIS level on CY 1988-2010 NASS-CDS files. All these analyses use essentially the same method. They start with logistic regressions on driver-passenger pairs, with the same independent variables, and the dependent variable being whether or not a specific type of injury occurred – and for all pairs included in the regression, either the driver, or the passenger, or both, had to have the specific type of injury in question. They apply the regression coefficients to the full database of 58,438 CY 2001-2010 FARS records to obtain, by double-pair comparison, estimates of the risk increase for aging or for females. Always using the same 58,438-record database for the computations allows direct comparison of the analysis results, even if the regressions were performed on different files.

TABLE 2-4: SUMMARY STATISTICS FOR ALL CARS AND LTVs, MY 1960-2011
INCREASE IN FATALITY RISK FOR EACH YEAR THAT AN OCCUPANT GETS OLDER
AND FOR A FEMALE RELATIVE TO A FEMALE OF THE SAME AGE
(Occupants 21 to 96, FARS 1975-2010)

Percent Fatality Increase per Year of Aging

	Males			Females			Average of Males & Females		
	Driver	RF Pass	Average	Driver	RF Pass	Average	Driver	RF Pass	Average
	3.06 ± .09	3.51 ± .09	3.28 ± .08	2.60 ± .10	3.27 ± .08	2.93 ± .08	2.83 ± .08	3.39 ± .07	3.11 ± .08

Percent Fatality Increase per Year of Aging By Occupant Age Group →→→→→→→

	Driver			RF Passenger		
	Avg 21-96	21-30	65-74	Avg 21-96	21-30	65-74
Males	3.06 ± .09	2.90	3.39	3.51 ± .09	2.90	4.58

Risk at Age 75 ÷ Risk at Age 21

	Males		Females	
	Drivers	RF Pass	Drivers	RF Pass
	5.04	6.70	3.87	5.67

Percent Fatality Increase for Females →→→→→→→

	Drivers			RF Passengers			Avg of Driver & RF
Occupant Age Group →→→→→→→	21-96	21-30	65-74	21-96	21-30	65-74	21-96
	13.4 ± 2.0	25.9	- 1.4	20.5 ± 2.2	29.2	11.4	17.0 ± 1.5

Subsequent chapters will describe in detail how injuries are defined and how the MCOD and CDS databases are created. The discussion here is limited to how the mechanics of the estimation algorithm needs to be modified.

The risk analysis of back-seat occupants on FARS starts with regressions on driver-passenger pairs, where the passenger is an occupant of either of the outboard rear seats – and where either the driver or that passenger, or possibly both, are fatalities. Unlike the front-seat analysis which is limited to one record per vehicle, multiple records may be created for one vehicle if there are multiple back-seat occupants. From this point onward, the analysis proceeds without additional change through the remainder of the algorithm: except that the passenger statistics will now be for the back seat rather than the RF seat, and the driver statistics will be of little interest.

The analyses of FARS-MCOD are limited to drivers and RF passengers. MCOD is supplementary injury information obtained by NHTSA from the National Center for Health Statistics for a subset of the person records on FARS (namely, the fatalities). The MCOD records can be merged to the FARS person-level records by ST_CASE, VEH_NO, and PER_NO. All the FARS-MCOD cases analyzed in this report will be a subset of the 154,467 FARS cases used in the regressions for drivers and RF passengers. Thus, the mechanics of the FARS-MCOD analyses are exactly the same as for FARS, with the caution that any regression analysis for a specific type of injury must be limited to the driver-RF pairs where at least one or possibly both occupants have that type of injury.

The CDS analyses are also limited to driver-RF pairs. The regression analyses will use unweighted CDS cases, as will be discussed in Section 5.1. With unweighted CDS data, the analyses used to obtain point estimates of statistics will be identical to the FARS analyses, again with the caution that at least one or possibly both of a driver-RF pair must have the type of injury that is the dependent variable in the regression. For confidence bounds, however, some modifications are in order. CDS does not have a ST_CASE variable to number the cases. Furthermore, CDS is not a census, but a national cluster sample of primary sampling units (PSU). During CY 1988-2010, there were a total of 36 different PSUs in NASS in various years, including 24 that furnished data in all of those years. Statistics from a cluster sample usually have larger sampling error than a simple random sample of the same N. For confidence bounds that take the cluster sampling into account, it is necessary to subdivide the PSUs, not the individual cases into systematic random subsamples, in a manner that each subsample contains approximately the same number of individual cases. Moreover, it is a bit easier to subdivide the PSUs into 8 rather than 10 systematic random subsamples of fairly similar size, because several of the PSUs each happen to furnish over 5 percent of the cases.

NASS identifies each PSU with a two-digit number. The numbers 0 to 99 are randomly re-ordered by a SAS random-number generator and listed in the new order, thereby scrambling the list of PSUs. The scrambled list is split into 8 groups of consecutive (by the new order) PSUs, each totaling as close as possible to 1/8 of the cases. The jackknife technique performs 8 regressions, each using the 7/8 of the CDS data that remain after one of the subsamples is removed. The 8 sets of alternative coefficients are each used in the remainder of the estimation algorithm to obtain 8 estimates of the key statistics. If the original key statistic is x and that statistic changes to $x + h$ with 7/8 of the data, a "pseudo-estimate" $x - 7h$ is generated for the subsample consisting of 1/8 of the data (because if a regression could have been run using only these cases, its coefficients would have generated the key statistic $x - 7h$ in order for it and the $x + h$ generated from the other 7/8 of the data to average out to x). The standard error of these 8 pseudo-estimates serves as the standard deviation of the original key statistic and it can be treated as a t-distribution with 7 degrees of freedom (df). Here, too, the subsampling process is iterated 11 times and the median of the 11 estimates of the standard deviations is used. If the point estimate is x and the median of the 11 estimates of the standard deviation is s, the confidence bounds for the key statistic are $x \pm 2.365\, s$, where 2.365 is the 97.5[th] percentile of a t-distribution with 7 df.[29]

---

[29] Kahane (2009), pp. 52-54 similarly groups the NASS-GES data, also a cluster sample, into systematic random subsamples of PSUs each containing approximately equal numbers of cases, iterates the error estimation 11 times, and takes the median of the 11 results.

CHAPTER 3

EFFECT OF OCCUPANT AGE AND GENDER ON OVERALL FATALITY RISK – ANALYSES OF 1975-2010 FARS DATA

3.0   Summary

Based on analyses of the relative fatality risk of two occupants of the same vehicle as a function of their ages and genders, risk is estimated to increase by an average of 2.83 ± .08 percent for each year that a driver gets older, 3.39 ± .07 percent for right-front passengers, and 3.46 ± .14 percent for back-seat outboard passengers. Fatality risk is an average of 13.4 ± 2.0 percent higher for a female driver than for a male driver of the same age exposed to similar physical insults; the corresponding increase for RF passengers is 20.5 ± 2.2 percent and for back-seat passengers, 15.7 ± 6.1 percent. The risk increase for aging one year intensifies with age; by contrast, the risk increase for females relative to males declines with age and, in the case of drivers, changes to a reduction beyond 65 to 74. The risk increases for aging and females may have both intensified slightly from vehicles of the 1960s up to about 1990; since then, the added risk for females has substantially diminished, probably to less than half, while the increase for aging may also have decreased, but by a much smaller amount. Air bags have mitigated the risk increases for females and for aging; belt use in the front seat has mitigated the risk increase for females. The effects of aging are similar across the various types of crashes, although perhaps slightly higher in frontals and nearside impacts than other crashes. The added risk for females is slightly higher in nearside impacts than in other crashes.

3.1   Average effect for all cars and LTVs, MY 1960-2011

Chapter 2 defined a database of 154,467 cars and LTVs in 1975-2010 FARS, occupied by a driver and a RF passenger whose age are both in the range of 21 to 96, and at least one or possibly both was a fatality (Section 2.2). Techniques combining logistic regression and double-pair comparison estimate the increase in fatality risk, given similar physical insults, as a male or female occupant ages one year and the fatality risk for a female relative to a male of the same age (Sections 2.1, 2.6, and 2.8), confidence bounds for the estimates (Section 2.9) and supplementary statistics (Section 2.10). The purpose of these tools is to compare the effects of aging, or of being female, for various types of vehicles, crashes, and occupant protection systems. But first, as a benchmark, let us estimate the average effects across 1975-2010 FARS, comprising all cars and LTVs of the past 52 model years.

Table 3-1a estimates the percent increase in fatality risk for a male or female driver or RF passenger aging one year, and its confidence bounds. The first row of Table 3-1a estimates that risk for male drivers increases by 3.06 ± .09 percent for each year of aging; that is the average increase per year for drivers ranging from 21 to 96. The increase for male RF passengers is higher, 3.51 ± .09 percent per year. As discussed in Section 2.10, it is unclear to what extent it is higher due to genuine differences in the crash environments faced by the driver and the RF passenger and to what extent it may reflect that when two people travel together, the one who is especially fragile or frail might tend to sit in the RF seat and let the healthier person drive.

49

TABLE 3-1a: INCREASE IN FATALITY RISK FOR EACH YEAR THAT AN OCCUPANT GETS OLDER **OVERALL AND BY MODEL YEAR** – CARS AND LTVs, DRIVERS AND RIGHT-FRONT PASSENGERS
(Given the same crash scenario, average for occupants 21 to 96, FARS 1975-2010)

Percent Fatality Increase per Year of Aging

	Males			Females			Average of Males & Females		
	Driver	RF Pass	Average	Driver	RF Pass	Average	Driver	RF Pass	Average
ALL CARS AND LTVs, MY 1960-2011	**3.06 ± .09**	**3.51 ± .09**	**3.28 ± .08**	**2.60 ± .10**	**3.27 ± .08**	**2.93 ± .08**	**2.83 ± .08**	**3.39 ± .07**	**3.11 ± .08**
Drivers wtd by RF age-gender distribution	3.10			2.68			2.89		
Cars only	3.04 ± .09	3.56 ± .09	3.30 ± .08	2.65 ± .09	3.31 ± .08	2.98 ± .08	2.85 ± .09	3.44 ± .07	3.14 ± .08
LTVs only	3.12 ± .16	3.28 ± .19	3.20 ± .16	2.41 ± .25	3.17 ± .18	2.79 ± .20	2.77 ± .18	3.22 ± .16	3.00 ± .17
Cars only, MY ≥ 2000	2.99 ± .34	3.14 ± .43	3.06 ± .34	2.40 ± .35	3.18 ± .30	2.79 ± .30	2.69 ± .30	3.16 ± .37	2.92 ± .30
LTVs only, MY ≥ 2000	3.10 ± .57	3.39 ± .49	3.25 ± .52	2.47 ± .49	3.24 ± .45	2.86 ± .44	2.79 ± .48	3.32 ± .47	3.05 ± .45
Cars and LTVs, by model year range									
1960-1966	2.71 ± .37	3.40 ± .40	3.05 ± .35	2.35 ± .60	3.11 ± .37	2.73 ± .42	2.53 ± .44	3.25 ± .35	2.89 ± .36
1967-1974	2.85 ± .17	3.28 ± .22	3.07 ± .16	2.61 ± .29	3.03 ± .18	2.82 ± .20	2.73 ± .18	3.16 ± .19	2.94 ± .17
1975-1979	3.04 ± .21	3.64 ± .15	3.34 ± .16	2.66 ± .25	3.35 ± .17	3.01 ± .18	2.85 ± .18	3.50 ± .17	3.17 ± .17
1980-1984	3.00 ± .21	3.52 ± .20	3.26 ± .20	2.42 ± .32	3.27 ± .21	2.84 ± .24	2.71 ± .24	3.40 ± .20	3.05 ± .20
1985-1989	3.19 ± .22	3.61 ± .20	3.40 ± .18	2.75 ± .25	3.43 ± .16	3.09 ± .17	2.97 ± .20	3.52 ± .16	3.24 ± .17
1990-1994	3.31 ± .29	3.65 ± .29	3.48 ± .27	2.81 ± .27	3.46 ± .28	3.13 ± .25	3.06 ± .27	3.55 ± .28	3.31 ± .25
1995-1999	2.98 ± .29	3.40 ± .31	3.19 ± .26	2.58 ± .33	3.03 ± .27	2.80 ± .29	2.78 ± .27	3.22 ± .27	3.00 ± .28
2000-2004	3.00 ± .42	3.18 ± .36	3.09 ± .38	2.31 ± .40	3.19 ± .32	2.75 ± .33	2.65 ± .38	3.18 ± .32	2.92 ± .33
2005-2011	3.19 ± .59	3.34 ± .59	3.27 ± .55	2.80 ± .60	3.27 ± .51	3.04 ± .53	3.00 ± .56	3.31 ± .53	3.15 ± .53

While it is possible that more safety features have been directed at drivers, because there are more driver than RF passenger fatalities, this would have tended to reduce drivers' absolute fatality risk at all ages and not necessarily influence the relative effects of aging one year. Furthermore, the difference between drivers and RF passengers in the effect of aging was already there in pre-1968 cars, before the existence of the FMVSS and has stayed about the same since then (as can be seen in the lower section of Table 3-1a and will be discussed in Section 3.3).

For males, the average of the effects for the driver and the RF passenger is a risk increase of 3.28 ± .08 percent per year. The year-to-year increase is lower for female occupants: 2.60 ± .10 percent for drivers and 3.27 ± .08 percent for RF passengers, averaging to 2.93 ± .08 percent. Essentially, females age slower than males. Or, as discussed in Section 2.10, the added risk for a female relative to a male diminishes with age, and this diminution may be deducted from the males' year-to-year risk increase. The right side of Table 3-1a averages the results for males and females: a risk increase of 2.83 ± .08 percent for drivers and 3.39 ± .07 percent for passengers. That averages out to 3.11 ± .08 percent for all front-outboard occupants, perhaps the best single indicator of the overall average increase in fatality risk for aging one year. All of these estimates are quite precise; the sampling errors (widths of the confidence bounds) are negligible relative to the point estimate. Those confidence bounds will become steadily wider as the analysis focuses on ever smaller subsets of crashes, vehicles, or occupants.

One guideline for the analysis, as discussed in Sections 2.6 and 2.8, was to make it as "portable" as possible, to allow direct side-by-side comparison of results for, say, vehicles of the 1960s and vehicles of today – even though the proportion of elderly and female drivers has increased in the general driving population. This was achieved by (1) having the regressions include second-order terms that estimated the effect of aging or gender as a function of age and (2) always weighting the results across ages 21-96 by the same population, namely the distribution of age and gender in CY 2001-2010 FARS. One exception to the "portable" approach was that all the results for drivers are weighted by the 2001-2010 age/gender distribution of drivers, while the results for RF passengers are weighted by the 2001-2010 distribution of RF passengers. In that sense, the driver and RF passenger results, themselves, are not directly comparable. The second row of Table 3-1a shows the averages obtained if the driver results are also averaged by the 2001-2010 age/gender distribution of RF passengers. It changes the results little, always less than the width of the confidence bounds: the estimated effect of a year of aging rises from 3.06 to 3.10 percent for male drivers, 2.60 to 2.68 percent for female drivers, and 2.83 to 2.89 percent for all drivers. The substantial difference between drivers and RF passengers in the first row of Table 3-1a (e.g., 3.06 versus 3.51 for males) is only to a slight extent due to the different weighting of drivers and passengers in that row. (Similarly, the second rows of Tables 3-1b, 3-1c, and 3-1d show that the choice of weight factors has little influence on the statistics in those tables.)

Table 3-1b shows how the effect of aging increases with age. For a male driver, the increase averages 3.06 percent per year of aging across ages 21-96. However, for young adults 21 to 30, the estimated increase is slightly less, just 2.90 percent per year. By 65 to 74, the increase has escalated somewhat, to 3.39 percent per year (see Figure 2-9 for the year-by-year trend). For RF passengers, however, the increase starts at the same rate, 2.90 percent per year at ages 21-30, but escalates considerably more, reaching 4.58 percent at 65 to 74. That results in a higher overall average: 3.51 percent.

TABLE 3-1b

INCREASE IN FATALITY RISK FOR EACH YEAR THAT A MALE OCCUPANT GETS OLDER
AT AGE 21 TO 30 VERSUS AGE 65 TO 74
(Given the same crash scenario, FARS 1975-2010)

OVERALL AND BY MODEL YEAR
CARS AND LTVs, DRIVERS AND RIGHT-FRONT PASSENGERS

	Percent Fatality Increase per Year of Aging					
Seat Position →→→→→→→→→→→	Driver			RF Passenger		
Occupant Age Group →→→→→→→→	Avg 21-96	21-30	65-74	Avg 21-96	21-30	65-74
ALL CARS & LTVs, MY 1960-2011	**3.06 ± .09**	**2.90**	**3.39**	**3.51 ± .09**	**2.90**	**4.58**
Drivers wtd by RF age-gender distrib	*3.10*	*2.91*	*3.42*			
Cars only	3.04 ± .09	2.83	3.45	3.56 ± .09	2.93	4.74
LTVs only	3.12 ± .16	2.96	3.43	3.28 ± .19	2.97	3.83
Cars only, MY ≥ 2000	2.99 ± .34	2.47	3.77	3.14 ± .43	2.12	4.70
LTVs only, MY ≥ 2000	3.10 ± .57	2.54	3.96	3.39 ± .49	2.93	4.14
Cars and LTVs, by model year range						
1960-1966	2.71 ± .37	2.37	3.26	3.40 ± .40	2.84	4.52
1967-1974	2.85 ± .17	2.89	2.91	3.28 ± .22	2.97	3.94
1975-1979	3.04 ± .21	2.99	3.24	3.64 ± .15	3.01	4.78
1980-1984	3.00 ± .21	3.10	2.99	3.52 ± .20	3.02	4.50
1985-1989	3.19 ± .22	2.87	3.73	3.61 ± .20	3.02	4.66
1990-1994	3.31 ± .29	3.24	3.54	3.65 ± .29	2.99	4.78
1995-1999	2.98 ± .29	2.57	3.62	3.40 ± .31	2.42	5.05
2000-2004	3.00 ± .42	2.47	3.78	3.18 ± .36	2.36	4.44
2005-2011	3.19 ± .59	3.00	3.52	3.34 ± .59	2.31	4.90

TABLE 3-1c

FATALITY RISK FOR A 75-YEAR-OLD OCCUPANT
RELATIVE TO A 21-YEAR-OLD OCCUPANT
(Given similar physical insults, FARS 1975-2010)

CARS AND LTVs, DRIVERS AND RIGHT-FRONT PASSENGERS
OVERALL AND BY MODEL YEAR

	Risk at 75 ÷ Risk at Age 21			
	Males		Females	
	Drivers	RF Pass	Drivers	RF Pass
ALL CARS AND LTVs, MY 1960-2011	**5.04**	**6.70**	**3.87**	**5.67**
Drivers wtd by RF age-gender distribution	*5.18*		*3.97*	
Cars only	4.99	6.94	3.98	5.75
LTVs only	5.22	5.86	3.47	5.46
Cars only, MY ≥ 2000	4.93	5.55	3.50	5.72
LTVs only, MY ≥ 2000	5.27	6.37	3.63	5.62
Cars and LTVs, by model year range				
1960-1966	4.21	6.41	3.43	5.22
1967-1974	4.50	5.85	3.90	4.92
1975-1979	4.97	7.23	4.02	5.96
1980-1984	4.86	6.80	3.49	5.67
1985-1989	5.40	7.06	4.20	6.25
1990-1994	5.75	7.26	4.32	6.29
1995-1999	4.86	6.46	3.85	4.83
2000-2004	4.98	5.68	3.23	5.71
2005-2011	5.46	6.25	4.38	5.85

Again, it is unknown to what extent the difference between drivers and RF passengers reflects "real" vehicle factors and to what extent, if any, a tendency for the frailest and most fragile occupants for any given age to ride as passengers rather than to drive.

Table 3-1c estimates the cumulative effect of all the year-to-year increases in risk. A 75-year-old male driver is 5.04 times as likely to die as a 21-year-old from similar physical insults. For male RF passengers, the 75-year-old has 6.70 times the fatality risk as the 21-year-old. The cumulative effects are smaller for females: the risk for a 75-year-old female driver is 3.87 times the risk at 21; for RF passengers, it is 5.67 times as high (see Figure 2-11 for year-by-year trends).

Table 3-1d estimates the increase in fatality risk for a female relative to a male of the same age, given a similar crash scenario or physical insult. For drivers, the average across all ages 21-96 is a 13.4 ± 2.0 percent higher risk for the female. But it varies substantially with age: among young adults 21 to 30, the female is at 25.9 percent higher risk than the male of the same age, but by 65 to 74, risk is nearly equal; in fact, the female's risk is an estimated 1.4 percent lower (see Figure 2-10 for the year-by-year trend). For RF passengers, the average additional risk for females is somewhat higher, 20.5 ± 2.2 percent. It starts just a bit higher than for drivers at 21 to 30 (29.2% versus 25.9%), but then drops off much slower than for drivers; even by 65 to 74, the female is still at 11.4 percent higher risk than the male. Here, too, it is unknown to what extent the differences of the driver and RF passengers results reflect the vehicle, or to what extent they may be due to elderly female drivers being healthier than female RF passengers of the same age. The average for drivers and RF passengers, across all ages, is 17.0 ± 1.5 percent higher risk for a female than for a male of the same age in the same seat. The sampling error is small relative to the point estimate, although not as small, in relative terms, as in the estimates of the effects of aging one year. In Tables 3-1b, 3-1c, and 3-1d, as in Table 3-1a, it makes little difference if the driver results are weighted by the 2001-2010 age/gender distribution of drivers (first row of each table) or RF passengers (second row).

3.2   Cars versus LTVs

The overall effect of aging is nearly equal in cars and LTVs: the right column of Table 3-1a shows the average risk increase per year for male and female drivers and RF passengers is 3.14 ± .08 percent in cars and 3.00 ± .17 percent in LTVs. The sampling error for the LTV estimate is higher because it is based on fewer crash cases. The effect of gender is also nearly the same in cars and LTVs: a 16.8 ± 1.8 percent increase for females relative to males in cars, according to the right column of Table 3-1d, versus LTVs' 17.3 ± 3.2 percent. The market for LTVs has changed over the years from predominantly pickup trucks to increasing shares for SUVs and, more recently, crossover utility vehicles (CUV). But the effects of aging and gender in LTVs have stayed similar to the corresponding effects in cars. When the data are limited to MY 2000 and later vehicles, the overall effect of aging one year is a 2.92 ± .30 percent increase in cars, 3.05 ± .45 percent in LTVs; the added risk for females is 7.2 ± 4.4 percent in cars, 7.5 ± 7.0 percent in LTVs.

TABLE 3-1d: INCREASE IN FATALITY RISK FOR A FEMALE RELATIVE TO A MALE OF THE SAME AGE
OVERALL AND BY MODEL YEAR – CARS AND LTVs, DRIVERS AND RIGHT-FRONT PASSENGERS
(Given the same crash scenario, average for occupants 21 to 96, FARS 1975-2010)

Percent Fatality Increase for Females

Seat Position →→→→→→→→→→	Drivers			RF Passengers			Avg of Driver & RF
Occupant Age Group →→→→→→→→→	21-96	21-30	65-74	21-96	21-30	65-74	21-96
ALL CARS AND LTVs, MY 1960-2011	**13.4 ± 2.0**	**25.9**	**- 1.4**	**20.5 ± 2.2**	**29.2**	**11.4**	**17.0 ± 1.5**
Drivers wtd by RF age-gender distribution	*12.9*	*26.0*	*- 1.4*				
Cars only	15.8 ± 2.7	26.9	2.5	17.8 ± 2.6	27.2	7.4	16.8 ± 1.8
LTVs only	9.2 ± 4.5	29.2	- 12.8	25.5 ± 4.7	30.7	20.9	17.3 ± 3.2
Cars only, MY ≥ 2000	10.4 ± 9.4	25.1	- 8.1	4.1 ± 7.4	3.8	4.7	7.2 ± 4.4
LTVs only, MY ≥ 2000	3.4 ± 12.4	19.3	- 15.9	11.6 ± 10.6	19.6	2.9	7.5 ± 7.0
Cars and LTVs, by model year range							
1960-1966	8.3 ± 13.6	18.7	- 3.8	25.1 ± 12.5	35.8	14.1	16.7 ± 9.6
1967-1974	10.9 ± 6.4	18.0	2.3	24.4 ± 6.1	33.9	15.0	17.7 ± 4.5
1975-1979	17.3 ± 5.2	27.8	5.2	18.3 ± 5.3	26.9	9.2	17.8 ± 3.5
1980-1984	16.9 ± 5.2	33.2	- 1.8	23.0 ± 4.1	32.9	12.9	20.0 ± 2.9
1985-1989	15.4 ± 5.2	28.0	.5	25.7 ± 4.7	32.1	19.4	20.5 ± 3.8
1990-1994	13.8 ± 7.3	26.4	- 1.2	23.1 ± 7.8	31.3	14.1	18.5 ± 5.2
1995-1999	8.8 ± 6.0	19.0	- 4.1	15.3 ± 5.5	29.8	- 1.0	12.0 ± 3.3
2000-2004	7.2 ± 7.6	24.2	- 13.7	6.8 ± 6.3	7.6	6.2	7.0 ± 4.2
2005-2011	6.9 ± 17.0	16.2	- 5.2	10.7 ± 14.0	15.7	4.9	8.8 ± 8.5

The full database, as stated above, contains 154,467 data points. But the N of cases quickly decreases when the data is split up into subgroups. Confidence bounds get wider (more or less inversely proportional to $\sqrt{N}$). Appendix F lists the number of data points for every regression analysis in Chapters 3-5. Analyses based on 10,000 cases or more generate precise estimates, but confidence bounds are rather wide at 2,000-5,000 cases and estimates are barely statistically meaningful at less than 1,000 cases.

There are some observed differences by seat position. Whereas the effect of male drivers' aging is almost equal in cars and LTVs (3.04 ± .09% versus 3.12 ± .16%), the effect for male RF passengers is somewhat higher in cars (3.56 ± .09%) than in LTVs (3.28 ± .19%), especially for older RF passengers (4.74% at 65 to 74 in cars, versus 3.83% in LTVs, according to Table 3-1b). The risk increase for females is lower in LTVs for drivers (15.8 ± 2.7% in cars versus 9.2 ± 4.5% for LTVs, according to Table 3-1d), but higher for RF passengers (17.8 ± 2.6% in cars versus 25.5 ± 4.7% in LTVs); moreover, the difference is observed primarily for the older occupants: at 65 to 74, female drivers in LTVs are at 12.8 percent lower risk than male drivers, while female RF passengers are at 20.9 percent higher risk than male RF passengers. NHTSA has no convincing explanation for the observed differences between drivers and RF passengers in LTVs; they might not be real differences, given the more limited data for LTVs.

3.3   Trends by model year

Surely, a principal question is whether the relative risk increases for aging and for females have diminished, stayed about the same, or intensified during the past 50 years. One approach is simply to group the vehicles into approximately five-model-year cohorts, starting with MY 1960-1966 – a nearly "pristine" state before the FMVSS with little belt use in fatal crashes – and ending with MY 2005-2011, when all light vehicles were equipped with frontal air bags, almost all with belt pretensioners and load limiters, and many with curtain- and/or side air bags. It is important to note that an intensification of the aging effect from one cohort to the next **does not mean that absolute risk has increased for elderly occupants**. It only means that the intervening safety improvements helped young people even more than they helped the elderly – and as a consequence, the relative gap between the elderly and the young increased. Likewise, if the aging effect stayed the same from one cohort to the next, it does not mean safety stayed the same, but that it improved about equally for young and old occupants. Figure 3-1, for example, based on a 2004 NHTSA report, shows that the overall, **absolute** risk index for occupants of all ages had decreased from 100 for vehicles on the road in CY 1955-1960 to 58 for vehicles on the road in CY 2002: a 42-percent reduction of overall fatality risk due attributed to increased belt use, air bags, and the other FMVSS up to that date.[30] This rising tide of safety benefited everyone: absolute risk decreased substantially for males and females of all ages, but not necessarily by exactly equal amounts.

---

[30] Based on Kahane (2004), Table 2 on p. xv.

FIGURE 3-1: OCCUPANT FATALITY-RISK INDEX BY CALENDAR YEAR
BASED ON LIVES SAVED BY THE FMVSS (1955 = 100)

Table 3-1a suggests the **relative** effect of aging one year possibly intensified from 1960 up to about 1990-1994 and has likely diminished since then. In the right column of Table 3-1a, the overall average effect started at 2.89 percent, increases with each successive cohort except one to a peak of 3.31 percent in MY 1990-1994, and drops back to 2.92 percent in MY 2000-2004, close to where it started. However, there is some uncertainty about the trend, because all estimates have sampling error, especially the first and the last, which are based on fewer crash cases (5,030 and 3,587 cases, respectively, according to Appendix F, as compared to 10,313 to 29,205 in the middle cohorts). Furthermore, even if there is a trend, it is not a strong one; there is little difference between the smallest (2.89) and largest (3.31) effects. Figure 3-2 graphs these effects, showing the point estimate and confidence bounds for each cohort, putting the trends in perspective. Whereas there is an increase in the point estimate in five of the first six cohorts, the increases are small both relative to the point estimates themselves and the confidence bounds. The decreases starting in 1995 are likewise small in absolute terms and relative to sampling error. It is unclear whether there has been a real change over time or whether the relative effect of aging a year has stayed basically the same over a time period when vehicles became much safer for occupants in absolute terms.

The cumulative impacts of these changes are illustrated by Table 3-1c, which shows that the risk for a 75-year-old male driver was 4.21 times as high as that for a 21-year-old male driver in MY 1960-1966, escalating to 5.75 times as high in MY 1990-1994, and then dropping again, and similarly for RF passengers. Table 3-1a shows similar patterns for drivers and RF passengers, but with consistently higher aging effects at the RF position. However, for male occupants, the difference between the effects for drivers and RF passengers has diminished from .6-.7 percentage points in the first three cohorts (e.g., 2.71 and 3.40, the two left columns of the 1960-1966 cohort in Table 3-1a) to just .2 percentage points in the last two cohorts (e.g., 3.19 and 3.34, the two left columns of the 2005 to 2011 cohort).

Table 3-1d shows a similar historical pattern in the added risk for females relative to males – but with a stronger diminution of the risk in the most recent vehicles. The point estimates of the added risk, averaged for driver and RF passenger, increased four times in a row, from 16.7 percent in MY 1960-1966 to 20.5 percent in MY 1985-1989 (but given the substantial sampling errors for each observation, it is hard to say how "real" the trend is). After MY 1990, however, the effect drops by more than half, down to 7.0 ± 4.2 percent in MY 2000-2004 and 8.8 ± 8.5 percent in MY 2005-2011. Figure 3-3 shows the drops after 1990 are large in absolute terms and also relative to confidence bounds. Unlike the effect of aging, which always stays close to 3 percent even as it fluctuates, the "gender gap" has been to a large extent mitigated in the latest vehicles.

The next set of tables will, in some ways, refine the trend analysis by grouping occupants by the types of occupant protection they had available and they used. However, even the trends based just on model-year cohorts are consistent with two hypotheses that often come up in this report:

- Safety technologies that especially benefit elderly occupants often (but not always) also especially benefit females, and vice-versa.

FIGURE 3-2: FATALITY INCREASE (%) FOR AGING ONE YEAR, BY MODEL YEAR COHORT

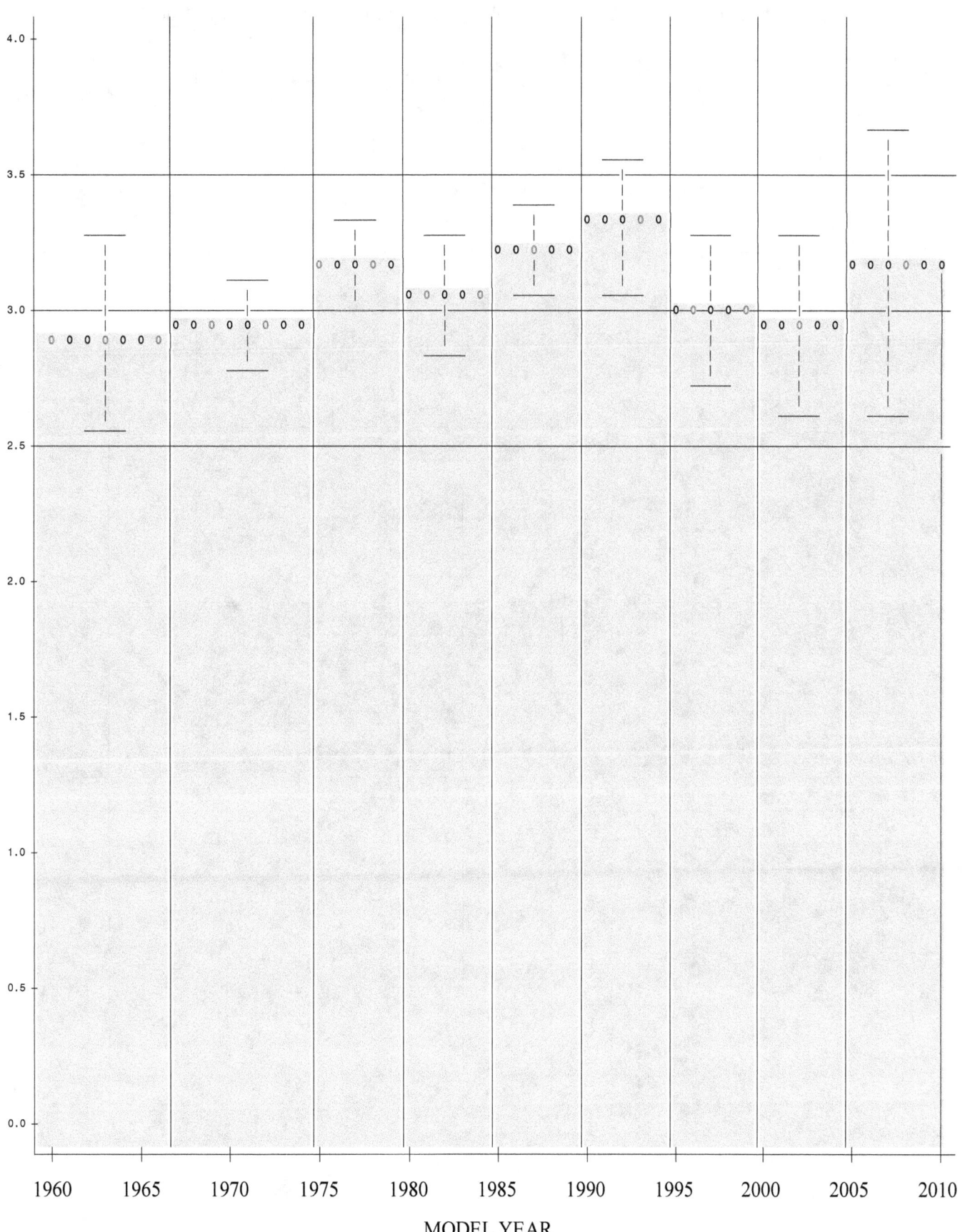

MODEL YEAR

Figure 3-3: Fatality Increase (%) for a Female Relative to a Male of the Same Age, by Model Year Cohort

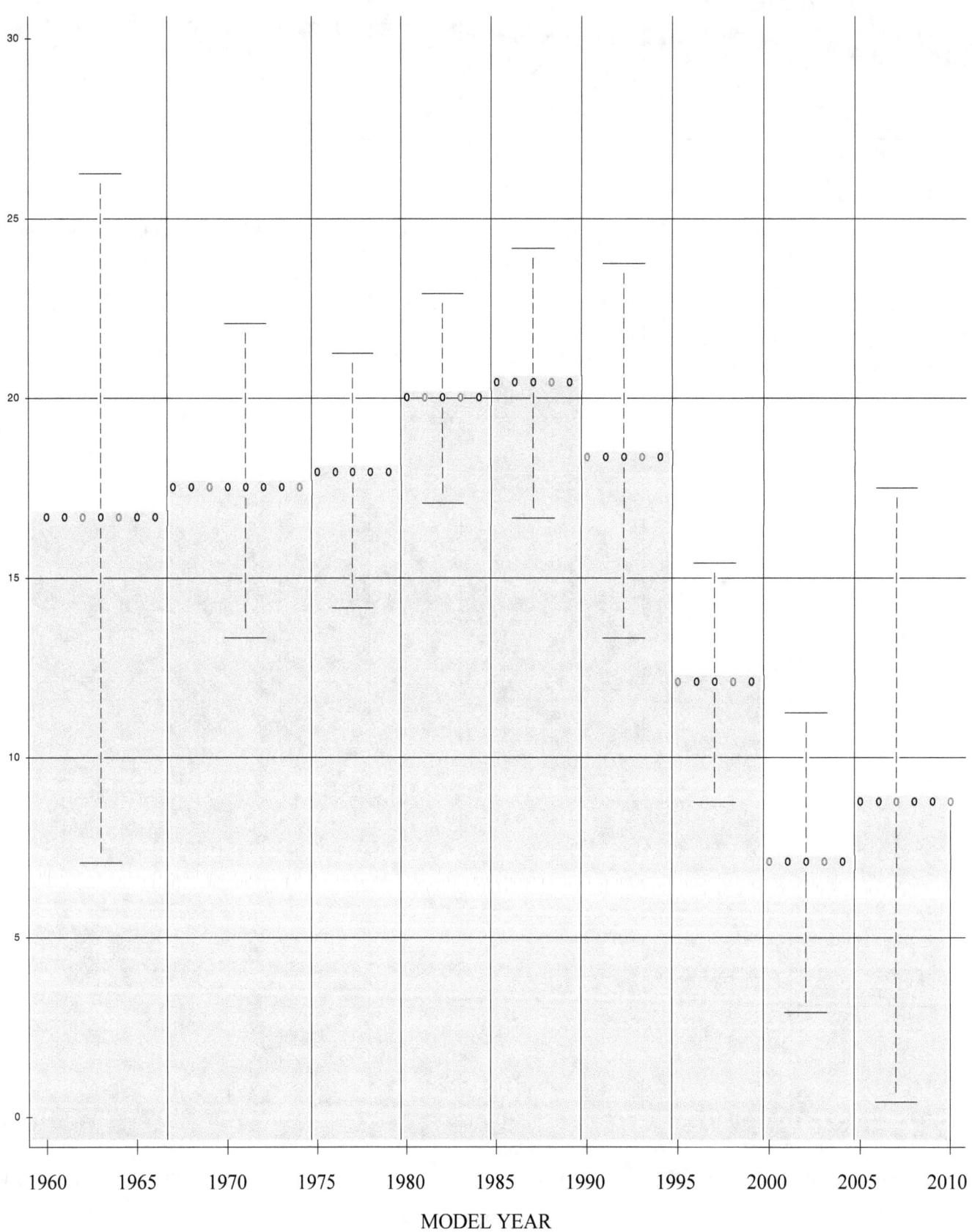

MODEL YEAR

- The initial safety technologies were especially beneficial for young male occupants and consequently may have intensified the relative effects of aging and the gender gap. But more recent technologies, informed by a deeper understanding of human tolerance and biomechanics, may have especially benefited older occupants and females, and reduced the relative effects of aging (perhaps) and gender.

3.4   Trends across "generations" of car and LTV occupants

Table 3-2a subdivides passenger-car occupants into eight "generational" groups based on a combination of the vehicle's model year, the occupants' belt use, and if the vehicle is equipped with dual frontal air bags (whether they deployed or not):

- Pre-1968, no energy-absorbing steering assemblies, unbelted;
- MY 1969-1982, unbelted;
- MY 1983-1996, no air bags, unbelted;
- 3-point belted occupants of cars without air bags;
- Automatic 2-point belted occupants of cars without air bags;
- Unbelted occupants of cars with dual frontal air bags;
- Dual frontal air bags, 3-point belted, no pretensioners and no load limiters; and
- Dual frontal air bags, 3-point belted, with pretensioners and load limiters.

Identification of the available occupant protection is based on NHTSA's VIN analysis programs (or is obvious from the model year of the vehicle). Cars/occupants that do not clearly fit into one of these categories are excluded from the analysis. As stated above, the last three groups include all vehicles equipped with dual frontal air bags, whether they deployed or not. They were not limited to deployment crashes because: (1) this would have limited them to primarily frontal impacts and made them not comparable to the other five groups; (2) the deployment variable on FARS has extensive missing data, especially in the earlier calendar years.

For six of the eight groups, the overall average risk increase for aging one year is quite close to 3 percent. The effect of aging is substantially higher, $3.92 \pm .35$ percent, with automatic 2-point belts, a technology that was phased out after 1996. Similarly, Table 3-2c indicates that with 2-point belts, risk increases 7.89 times for 75-year-old male drivers relative to 21-year-old, and 10.80 times for passengers: risk multipliers considerably higher than any number previously seen in the tables. The effect of aging is likely somewhat lower than 3 percent, although this is not certain in view of the sampling error, for unbelted occupants of cars with dual air bags: $2.62 \pm .35$ percent: this configuration might be **relatively** more advantageous for older occupants because the air bags distribute the load over a wide area of the body without belts' concentrated load on the ribs or the focused blunt-impact trauma of completely unrestrained occupants. (The keyword is "relatively": in absolute terms, of course, belted occupants of any age are at less risk than unbelted occupants of the same age.)

TABLE 3-2a: INCREASE IN FATALITY RISK FOR EACH YEAR THAT AN OCCUPANT GETS OLDER BY TYPE OF OCCUPANT PROTECTION – CARS AND LTVs, DRIVERS AND RIGHT-FRONT PASSENGERS
(Given the same crash scenario, average for occupants 21 to 96, FARS 1975-2010)

Percent Fatality Increase per Year of Aging

	Males			Females			Average of Males & Females		
	Driver	RF Pass	Average	Driver	RF Pass	Average	Driver	RF Pass	Average
EIGHT "GENERATIONS" OF CAR OCCUPANTS									
Unbelted, pre-MY 1968, no EA columns	2.77 ± .46	3.41 ± .58	3.09 ± .47	2.40 ± .64	3.08 ± .39	2.74 ± .46	2.58 ± .46	3.24 ± .45	2.91 ± .45
Unbelted occupants of MY 1969-1982 cars	2.92 ± .15	3.52 ± .16	3.22 ± .12	2.69 ± .21	3.24 ± .14	2.96 ± .16	2.81 ± .15	3.38 ± .13	3.09 ± .14
Unbelted, MY 1983-1996, no air bags	3.12 ± .28	3.57 ± .22	3.35 ± .23	2.72 ± .21	3.25 ± .22	2.98 ± .20	2.92 ± .23	3.41 ± .21	3.16 ± .22
3-pt. belted occupants of cars w/o air bags	2.99 ± .22	3.63 ± .21	3.31 ± .19	2.59 ± .24	3.44 ± .21	3.01 ± .19	2.79 ± .20	3.54 ± .19	3.16 ± .19
2-pt. belted occupants of cars w/o air bags	3.93 ± .40	4.38 ± .38	4.16 ± .39	3.24 ± .49	4.13 ± .41	3.68 ± .39	3.58 ± .38	4.26 ± .36	3.92 ± .35
Unbelted occupants of cars with dual air bags	2.53 ± .40	2.98 ± .41	2.75 ± .33	2.22 ± .51	2.74 ± .40	2.48 ± .43	2.37 ± .42	2.86 ± .36	2.62 ± .35
Belted, dual air bags, no pretens/load lim	2.92 ± .41	3.52 ± .39	3.22 ± .39	2.58 ± .44	3.07 ± .37	2.83 ± .37	2.75 ± .39	3.30 ± .37	3.02 ± .36
Belted, dual air bags, pretensioners, load lim	3.10 ± .50	3.16 ± .55	3.13 ± .44	2.42 ± .58	3.24 ± .45	2.83 ± .47	2.76 ± .48	3.20 ± .46	2.98 ± .44
All cars	3.04 ± .09	3.56 ± .09	3.30 ± .08	2.65 ± .10	3.31 ± .08	2.98 ± .08	2.85 ± .09	3.44 ± .07	3.14 ± .08
FOUR "GENERATIONS" OF LTV OCCUPANTS									
Unbelted occupants of LTVs w/o air bags	2.93 ± .18	3.12 ± .22	3.02 ± .18	2.24 ± .52	2.98 ± .22	2.61 ± .33	2.58 ± .29	3.05 ± .19	2.82 ± .21
Belted occupants of LTVs w/o air bags	3.42 ± .39	3.53 ± .37	3.47 ± .34	2.98 ± .47	3.63 ± .39	3.30 ± .36	3.20 ± .38	3.58 ± .37	3.39 ± .36
Unbelted, dual air bags (no on-off switches)	2.75 ± .62	3.15 ± .64	2.95 ± .56	2.22 ± .92	2.75 ± .60	2.49 ± .69	2.49 ± .70	2.95 ± .60	2.72 ± .63
Belted, dual air bags (no on-off switches)	3.35 ± .56	3.31 ± .43	3.33 ± .48	2.54 ± .51	3.26 ± .46	2.90 ± .46	2.95 ± .48	3.28 ± .42	3.11 ± .46
All LTVs	3.12 ± .16	3.28 ± .19	3.20 ± .16	2.41 ± .25	3.17 ± .18	2.79 ± .20	2.77 ± .18	3.22 ± .16	3.00 ± .17

TABLE 3-2b

INCREASE IN FATALITY RISK FOR EACH YEAR THAT A MALE OCCUPANT GETS OLDER
AT AGE 21-30 VERSUS AGE 65-74
(Given the same crash scenario, FARS 1975-2010)

BY TYPE OF OCCUPANT PROTECTION
CARS AND LTVs, DRIVERS AND RIGHT-FRONT PASSENGERS

	Percent Fatality Increase per Year of Aging					
Seat Position →→→→→→→→→→→	Driver			RF Passenger		
Occupant Age Group →→→→→→→→	Avg 21-96	21-30	65-74	Avg 21-96	21-30	65-74
EIGHT "GENERATIONS" OF CAR OCCUPANTS						
Unbelted, pre-MY 1968, no EA columns	2.77 ± .46	2.64	3.03	3.41 ± .57	2.94	4.43
Unbelted occupants of MY 1969-1982 cars	2.92 ± .15	2.98	2.99	3.52 ± .16	3.08	4.42
Unbelted, MY 1983-1996, no air bags	3.12 ± .28	3.29	3.02	3.57 ± .22	3.21	4.37
3-pt. belted occupants of cars w/o air bags	2.99 ± .22	2.46	3.82	3.63 ± .21	2.73	5.16
2-pt. belted occupants of cars w/o air bags	3.93 ± .40	3.42	4.77	4.39 ± .38	4.08	5.24
Unbelted occupants of cars with dual air bags	2.53 ± .41	2.14	3.12	2.98 ± .41	2.35	4.03
Belted, dual air bags, no pretens/load lim	2.92 ± .41	2.57	3.49	3.52 ± .39	2.53	5.21
Belted, dual air bags, pretensioners, load lim	3.10 ± .50	2.62	3.87	3.16 ± .55	2.05	4.85
All cars	3.04 ± .09	2.83	3.45	3.56 ± .09	2.93	4.74
FOUR "GENERATIONS" OF LTV OCCUPANTS						
Unbelted occupants of LTVs w/o air bags	2.93 ± .18	3.11	2.80	3.12 ± .22	3.06	3.34
Belted occupants of LTVs w/o air bags	3.42 ± .39	2.81	4.34	3.53 ± .37	3.01	4.32
Unbelted, dual air bags (no on-off switches)	2.75 ± .62	2.54	3.12	3.15 ± .64	3.34	3.20
Belted, dual air bags (no on-off switches)	3.35 ± .56	2.94	4.01	3.31 ± .43	2.51	4.45
All LTVs	3.12 ± .16	2.96	3.43	3.28 ± .19	2.97	3.83

TABLE 3-2c

FATALITY RISK FOR A 75-YEAR-OLD OCCUPANT
RELATIVE TO A 21-YEAR-OLD OCCUPANT
(Given similar physical insults, FARS 1975-2010)

CARS AND LTVs, DRIVERS AND RIGHT-FRONT PASSENGERS
BY TYPE OF OCCUPANT PROTECTION

Risk at Age 75 ÷ Risk at Age 21

	Males		Females	
	Drivers	RF Pass	Drivers	RF Pass
EIGHT "GENERATIONS" OF CAR OCCUPANTS				
Unbelted, pre-MY 1968, no EA columns	4.33	6.45	3.52	5.02
Unbelted occupants of MY 1969-1982 cars	4.67	6.73	4.08	5.50
Unbelted, MY 1983-1996, no air bags	5.19	6.93	4.11	5.47
3-pt. belted occupants of cars w/o air bags	4.90	7.32	3.90	6.38
2-pt. belted occupants of cars w/o air bags	7.89	10.80	5.35	8.84
Unbelted occupants of cars with dual air bags	3.83	5.00	3.20	4.18
Belted, dual air bags, no pretens/load lim	4.74	7.03	3.87	4.88
Belted, dual air bags, pretensioners, load lim	5.21	5.59	3.52	5.91
All cars	4.99	6.94	3.98	5.75
FOUR "GENERATIONS" OF LTV OCCUPANTS				
Unbelted occupants of LTVs w/o air bags	4.72	5.41	3.17	4.95
Belted occupants of LTVs w/o air bags	6.16	6.64	4.78	7.30
Unbelted, dual air bags (no on-off switches)	4.36	5.56	3.20	4.15
Belted, dual air bags (no on-off switches)	5.97	6.03	3.71	5.87
All LTVs	5.22	5.86	3.47	5.46

TABLE 3-2d: INCREASE IN FATALITY RISK FOR A FEMALE RELATIVE TO A MALE OF THE SAME AGE
BY TYPE OF OCCUPANT PROTECTION – CARS AND LTVs, DRIVERS AND RIGHT-FRONT PASSENGERS
(Given the same crash scenario, average for occupants 21 to 96, FARS 1975-2010)

Percent Fatality Increase for Females

Seat Position →→→→→→→→→→	Drivers			RF Passengers			Avg of Driver & RF
Occupant Age Group →→→→→→→→→	21-96	21-30	65-74	21-96	21-30	65-74	21-96
EIGHT "GENERATIONS" OF CAR OCCUPANTS							
Unbelted, pre-MY 1968, no EA columns	5.8 ± 13.4	15.0	- 5.5	25.0 ± 12.5	38.4	10.7	15.4 ± 9.4
Unbelted occupants of MY 1969-1982 cars	16.6 ± 4.3	24.4	7.4	17.6 ± 4.2	27.5	7.2	17.1 ± 2.9
Unbelted, MY 1983-1996, no air bags	20.5 ± 6.2	34.2	5.1	19.1 ± 5.5	30.5	6.7	19.8 ± 4.2
3-pt. belted occupants of cars w/o air bags	16.7 ± 7.2	26.1	4.9	19.4 ± 7.2	25.7	12.2	18.0 ± 4.4
2-pt. belted occupants of cars w/o air bags	20.4 ± 14.1	39.6	- 1.1	35.0 ± 17.9	47.3	21.8	27.7 ± 10.0
Unbelted occs of cars with dual air bags	18.6 ± 9.8	28.8	6.8	8.6 ± 11.2	17.4	- 1.2	13.6 ± 6.0
Belted, dual air bags, no pretens/load lim	13.8 ± 9.7	22.1	3.2	8.8 ± 11.1	25.4	- 10.4	11.3 ± 6.8
Belted, dual air bags, pretensioners, load lim	6.6 ± 10.6	21.5	- 12.0	4.0 ± 11.9	3.0	5.6	5.3 ± 9.0
All cars	15.8 ± 2.7	26.9	2.5	17.8 ± 2.6	27.2	7.4	16.8 ± 1.8
FOUR "GENERATIONS" OF LTV OCCUPANTS							
Unbelted occupants of LTVs w/o air bags	15.7 ± 8.6	37.3	- 7.0	30.0 ± 6.5	35.9	25.0	22.9 ± 5.5
Belted occupants of LTVs w/o air bags	1.7 ± 11.3	10.7	- 9.5	30.7 ± 14.9	27.1	37.1	16.2 ± 9.1
Unbelted, dual air bags (no on-off switches)	8.1 ± 15.8	22.2	- 7.4	12.0 ± 13.0	28.3	- 3.0	10.0 ± 8.8
Belted, dual air bags (no on-off switches)	1.4 ± 11.4	20.8	- 20.8	10.8 ± 12.7	12.2	9.7	6.1 ± 7.6
All LTVs	9.2 ± 4.5	29.2	- 12.8	25.5 ± 4.7	30.7	20.9	17.3 ± 3.2

65

LTVs, where fewer crash cases are available, are allocated among four "generations":

- No air bags, unbelted;
- No air bags, 3-point belted;
- Dual frontal air bags, unbelted; and
- Dual frontal air bags, 3-point belted.

Table 3-2a indicates that the observed effect of aging is slightly higher when occupants are belted, and slightly lower when the LTV is equipped with dual air bags; thus, the highest observed effect is for 3-point belted without air bags, $3.39 \pm .36$ percent, and the lowest for unbelted with dual air bags, $2.72 \pm .63$ percent; however, the fairly wide confidence bounds make it difficult to draw definitive conclusions. Table 3-2b estimates the effects of aging one year at 21 to 30 and contrasts them with the effects at 65 to 74. Most of the results parallel the overall effects, but one group of findings stands out: for unbelted drivers of cars and LTVs without air bags, the effect is almost the same at 65-74 as at 21-30, or even lower (unlike the usual pattern of escalating effects). This is consistent with Evans' findings from the 1980's, based on vehicles without air bags in which 97 percent of the occupant fatalities were unbelted: fatality risk for drivers increased at a slower rate in the higher age brackets.[31] For all RF passengers and for drivers protected by belts and/or air bags, on the other hand, Table 3-2b consistently shows a higher effect of aging one year at 65 to 74 than at 21-30 – consistent with Zhou's findings that aging has a stronger effect on belt-loading tolerance than on blunt-impact tolerance.[32]

Table 3-2d estimates the risk for females relative to males of the same age. The right column shows the average effect for drivers and RF passengers across ages 21-96. Two findings resemble the results on aging: (1) 2-point automatic belts pose relatively the highest extra risk for females ($27.7 \pm 10.0\%$); (2) frontal air bags reduce the relative extra risk for females in both cars and LTVs. However, unlike the results on aging, 3-point belt use also tends to reduce the extra risk for females, especially in cars and LTVs with dual air bags and even more so when the belts are equipped with pretensioners and load limiters. As a result, the added risk for females in the latest generation of cars (dual air bags, belted, with pretensioners and load limiters) has dropped to a non-significant level: 5.3 ± 9.0 percent. The gender gap is likewise non-significant ($6.1 \pm 7.6\%$) in the latest generation of LTVs and it may be close to zero ($1.4 \pm 11.4\%$) for their drivers. Fundamentally, the latest air bag and belt systems are apparently somewhat more effective for females than for males and this new factor benefitting females appears to have greatly reduced or possibly even eliminated the original gender gap seen in the many preceding decades. Several more years of data will be needed, however, before there can be a firm conclusion. Because the currently available data show fairly small but non-significant risk increases for females ($5.3 \pm 9.0\%$ in cars, $6.1 \pm 7.6\%$ in LTVs), it is not yet clear whether the gender gap has been eliminated or merely reduced.

---

[31] Evans (1991), p. 27.
[32] Zhou, Rouhana, and Melvin (1996).

3.5   Variations by impact location

Table 3-3a separately estimates the fatality-risk increase when an occupant ages one year for five types of impact: frontal, nearside, far-side, first-event rollover, and rear/other. Based on FARS data, a "first-event rollover" includes vehicles in single-vehicle crashes where the first harmful event is a rollover, or where the first harmful event is essentially a non-event such as hitting a curb and the most harmful event is a rollover. The "rear/other" category includes specific first-event non-collisions such as running into the water or falling out of a moving vehicle, rear impacts (the principal impact, IMPACT2 = 5-7), and "other non-collisions" for which no impact location is specified.

All other impacts are classified by the principal impact (IMPACT2): 11, 12, or 1 for frontals; 2-4 or 81-83 for right-side; and 8-10 or 61-63 for left-side. If IMPACT2 is unknown, the initial impact (IMPACT1) is used. A "nearside" impact is a left-side impact for the driver or a right-side impact for the RF passenger. A "far-side" impact is a right-side impact for the driver or a left-side impact for the RF passenger. If FARS provides no impact location but says the vehicle was involved in a head-on crash, the impact is assumed to be frontal. Post-impact rollovers, unless the first "event" was basically just a tripping mechanism such as hitting a curb, are not included with the rollovers but are classified according to the location of the principal impact.

Thus, in Table 3-3a, the "nearside" estimates for the driver are based on left-side impacts and for the RF passenger, right-side impacts. The estimates for "average of driver and RF" are the arithmetic mean of the driver and RF-passenger estimates, with confidence bounds based on assuming the driver and RF-passenger estimates have no covariance (because they are based on entirely different data); likewise for the "far-side" estimates.

The top half of Table 3-3a includes cars and LTVs; the lower half, cars only. Results are quite similar. The overall average effect of aging one year is slightly higher in frontal impacts ($3.20 \pm .09\%$) and nearside impacts ($3.19 \pm .14\%$) than in far-side impacts ($2.84 \pm .14\%$) and rear/other crashes ($2.74 \pm .27\%$), with first-event rollovers in between ($3.12 \pm .25\%$). Table 3-3b suggests that the effect of aging in frontal and nearside impacts is especially high for the older occupants, whereas in rollovers, it is high for the younger occupants but escalates little at the older ages. But Table 3-3c suggests that the risk ratio for a 75-year-old relative to a 21-year-old is fairly similar for the various impact types. On the whole, the effects of aging show a lot more similarities than differences across impact types.

Table 3-3d suggests that the added risk for a female relative to a male of the same age is, on the average, highest in nearside impacts ($24.4 \pm 4.7\%$ higher risk for females) and rollovers ($22.2 \pm 4.9\%$). The effect is especially high for female RF passengers in nearside impacts and rollovers, both for young passengers and older ones. Added risk for females is lower in frontals ($14.1 \pm 2.2\%$), far-side impacts ($13.0 \pm 4.1\%$) and, probably, rear/other crashes ($17.7 \pm 5.4\%$). As will be discussed shortly, air bags are a technology that especially protects females; whereas many vehicles in the database are equipped with frontal air bags, a smaller percentage is equipped with curtains and/or torso air bags that would help in the nearside impacts. . .

TABLE 3-3a: INCREASE IN FATALITY RISK FOR EACH YEAR THAT AN OCCUPANT GETS OLDER **BY IMPACT TYPE** – CARS AND LTVs, DRIVERS AND RIGHT-FRONT PASSENGERS
(Given the same crash scenario, average for occupants 21 to 96, FARS 1975-2010)

Percent Fatality Increase per Year of Aging

	Males			Females			Average of Males & Females		
	Driver	RF Pass	Average	Driver	RF Pass	Average	Driver	RF Pass	Average
CARS PLUS LTVs									
All impacts	3.06 ± .09	3.51 ± .09	3.28 ± .08	2.60 ± .10	3.27 ± .08	2.93 ± .08	2.83 ± .08	3.39 ± .07	3.11 ± .08
Frontal impacts	3.17 ± .11	3.55 ± .11	3.36 ± .09	2.64 ± .15	3.44 ± .10	3.04 ± .10	2.91 ± .11	3.49 ± .09	3.20 ± .09
Nearside impacts	3.21 ± .23	3.36 ± .21	3.28 ± .16	2.84 ± .24	3.15 ± .23	2.99 ± .17	3.02 ± .21	3.36 ± .18	3.19 ± .14
Far-side impacts	2.74 ± .27	3.48 ± .23	3.11 ± .17	2.13 ± .21	3.24 ± .17	2.68 ± .14	2.43 ± .20	3.26 ± .20	2.84 ± .14
First-event rollovers	3.16 ± .31	3.47 ± .34	3.31 ± .27	2.48 ± .42	3.35 ± .28	2.92 ± .28	2.82 ± .29	3.41 ± .26	3.12 ± .25
Rear impacts & other crashes	2.65 ± .37	3.18 ± .23	2.92 ± .27	2.33 ± .34	2.79 ± .32	2.56 ± .30	2.49 ± .31	2.99 ± .27	2.74 ± .27
CARS ONLY									
All impacts	3.04 ± .09	3.56 ± .09	3.30 ± .08	2.65 ± .10	3.31 ± .08	2.98 ± .08	2.85 ± .09	3.44 ± .07	3.14 ± .08
Frontal impacts	3.17 ± .12	3.64 ± .13	3.41 ± .11	2.71 ± .16	3.50 ± .11	3.11 ± .13	2.94 ± .12	3.58 ± .11	3.26 ± .11
Nearside impacts	3.27 ± .27	3.38 ± .27	3.33 ± .19	2.87 ± .22	3.17 ± .27	3.02 ± .18	3.07 ± .22	3.27 ± .26	3.17 ± .17
Far-side impacts	2.62 ± .27	3.47 ± .23	3.05 ± .18	2.13 ± .24	3.27 ± .19	2.70 ± .15	2.38 ± .23	3.37 ± .16	2.87 ± .14
First-event rollovers	3.17 ± .42	3.55 ± .44	3.35 ± .39	2.47 ± .56	3.32 ± .42	2.89 ± .44	2.82 ± .42	3.43 ± .37	3.13 ± .39
Rear impacts & other crashes	2.47 ± .35	3.23 ± .32	2.85 ± .34	2.34 ± .39	2.85 ± .35	2.60 ± .32	2.40 ± .34	3.05 ± .30	2.72 ± .29

TABLE 3-3b

INCREASE IN FATALITY RISK FOR EACH YEAR THAT A MALE OCCUPANT GETS OLDER
AT AGE 21-30 VERSUS AGE 65-74
(Given the same crash scenario, FARS 1975-2010)

BY IMPACT TYPE
CARS AND LTVs, DRIVERS AND RIGHT-FRONT PASSENGERS

	Percent Fatality Increase per Year of Aging					
Seat Position →→→→→→→→→→→	Driver			RF Passenger		
Occupant Age Group →→→→→→→	Avg 21-96	21-30	65-74	Avg 21-96	21-30	65-74
CARS PLUS LTVs						
All impacts	3.06 ± .09	2.90	3.39	3.51 ± .09	2.90	4.58
Frontal impacts	3.17 ± .11	3.11	3.38	3.55 ± .11	2.91	4.68
Nearside impacts	3.21 ± .23	2.95	3.72	3.36 ± .21	2.86	4.53
Far-side impacts	2.74 ± .27	2.34	3.36	3.49 ± .23	2.92	4.24
First-event rollovers	3.16 ± .31	3.40	2.96	3.46 ± .34	3.40	3.71
Rear impacts & other crashes	2.65 ± .37	2.16	3.38	3.18 ± .23	2.40	4.49
CARS ONLY						
All impacts	3.04 ± .09	2.83	3.45	3.56 ± .09	2.93	4.74
Frontal impacts	3.17 ± .12	3.11	3.39	3.64 ± .13	2.99	4.87
Nearside impacts	3.27 ± .27	3.04	3.76	3.38 ± .27	2.86	4.63
Far-side impacts	2.62 ± .27	2.08	3.45	3.47 ± .23	2.83	4.31
First-event rollovers	3.17 ± .42	3.42	2.96	3.55 ± .44	3.36	3.99
Rear impacts & other crashes	2.47 ± .35	2.15	2.98	3.23 ± .32	2.48	4.60

TABLE 3-3c

FATALITY RISK FOR A 75-YEAR-OLD OCCUPANT
RELATIVE TO A 21-YEAR-OLD OCCUPANT
(Given similar physical insults, FARS 1975-2010)

CARS AND LTVs, DRIVERS AND RIGHT-FRONT PASSENGERS
BY IMPACT TYPE

| | Risk at Age 75 ÷ Risk at Age 21 | | | |
| | Males | | Females | |
	Drivers	RF Pass	Drivers	RF Pass
CARS PLUS LTVs				
All impacts	5.04	6.70	3.87	5.67
Frontal impacts	5.33	6.90	3.94	6.41
Nearside impacts	5.50	6.56	4.38	5.34
Far-side impacts	4.30	6.26	3.07	5.38
First-event rollovers	5.29	6.48	3.57	6.01
Rear impacts & other crashes	4.08	5.59	3.41	4.22
CARS ONLY				
All impacts	4.99	6.94	3.98	5.75
Frontal impacts	5.33	7.30	4.10	6.58
Nearside impacts	5.66	6.70	4.41	5.35
Far-side impacts	4.09	6.17	3.11	5.49
First-event rollovers	5.29	6.74	3.54	5.77
Rear impacts & other crashes	3.69	5.83	3.44	4.38

TABLE 3-3d: INCREASE IN FATALITY RISK FOR A FEMALE RELATIVE TO A MALE OF THE SAME AGE **BY IMPACT TYPE** – CARS AND LTVs, DRIVERS AND RIGHT-FRONT PASSENGERS
(Given the same crash scenario, average for occupants 21 to 96, FARS 1975-2010)

Percent Fatality Increase for Females

Seat Position →	Drivers			RF Passengers			Avg of Driver & RF
Occupant Age Group →	21-96	21-30	65-74	21-96	21-30	65-74	21-96
CARS PLUS LTVs							
All impacts	13.4 ± 2.0	25.9	- 1.4	20.5 ± 2.2	29.2	11.4	17.0 ± 1.5
Frontal impacts	11.4 ± 3.0	25.7	- 5.4	16.7 ± 3.3	20.9	13.0	14.1 ± 2.2
Nearside impacts	18.1 ± 7.1	30.2	3.7	30.7 ± 6.2	41.9	14.5	24.4 ± 4.7
Far-side impacts	16.9 ± 5.6	28.7	.6	9.0 ± 6.1	15.7	2.9	13.0 ± 4.1
First-event rollovers	10.7 ± 6.7	30.8	- 11.7	33.8 ± 10.3	38.6	30.1	22.2 ± 4.9
Rear impacts & other crashes	10.3 ± 9.4	18.4	.0	25.2 ± 8.0	39.7	9.0	17.7 ± 5.4
CARS ONLY							
All impacts	15.8 ± 2.7	26.9	2.5	17.8 ± 2.6	27.2	7.4	16.8 ± 1.8
Frontal impacts	13.9 ± 3.5	27.2	- 1.8	14.2 ± 3.5	19.6	8.5	14.0 ± 2.2
Nearside impacts	16.2 ± 8.2	28.9	.9	28.7 ± 7.6	40.3	11.2	22.4 ± 5.6
Far-side impacts	21.7 ± 7.4	30.9	8.7	10.7 ± 6.9	15.5	6.5	16.2 ± 5.1
First-event rollovers	15.2 ± 11.2	37.6	- 9.4	37.7 ± 15.1	47.4	28.3	26.4 ± 7.0
Rear impacts & other crashes	8.3 ± 10.3	12.8	2.5	16.4 ± 7.7	30.1	.6	12.3 ± 5.8

3.6   Variations by impact location and type of occupant protection

Table 3-4a builds on the analysis of passenger cars in the lower half of Table 3-3a. The five impact locations are further subdivided by the type of occupant protection available and in use. In frontal impacts, where there are many crash cases, it is possible to consider five types of occupant protection, based on belt use and frontal-air-bag availability (and, these being frontal crashes severe enough to result in fatalities, it is likely that the air bags deployed in most of the cases). The effect of aging one year is, on the average, fairly similar for those five types (3.22 ± .15% for unbelted occupants without air bags, 3.17 ± .33% for 3-point belted occupants without air bags; a possibly lower 2.85 ± .45% for unbelted occupants with air bags; 3.19 ± .45% with belts and air bags, but no pretensioners or load limiters on the belts; and 3.22 ± .61% with pretensioning, load-limiting belts and air bags) – at least, with these somewhat wide confidence bounds, it is not really possible to claim differences among the groups.

For the non-frontal impact types, the availability of frontal air bags is usually not relevant (because they do not deploy or, if they deployed, were less influential than in frontal impacts). It is sufficient to just compare belted and unbelted occupants. However, for side impacts, the influence of curtain and side air bags is of great interest. Unfortunately, there are relatively few vehicles equipped with them in the current database. Even combining nearside and far-side impacts and including the LTV cases yields a total of only 744 side impacts with curtain and side air bags (as compared to a range of 6,192 to 12,148 cases in the other four side-impact groups – see Appendix F); 744 cases is barely enough to make the estimates of the overall effects of aging and gender statistically meaningful, let alone more detailed analyses, such as those in Table 3-4b, of how the effect changes with age.[33]

Table 3-4a shows that, in cars without curtain or side air bags, the effect of aging one year is about the same for unbelted occupants (3.08 ± .22%) and belted occupants (3.19 ± .32%) in nearside impacts; likewise in far-side impacts (2.90 ± .20% versus 2.94 ± .30%). But in side impacts of vehicles equipped with separate curtain and torso bags or with combination bags, the observed effect of aging one year is considerably less: 2.28 ± 1.22 percent. In fact, the lowest point estimate is half a percentage point lower than any other point estimate of the effect of aging (average for male and female driver and RF passenger) so far. But due to the high sampling error, it is premature to draw firm conclusions; the analysis would need to be repeated in about three or four years to see if the result holds up. Similarly, Table 3-4c shows an exceptionally low cumulative risk increase from 21 to 75 in side impacts of vehicles with curtain-plus-torso or combination air bags.

The analyses of first-event rollovers and rear/other crashes show similar or at least not significantly different effects of aging one year for belted and unbelted occupants.

---

[33] Table 3-4b shows an exceptional escalation of the aging effect at 65 to 74 relative to at 21-30 with curtain and side air bags. But these estimates are unlikely to be accurate, because the effectiveness analysis for curtain and side air bags in Section 9.7 shows that they are actually especially beneficial for older occupants.

TABLE 3-4a: INCREASE IN FATALITY RISK FOR EACH YEAR THAT AN OCCUPANT GETS OLDER
BY TYPE OF IMPACT AND OCCUPANT PROTECTION – CARS ONLY, DRIVERS AND RIGHT-FRONT PASSENGERS

Percent Fatality Increase per Year of Aging

	Males			Females			Average of Males & Females		
	Driver	RF Pass	Average	Driver	RF Pass	Average	Driver	RF Pass	Average
FRONTAL IMPACTS									
Unbelted occupants of cars without air bags	3.13 ± .17	3.64 ± .24	3.39 ± .17	2.73 ± .19	3.39 ± .14	3.06 ± .17	2.93 ± .17	3.52 ± .18	3.22 ± .15
3-pt. belted occupants of cars w/o air bags	2.83 ± .41	3.78 ± .42	3.30 ± .36	2.39 ± .44	3.69 ± .37	3.04 ± .34	2.61 ± .37	3.74 ± .39	3.17 ± .33
Unbelted occupants of cars with dual air bags	2.45 ± .53	3.30 ± .58	2.88 ± .46	2.77 ± .58	2.88 ± .54	2.82 ± .46	2.61 ± .52	3.09 ± .51	2.85 ± .45
Belted, dual air bags, no pretens/load lim	3.28 ± .55	3.44 ± .63	3.36 ± .52	2.56 ± .68	3.47 ± .55	3.02 ± .49	2.92 ± .50	3.45 ± .58	3.19 ± .45
Belted, dual air bags, pretensioners, load lim	3.34 ± .80	3.31 ± .69	3.32 ± .66	2.70 ± .81	3.54 ± .65	3.12 ± .66	3.02 ± .73	3.43 ± .66	3.22 ± .61
NEARSIDE IMPACTS									
Unbelted occs of cars w/o side/curtain bags	3.02 ± .31	3.24 ± .36	3.13 ± .24	3.00 ± .38	3.07 ± .32	3.03 ± .25	3.01 ± .28	3.15 ± .33	3.08 ± .22
3-pt belted occs of cars w/o side/curtain bags	3.34 ± .46	3.51 ± .53	3.43 ± .35	2.75 ± .46	3.17 ± .51	2.96 ± .34	3.05 ± .42	3.34 ± .48	3.19 ± .32
FAR-SIDE IMPACTS									
Unbelted occs of cars w/o side/curtain bags	2.78 ± .34	3.38 ± .34	3.08 ± .24	2.29 ± .30	3.15 ± .26	2.72 ± .20	2.53 ± .27	3.26 ± .28	2.90 ± .20
3-pt belted occs of cars w/o side/curtain bags	2.52 ± .60	3.75 ± .42	3.14 ± .37	1.90 ± .44	3.58 ± .47	2.74 ± .32	2.21 ± .44	3.67 ± .41	2.94 ± .30
ALL SIDE IMPACTS									
Cars/LTVs w curtain+torso or combo bags	2.45 ±1.31	2.30 ± 1.43	2.38 ±1.24	2.01 ±1.47	2.36 ±1.28	2.18 ±1.18	2.23 ±1.17	2.33 ±1.36	2.28 ±1.22
FIRST-EVENT ROLLOVERS									
Unbelted occupants	2.64 ± .50	3.34 ± .52	2.99 ± .45	2.25 ± .64	3.13 ± .48	2.69 ± .45	2.45 ± .46	3.24 ± .44	2.84 ± .40
3-point belted occupants	3.66 ±1.10	3.39 ±1.10	3.52 ±1.01	2.98 ±1.21	3.29 ± .93	3.13 ±1.05	3.32 ±1.09	3.34 ± .99	3.33 ±1.06
REAR IMPACTS & OTHER CRASHES									
Unbelted occupants	2.36 ± .41	3.03 ± .49	2.69 ± .40	2.17 ± .58	2.68 ± .43	2.42 ± .43	2.26 ± .48	2.85 ± .40	2.56 ± .42
3-point belted occupants	2.24 ± .59	3.15 ± .69	2.69 ± .59	2.28 ± .70	2.71 ± .69	2.50 ± .61	2.26 ± .63	2.93 ± .68	2.59 ± .62

TABLE 3-4b: INCREASE IN FATALITY RISK FOR EACH YEAR
THAT A MALE OCCUPANT GETS OLDER
AT AGE 21-30 VERSUS AGE 65-74
(Given the same crash scenario, FARS 1975-2010)

BY TYPE OF IMPACT AND OCCUPANT PROTECTION
CARS ONLY, DRIVERS AND RIGHT-FRONT PASSENGERS

Percent Fatality Increase per Year of Aging

Seat Position →→→→→→→→→→→	Driver			RF Passenger		
Occupant Age Group →→→→→→→→	Avg 21-96	21-30	65-74	Avg 21-96	21-30	65-74
FRONTAL IMPACTS						
Unbelted occupants of cars without air bags	3.13 ± .17	3.38	2.97	3.64 ± .24	3.13	4.68
3-pt. belted occupants of cars w/o air bags	2.83 ± .41	2.24	3.74	3.78 ± .42	2.82	5.52
Unbelted occupants of cars with dual air bags	2.45 ± .53	2.09	3.00	3.30 ± .58	2.79	4.39
Belted, dual air bags, no pretens/load lim	3.28 ± .55	3.20	3.50	3.44 ± .63	2.54	4.88
Belted, dual air bags, pretensioners, load lim	3.34 ± .80	3.33	3.46	3.31 ± .69	2.00	5.30
NEARSIDE IMPACTS						
Unbelted occs of cars w/o side/curtain bags	3.02 ± .31	3.15	2.92	3.24 ± .36	2.63	4.51
3-pt belted occs of cars w/o side/curtain bags	3.34 ± .46	3.00	4.01	3.51 ± .53	3.05	4.56
FAR-SIDE IMPACTS						
Unbelted occs of cars w/o side/curtain bags	2.78 ± .34	2.35	3.42	3.38 ± .34	3.13	3.80
3-pt belted occs of cars w/o side/curtain bags	2.52 ± .60	1.56	3.94	3.75 ± .42	2.94	4.72
ALL SIDE IMPACTS						
Cars/LTVs w curtain+torso or combo bags	2.45 ±1.30	1.29	4.18	2.31 ±1.43	.45	5.21
FIRST-EVENT ROLLOVERS						
Unbelted occupants	2.64 ± .50	3.56	1.55	3.34 ± .52	3.56	3.35
3-point belted occupants	3.66 ±1.10	3.21	4.46	3.39 ±1.10	2.41	4.74
REAR IMPACTS & OTHER CRASHES						
Unbelted occupants	2.36 ± .41	2.48	2.29	3.03 ± .49	2.78	3.61
3-point belted occupants	2.24 ± .59	1.42	3.45	3.15 ± .69	1.63	5.86

TABLE 3-4c: FATALITY RISK FOR A 75-YEAR-OLD OCCUPANT RELATIVE TO A 21-YEAR-OLD OCCUPANT
(Given similar physical insults, FARS 1975-2010)

CARS ONLY, DRIVERS AND RIGHT-FRONT PASSENGERS
BY TYPE OF IMPACT AND OCCUPANT PROTECTION

	Risk at Age 75 ÷ Risk at Age 21			
	Males		Females	
	Drivers	RF Pass	Drivers	RF Pass
FRONTAL IMPACTS				
Unbelted occs of cars without air bags	5.20	7.31	4.12	6.08
3-pt. belted occs of cars w/o air bags	4.52	8.20	3.52	7.72
Unbelted occs of cars with dual air bags	3.72	6.21	4.46	4.30
Belted, dual air bags, no pretens/load lim	5.66	6.47	3.75	6.61
Belted, dual air bags, pretensioners, load lim	5.85	6.10	4.07	7.43
NEARSIDE IMPACTS				
Unbelted occs of cars w/o side/curtain bags	4.94	6.06	4.87	5.09
3-pt belted occs of cars w/o side/curtain bags	5.86	7.21	4.11	5.18
FAR-SIDE IMPACTS				
Unbelted occs of cars w/o side/curtain bags	4.39	6.00	3.35	5.23
3-pt belted occs of cars w/o side/curtain bags	3.92	7.10	2.83	6.44
ALL SIDE IMPACTS				
Cars/LTVs w curtain+torso or combo bags	3.78	3.63	2.94	3.80
FIRST-EVENT ROLLOVERS				
Unbelted occupants	4.02	6.17	3.25	5.26
3-point belted occupants	7.04	6.22	4.62	5.92
REAR IMPACTS & OTHER CRASHES				
Unbelted occupants	3.48	5.10	3.10	3.99
3-point belted occupants	3.35	6.02	3.46	4.05

TABLE 3-4d: INCREASE IN FATALITY RISK FOR A FEMALE RELATIVE TO A MALE OF THE SAME AGE
BY TYPE OF IMPACT AND OCCUPANT PROTECTION – CARS ONLY, DRIVERS AND RIGHT-FRONT PASSENGERS

Percent Fatality Increase for Females

Seat Position →→→→→→→→→→	Drivers			RF Passengers			Avg of Driver & RF
Occupant Age Group →→→→→→→→	21-96	21-30	65-74	21-96	21-30	65-74	21-96
FRONTAL IMPACTS							
Unbelted occupants of cars without air bags	15.3 ± 5.0	27.9	1.0	12.5 ± 5.8	20.9	3.5	13.9 ± 3.7
3-pt. belted occupants of cars w/o air bags	21.6 ± 9.4	34.2	6.8	12.2 ± 9.3	15.5	9.4	16.9 ± 6.3
Unbelted occupants of cars with dual air bags	11.1 ± 16.8	6.6	17.9	2.7 ± 16.4	21.2	-16.4	6.9 ± 7.2
Belted, dual air bags, no pretens/load lim	7.6 ± 17.0	24.6	-12.5	7.3 ± 16.4	6.7	8.0	7.5 ± 12.5
Belted, dual air bags, pretensioners, load lim	5.6 ± 17.2	22.4	-14.9	13.1 ± 19.0	7.6	22.3	9.3 ± 10.6
NEARSIDE IMPACTS							
Unbelted occs of cars w/o side/curtain bags	17.0 ± 9.2	19.7	13.9	26.2 ± 8.8	35.5	12.8	21.6 ± 6.3
3-pt belted occs of cars w/o side/curtain bags	7.7 ± 15.4	22.0	-8.9	23.3 ± 15.2	37.3	.7	15.5 ± 10.8
FAR-SIDE IMPACTS							
Unbelted occs of cars w/o side/curtain bags	23.6 ± 9.5	33.6	10.7	12.1 ± 7.3	17.6	7.3	17.8 ± 6.0
3-pt belted occs of cars w/o side/curtain bags	29.8 ± 15.5	40.4	13.7	7.3 ± 15.2	12.4	2.9	18.5 ± 10.9
ALL SIDE IMPACTS							
Cars/LTVs w curtain+torso or combo bags	17.8 ± 31.7	26.3	6.5	-12.7 ± 22.1	-14.6	-9.7	2.6 ± 21.5
FIRST-EVENT ROLLOVERS							
Unbelted occupants	26.9 ± 15.3	43.3	9.5	39.7 ± 16.9	51.0	30.0	33.3 ± 9.2
3-point belted occupants	-8.1 ± 18.4	8.5	-28.2	19.2 ± 25.7	18.9	19.8	5.5 ± 10.3
REAR IMPACTS & OTHER CRASHES							
Unbelted occupants	12.3 ± 15.2	19.9	3.2	17.9 ± 9.7	30.0	4.5	15.1 ± 7.7
3-point belted occupants	-.8 ± 16.3	-1.3	1.7	13.0 ± 19.7	30.4	-8.4	6.5 ± 10.2

Table 3-4b shows that the estimated effect of aging one year for unbelted drivers is actually a bit lower at 65 to 74 than at 21 to 30 in four of the five impact types (all except far-side). Again, this is consistent with Evans' results for mostly (97%) unbelted drivers, which showed the effect of aging leveling off rather than intensifying at the higher ages.

Table 3-4d shows that the risk increase for females has diminished in frontal impacts when cars are equipped with frontal air bags: all three point estimates for the average of driver and RF passenger are lower than 10 percent. A specific question is whether the certified advanced compliant (CAC) air bags phased in during MY 2004 to 2007, which must meet a barrier test using 5th-percentile-female dummies, are especially beneficial for female occupants.[34] It cannot directly be answered with the current database, because even in the most recent group of cars in Table 3-4d (with pretensioners and load limiters), only 34 percent of the cars are equipped with CAC air bags. However, it should be noted even without pretensioners and load limiters (where none of the cars have CAC air bags) and for unbelted occupants of cars with dual air bags (only 5% of which are CAC air bags) the risk increase for females is more or less equally low. This suggests air bags were already designed with attention to protecting female occupants even before the MY 2004-2007 phase-in of the test with the small-adult dummy.

In side impacts of vehicles with curtain-plus-torso or combination bags, the observed risk increase for females is likewise close to zero, 2.6 ± 21.5 percent, although there is much uncertainty due to the limited sample.

First-event rollovers show the greatest contrast between unbelted and belted females. The risk increase for unbelted females relative to unbelted males is exceptionally great: 33.3 ± 9.2 percent. But belted females only have 5.5 ± 10.3 percent higher risk than belted males. A hypothesis is that the results reflect differences in anatomy rather than the vehicle's crashworthiness technologies. Approximately 73 percent of the unrestrained fatalities in rollover crashes are ejected.[35] Females, being on the average smaller than males, can pass more easily through ejection portals when they are unrestrained: specifically the side-window area, a relatively small, but frequent ejection portal. The hypothesis is supported by a double-pair comparison analysis of CY 1991-2010 FARS rollovers, which shows that an unbelted female is 17 percent more likely to be ejected than an unbelted male, given the same type of rollover crash.[36] Belted fatalities in rollovers, on the other hand, may involve contact with intruding structures, such as roof crush. Here, small stature could be a possible advantage for belted females if it keeps them further away from the roof.

---

[34] Kahane, C. J. (2006, August). *An Evaluation of the 1998-1999 Redesign of Frontal Air Bags.* (Report No. DOT HS 810 685, p. 20). Washington, DC: National Highway Traffic Safety Administration. Available at www-nrd.nhtsa.dot.gov/Pubs/810685.PDF

[35] Kahane, C. J. (2000). *Fatality Reduction by Safety Belts for Front-Seat Occupants of Cars and Light Trucks: Updated and Expanded Estimates Based on 1986-99 FARS Data.* (Report No. DOT HS 809 199, p. 31). Washington, DC: National Highway Traffic Safety Administration. Available at www-nrd.nhtsa.dot.gov/Pubs/809199.PDF

[36] The analysis is similar to the basic double-pair comparison to estimate belt effectiveness, described at the beginning of Section 9.3 of this report (before the discussion of CATMOD), except that ejected fatalities (rather than all fatalities) are tabulated by the drivers' and RF passengers' gender (rather than by their belt use).

3.7   Effects of aging and gender for back-seat outboard occupants

FARS data for 1975-2010 contains 22,502 pairs consisting of a driver and a back-seat outboard occupant riding in the same vehicle, a MY 1960-2011 passenger car, where the driver's and the passenger's age are both in the range of 21 to 96 and at least one of them was a fatality. It is far fewer than the 154,467 driver-RF pairs that were the basis for the preceding analyses. The smaller number of cases will increase sampling error and limit the analysis of back-seat occupants to a few basic subsets of the data. There are fewer cases because occupancy of the back seat is lower than the RF seat and many of the back-seat occupants are younger than 21; also, LTVs have been excluded because their back seats vary so much – from no back seat or a limited seat in many pickup trucks to multiple rows in passenger vans. Slightly offsetting these factors that limit the number of cases, multiple records may be created for a single vehicle if there are multiple adult back-seat occupants. Appendix F states the N of cases for each analysis.

Table 3-5a estimates the increases in fatality risk when back-seat occupants age one year. The analysis method is identical to the earlier sections of this chapter and produces estimates for the driver as well as the back-seat occupant, but only the results for the back-seat occupants are shown in Table 3-5a. Overall, the increase in fatality risk for aging a year, averaged for males and females is 3.46 ± .14 percent; it is 3.57 ± .18 percent for males and 3.35 ± .14 percent for females. These estimates are virtually identical to the corresponding effects for right-front passengers of passenger cars (3.44 ± .07%, 3.56 ± .09%, and 3.31 ± .08%, according the third row of Table 3-1a, labeled "cars only"). Table 3-5b shows the risk increase intensifies with age, from an estimated 2.87 percent per year at ages 21-30 to 4.96 percent at 65 to 74; again, these are close to the effects for RF passengers (2.93% and 4.74% according to Table 3-1b). From 21 to 75, fatality risk has increased by a factor of 7.12 for males and 6.03 for females, according to Table 3-5c (the cumulative increases for RF passengers are 6.94 and 5.75, respectively, in Table 3-1c).

However, unlike the RF passengers, the effects of aging are different for belted and unbelted back-seat occupants. The risk increase for unbelted occupants aging a year is 3.31 ± .19 percent in the generation of cars without driver air bags and 2.38 ± .58 percent in the more recent generation of cars equipped with driver or dual air bags (based on relatively few cases, because most people are belted now) – averaging out to 3.18 ± .18 percent, slightly less than corresponding results for unbelted RF passengers (see Table 3-2a).[37] But for lap-belted back-seat passengers, the increase is 3.98 ± .82 percent; for 3-point belted passengers, it is an average of 3.94 ± .47 percent. These are among the highest point estimates we have seen so far, exceeded only by the 4.26 ± .36 percent for RF passengers with 2-point automatic belts in Table 3-2a.

---

[37] The estimates for the averages of the two generations of unbelted back-seat occupants are derived as follows. Point estimate: mean of the point estimates, weighted by N of crash cases. Confidence bounds: assumes the estimates for the two different generations have no covariance (because they are based on entirely different data). Likewise for the two generations of 3-point belted occupants.

TABLE 3-5a

BACK-SEAT OUTBOARD OCCUPANTS OF PASSENGER CARS
INCREASE IN FATALITY RISK FOR EACH YEAR THAT AN OCCUPANT GETS OLDER
BY TYPE OF OCCUPANT PROTECTION
(Given the same crash scenario, average for occupants 21 to 96, FARS 1975-2010)

	Percent Fatality Increase per Year of Aging		
	Males	Females	Average of M and F
All back-seat occupants of cars, MY 1960-2011	**3.57 ± .18**	**3.35 ± .14**	**3.46 ± .14**
FIVE "GENERATIONS" OF BACK-SEAT OCCUPANTS OF CARS			
Unbelted back-seat occs of cars w/o air bags	3.47 ± .26	3.15 ± .20	3.31 ± .19
Lap-belted back-seat occs of pre-1990 cars w/o air bags	4.14 ± .83	3.81 ± .83	3.98 ± .82
3-point belted back-seat occs of cars w/o air bags	4.38 ± .89	3.91 ± .88	4.15 ± .84
Unbelted back-seat occs of cars w driver/dual air bags	2.42 ± .72	2.34 ± .68	2.38 ± .58
3-pt belted back-seat occs of cars w driver/dual air bags	4.04 ± .60	3.74 ± .49	3.89 ± .54
Weighted average: unbelted back-seat occupants	3.32 ± .24	3.03 ± .19	3.18 + .18
Weighted average: 3-point belted back-seat occupants	4.10 ± .52	3.77 ± .44	3.94 ± .47

TABLE 3-5b

BACK-SEAT OUTBOARD OCCUPANTS OF PASSENGER CARS
INCREASE IN FATALITY RISK FOR EACH YEAR THAT A MALE OCCUPANT GETS OLDER
AT AGE 21-30 VERSUS AGE 65-74
BY TYPE OF OCCUPANT PROTECTION
(Given the same crash scenario, FARS 1975-2010)

	Percent Fatality Increase per Year of Aging		
Occupant Age Group →→→→→→→→→→→→→→→	21-96	21-30	65-74
All back-seat occupants of cars, MY 1960-2011	**3.57 ± .18**	**2.87**	**4.96**
FIVE "GENERATIONS" OF BACK-SEAT OCCUPANTS OF CARS			
Unbelted back-seat occupants of cars w/o air bags	3.47 ± .26	3.22	4.13
Lap-belted back-seat occs of pre-1990 cars w/o air bags	4.15 ± .84	3.14	6.27
3-point belted back-seat occs of cars w/o air bags	4.38 ± .89	3.39	7.02
Unbelted back-seat occs of cars with driver/dual air bags	2.42 ± .72	2.01	3.16
3-pt belted back-seat occs of cars with driver/dual air bags	4.04 ± .60	3.07	6.19
Weighted average: unbelted back-seat occupants	3.32 ± .24	3.04	3.99
Weighted average: 3-point belted back-seat occupants	4.10 ± .52	3.13	6.34

80

TABLE 3-5c

BACK-SEAT OUTBOARD OCCUPANTS OF PASSENGER CARS
FATALITY RISK FOR A 75-YEAR-OLD OCCUPANT
RELATIVE TO A 21-YEAR-OLD OCCUPANT
BY TYPE OF OCCUPANT PROTECTION
(Given similar physical insults, FARS 1975-2010)

OVERALL AND BY TYPE OF OCCUPANT PROTECTION

	Risk at Age 75 ÷ Risk at Age 21	
	Males	Females
All back-seat occupants of cars, MY 1960-2011	**7.12**	**6.03**

FIVE "GENERATIONS" OF BACK-SEAT OCCUPANTS OF CARS

	Males	Females
Unbelted back-seat occupants of cars w/o air bags	6.57	5.19
Lap-belted back-seat occs of pre-1990 cars w/o air bags	10.50	7.66
3-point belted back-seat occs of cars w/o air bags	13.80	9.02
Unbelted back-seat occs of cars with driver/dual air bags	3.76	3.62
3-pt belted back-seat occs of cars with driver/dual air bags	9.83	7.44
Weighted average: unbelted back-seat occupants	6.07	4.93
Weighted average: 3-point belted back-seat occupants	10.45	7.70

TABLE 3-5d

BACK-SEAT OUTBOARD OCCUPANTS OF PASSENGER CARS
INCREASE IN FATALITY RISK FOR A FEMALE RELATIVE TO A MALE OF THE SAME AGE
BY TYPE OF OCCUPANT PROTECTION
(Given the same crash scenario, average for occupants 21 to 96, FARS 1975-2010)

	Percent Fatality Increase for Females		
Occupant Age Group →→→→→→→→→→→→→→→→→→→→→	21-96	21-30	65-74
All back-seat occupants of cars, MY 1960-2011	**15.7 ± 6.1**	**23.8**	**7.1**
FIVE "GENERATIONS" OF BACK-SEAT OCCUPANTS OF CARS			
Unbelted back-seat occs of cars w/o air bags	12.2 ± 8.0	23.7	.6
Lap-belted back-seat occs of pre-1990 cars w/o air bags	31.9 ± 28.3	49.7	11.5
3-point belted back-seat occs of cars w/o air bags	30.4 ± 53.2	49.5	9.0
Unbelted back-seat occs of cars with driver/dual air bags	5.8 ± 15.7	5.5	5.0
3-point belted back-seat occs of cars with driver/dual air bags	27.5 ± 21.4	42.6	9.6
Weighted average: unbelted back-seat occupants	11.3 ± 7.2	21.1	1.3
Weighted average: 3-point belted back-seat occupants	28.0 ± 20.0	43.8	9.5

Furthermore, Table 3-5b shows the effect of aging intensifies for belted occupants as they get older, reaching 6.27 percent at 65 to 74 with lap belts and 6.34 percent with 3-point belts. Table 3-5c shows that the risk for a belted 75-year-old male is over 10 times as high as for a belted 21-year-old; the only other 10-fold increase so far in these tables was for 2-point automatic belts in the RF seat. Belts are not as effective for senior citizens as for young people in the back seat. Specifically, Table 9-5 of this report estimates effectiveness for four "generations" of 3-point belts in passenger cars, separately for drivers and RF passengers. These eight estimates range from 44 to 54 percent fatality reduction for people age 49 and younger, and each of these respective estimates ranges from 1 to 13 percentage points lower for people age 55 and older. But for back-seat occupants, Table 9-9 estimates a 61 percent fatality reduction for people age 49 and younger, but only a 30 percent reduction for age 55 and older.[38] These contrasting results help explain why the **relative** effect of aging is higher for back-seat occupants. Nevertheless, in **absolute** terms, belts are still effective for senior citizens in the back seat; a belted occupant is safer than an unbelted occupant of the same age.

Table 3-5d indicates risk increases for females that parallel the effects of aging. Overall, female back-seat occupants have an average of 15.7 ± 6.1 percent higher fatality risk than males of the same age, given similar physical insults. That is similar to the 17.8 ± 2.6 percent effect for RF passengers of cars (Table 3-1d). However, the increase is only 11.3 ± 7.2 percent for unbelted females relative to unbelted males, whereas with lap belts, females are at 31.9 ± 28.3 percent higher risk than males, and with 3-point belts, 28.0 ± 20.0 percent higher risk. In the front seat, belt use in most cases helped close the gender gap even while it sometimes possibly intensified the aging effect. In the back seat, belt use intensifies the added risk for aging and likely also for females relative to males.

---

[38] See also: Morgan, C. (1999, June). *Effectiveness of Lap/Shoulder Belts in the Back Outboard Seating Positions.* (Report No. DOT HS 808 945, pp 47-51). Washington, DC: National Highway Traffic Safety Administration. Available at www-nrd.nhtsa.dot.gov/Pubs/808945.PDF

CHAPTER 4

EFFECT OF OCCUPANT AGE AND GENDER ON INJURIES THAT "CONTRIBUTED TO DEATH" – ANALYSES OF 1987-2007 FARS-MCOD DATA

4.0   Summary

The FARS-MCOD file lists a fatally injured occupant's injuries that "contributed to death." Based on analyses of the relative injury risk of two occupants of the same vehicle as a function of their ages and genders, head-injury risk is estimated to increase by an average of $2.49 \pm .17$ percent for each year that a front-seat occupant gets older, torso-injury risk by $3.80 \pm .17$ percent (specifically, chest injury increases by $4.06 \pm .21\%$ and abdominal injury by $3.43 \pm .32\%$), and neck-injury risk by $3.19 \pm .37$ percent. Head-injury risk is an average of 14.6 ± 3.1 percent higher for a female than for a male of the same age, in the same seat position, exposed to similar physical insults; the corresponding increase for torso injuries is 13.3 ± 3.6 percent ($8.8 \pm 4.6\%$ for chest injuries and $31.9 \pm 8.3\%$ for abdominal injuries), but for neck injuries, a substantially higher 39.4 ± 9.4 percent. In other words, older occupants are exceptionally vulnerable to torso injuries, especially chest injuries, but females to neck and abdominal injuries. These effects are fairly similar across the various types of crashes: frontals, side impacts, and rollovers.

4.1   Preparation of a FARS-MCOD analysis file

FARS is a census of fatal crashes in the United States since 1975. The basic FARS data does not furnish information on the specific injuries of people involved in the crashes. The National Center for Health Statistics, however, has assembled a census of death certificates for people who died in the United States since 1968, from any causes.[39] Its data is called the Multiple Cause of Death file.[40] Death certificates list diseases, injuries, conditions and external factors that "contributed" to a person's death. Beginning with the 1987 data, NHTSA and NCHS have worked together to link records of fatalities on FARS to their corresponding death-certificate data on MCOD. These supplemental FARS-MCOD files can be merged with the basic FARS person-level data by ST_CASE, VEH_NO and PER_NO, but only for the fatally injured people. They list the injuries of the fatally injured people – not necessarily all their injuries, but only those that "contributed to the fatality" in the opinion of whoever filled out the death certificate. The MCOD data does not list the sources of the injuries (vehicle components contacted).

Injury data: MCOD uses the International Classification of Diseases to classify injuries. From 1987 to 1998, MCOD used the 9th revision of that system (ICD-9).[41] Since 1999, MCOD has used the 10th revision of that system (ICD-10).[42] The two versions differ substantially. In

---

[39] CDC. *Multiple Cause of Death, 1999-2006*. http://wonder.cdc.gov/wonder/help/mcd.html.

[40] NCHS. (2006). *Multiple Causes of Mortality, 2003; Documentation of the Mortality Tape File for 2003 Data*. Hyattsville, MD: National Center for Health Statistics. http://wonder.cdc.gov/wonder/sci_data/mort/mcmort/type_txt/mcmort03/mcmort03.asp.

[41] CDC (a) *Classification of Diseases and Injuries – ICD-9-CM Tabular List of Diseases (FY07)*. Dtab07.zip at ftp://ftp.cdc.gov/pub/Health_Statistics/NCHS/Publications/ICD9-CM/2006/, pp. 554-659.

[42] WHO (2005). *International Statistical Classification of Diseases and Health Related Problems, Tenth Revision – ICD-10, Second Edition*. Geneva: World Health Organization. www.who.int/classifications/icd/en/.

general, there is no simple one-to-one correspondence between specific injury codes. However, for the analyses of this chapter it is only necessary to identify the body region of the injury and weed out obviously minor injuries. That can be accomplished unambiguously with either version and it allows combining data from both versions. Because acquisition and processing of death-certificate data takes time, 2007 is the latest year of FARS-MCOD data as of October 2012. The analysis is based on 21 years of data, 1987 to 2007.

An ICD-9 code consists of a three-digit number and, possibly, a decimal point and another digit (for additional detail). For example, 800 is a fracture of the vault of the skull and 800.0 is a closed fracture of the vault of the skull without mention of intracranial injury. Relevant to this study are the codes 800-999, comprising all types of injuries as well as poisonings and consequences of trauma. Not relevant are the codes 001-799, comprising diseases and any codes preceded by a letter, such as the E codes for external causes including "motor vehicle crash."

The FARS-MCOD files for 1987-1998 may list up to 15 "record-axis" ICD-9 codes per fatally injured person, mostly 800-999 codes. The record-axis codes recapitulate whatever is on the death certificate, except that NCHS has screened them to eliminate contradictions and duplicate codes. The 1987-1998 FARS-MCOD files have 844,923 800-900 codes for 488,116 fatally injured people, an average of 1.7 injuries per person. The majority of people, 278,888 have only one injury listed; 126,246 have two; 48,020 have three; and 34,962, four or more. In other words, most of the death certificates appear to be coded appropriately, listing only the one, two, or occasionally three or four injuries that clearly contributed to the fatality. Far less often is there an extensive list of injuries that likely does not discriminate their importance.

However, injury information on death certificates does not necessarily derive from autopsies or hospital records and it may lack specifics needed even for basic analysis. The priority for many of the officials who fill out original death certificates is to register whether a person died from an unintentional crash as opposed to homicide, suicide, or a heart attack – not what particular injury caused the fatality. A common code, accounting for 19 percent of the reported injuries (156,483 of 844,923) is simply 959.8, "other or multiple unspecified injuries." Also prevalent are 959.9, "unknown injuries" (48,585 cases) and 869, "injury to unspecified organs" (31,370 cases). These cases are lost to the analyses.

Even when the codes specify a body region and can be used for analyses, they often say little else. For example, code 959.1, "unknown trunk injuries" (26,453 cases), although known to be a torso injury, is of unknown severity. Only 31 percent of the injuries are reported in enough detail to make it evident they are severe, such as "cerebral laceration."

Appendix A lists all the 800-999 codes, specifying which ones are included in the analyses as head injuries (blue print), torso injuries (red print), or neck injuries (green print) and which ones are excluded from the analyses (black print). Head injuries include brain, skull, and face, excluding evidently minor injuries such as "superficial head/face injury" (code 910) that likely did not contribute much to the fatality. Torso injuries include chest, abdomen, thoracic or lumbar spine, pelvis, clavicle, and scapula, excluding evidently minor injuries. Neck injuries include the cervical spine and throat. Excluded from the analyses are injuries to the arms or legs, injuries to multiple or unspecified body regions, evidently minor injuries, burns regardless of the body region, poisoning, and complications after an injury. Appendix A uses bold italics to indicate evidently severe head, torso, or neck injuries; regular red, blue, or green print indicates injuries of unknown severity.

An ICD-10 code consists of a letter, a two-digit number, and, possibly, a decimal point and another digit. For example, S02.0 is a fracture of the vault of the skull. Relevant to this study are the codes starting with S and T, comprising all types of injuries as well as poisonings and consequences of trauma. Not relevant are the codes starting with other letters, such as the A-R codes for diseases or the V codes for external causes including "motor vehicle crash."

The FARS-MCOD file may list up to 15 "record-axis" ICD-10 codes per fatally injured person, mostly S and T codes. The 1999-2007 FARS-MCOD files have 573,010 S and T codes for 353,563 fatally injured people, an average of 1.6 injuries per person. The majority of people, 222,793 have only one injury listed; 81,269 have two; 29,401 have three; and 20,100, four or more. The distribution is similar to the ICD-9 codes, with just slightly fewer injuries per person (1.6 versus 1.7). Here, too, many of the codes are not specific and are lost to the analyses. The most common code, accounting for 21 percent of the reported injuries (118,504 of 573,010) is simply T07, "unspecified multiple injuries." Also prevalent are T14.9, "injury, unspecified" (57,735 cases) and T14.8, "other injuries of unspecified body region" (8,792 cases). Even when the codes specify a body region and may be included in the analyses, they often say little else. For example, code S09.9, "unspecified injury of head" (110,849 cases) exceeds all other head injuries, combined. Only 23 percent of the injuries (less than the 31% with ICD-9) are reported in enough detail to make it evident they are severe. Appendix B lists all the S and T codes, using the same classification scheme and color-coding as Appendix A.

The ICD-9 and ICD-10 data have similar injury distributions by body region. For the injuries that were identified as head, neck, or torso, the ICD-9 distribution was 51 percent head-injury, 8 percent neck-injury, and 41 percent torso. The corresponding numbers for the ICD-10 data is 51, 10, and 39 percent. It is unknown whether the modest shift from torso to neck injuries is due to differences in the coding systems or a genuine shift in the vehicles, crashes, or occupants. Given the similarities in their numbers of injuries per person and distribution of injury body regions, it seems appropriate to combine the ICD-9 and ICD-10 data to obtain a single large database.

FARS-MCOD analysis file: The information on a person's individual injuries is used to define six variables at the person level:

- HEAD1 = the number of evidently severe head injuries for that person (range 0 to 15)
- HEAD2 = the number of evidently severe or unknown-severity head injuries
- TORSO1 = the number of evidently severe torso injuries
- TORSO2 = the number of evidently severe or unknown-severity torso injuries
- NECK1 = the number of evidently severe neck injuries
- NECK2 = the number of evidently severe or unknown-severity neck injuries

These person-level records are used to create a database of all FARS cases of cars or LTVs (MY ≥ 1960) involved in fatal crashes and in which the driver and RF passenger seats were occupied; where either the driver, or the RF passenger, or both were fatalities; where both occupants' age and gender must be known, and ages must be in the range of 21 to 96. Furthermore, these cases must also be on the FARS file analyzed in Chapters 2 and 3: specifically, the driver and RF passenger should have similar belt use and air bag availability (see Section 2.2) – i.e., either both occupants are belted or neither; the vehicle must either have dual air bags or have no air bags at all.

The database contains 90,179 records of driver-RF pairs riding in the same vehicle, namely all of the 154,467 records of the CY 1975-2010 database in Chapters 2 and 3 for which the CY is in the 1987-2007 range where MCOD data is available on the fatally injured person(s). But for analysis purposes, the database is effectively much smaller than 90,179 records. That is because, in any regression analysis, the data must be limited to cases where either the driver, or the RF passenger, or both have the type of injury in question. For example, there are many vehicles where neither occupant has a head injury reported in MCOD. These vehicles cannot be included in the basic regression analysis of head-injury risk; the regression is limited to the 33,998 cases where either the driver, or the RF passenger, or both had an evidently severe or unknown-severity head injury reported in MCOD. Similarly, the basic analysis of torso injury is limited to the 30,378 cases where one or both occupants had a torso injury reported and the basic analysis of neck injury is limited to the 8,049 cases where at least one of the occupants had a neck injury. Appendix F lists the N of cases for each of the individual regression analyses.

The dependent variable in the logistic regressions for drivers is whether the driver had any injuries of a specific type – e.g., if HEAD1 > 0 for the driver. It does not matter how many of these injuries, as long as there is at least one.

4.2   Effects of aging and gender on head, torso, and neck injury risk

Techniques combining the logistic regression coefficients with a double-pair comparison analysis generate estimates of the increase in injury risk, given similar physical insults, as a male or female occupant ages one year and the injury risk for a female relative to a male of the same age. Table 4-1a compares the effects of aging on head, torso, and neck injuries that, according to MCOD, "contributed to death" overall and separately for unbelted and 3-point belted occupants.

The first row of Table 4-1a copies the first row from Table 3-1a: the overall increase in fatality risk. The average for males and females is an increase of 2.83 ± .08 percent for each year that the driver ages and a 3.39 ± .07 percent increase for the RF passenger, averaging to 3.11 ± .08 percent for the driver and RF passenger together. These numbers as well as the separate estimates for males and females are the benchmarks for comparison with the results for the individual body regions.

TABLE 4-1a: INCREASE IN RISK OF "INJURIES THAT CONTRIBUTE TO DEATH" FOR EACH YEAR THAT AN OCCUPANT GETS OLDER **BY BODY REGION** – CARS AND LTVs (MY 1960-2008), DRIVERS AND RIGHT-FRONT PASSENGERS
(Given the same crash scenario, average for occupants 21 to 96, FARS MCOD 1987-2007)

Percent Injury Increase per Year of Aging

	Males			Females			Average of Males & Females		
	Driver	RF Pass	Average	Driver	RF Pass	Average	Driver	RF Pass	Average
OVERALL FATALITY RISK	**3.06 ± .09**	**3.51 ± .09**	**3.28 ± .08**	**2.60 ± .10**	**3.27 ± .08**	**2.93 ± .08**	**2.83 ± .08**	**3.39 ± .07**	**3.11 ± .08**
Evidently severe or unknown-severity head injury									
All occupants	2.42 ± .19	2.93 ± .21	2.68 ± .18	2.03 ± .18	2.57 ± .20	2.30 ± .16	2.23 ± .16	2.75 ± .20	2.49 ± .17
Unbelted occupants	2.20 ± .22	2.92 ± .28	2.56 ± .23	1.83 ± .34	2.57 ± .29	2.20 ± .25	2.02 ± .25	2.74 ± .25	2.38 ± .22
3-point belted occupants	2.51 ± .35	2.87 ± .36	2.69 ± .29	2.16 ± .29	2.59 ± .25	2.38 ± .25	2.33 ± .28	2.73 ± .30	2.53 ± .25
Evidently severe head injury									
All occupants	2.41 ± .43	2.78 ± .44	2.60 ± .36	1.87 ± .46	2.39 ± .42	2.13 ± .39	2.14 ± .39	2.58 ± .39	2.36 ± .35
Unbelted occupants	2.26 ± .51	2.98 ± .55	2.63 ± .48	2.26 ± .74	2.49 ± .61	2.37 ± .62	2.26 ± .60	2.73 ± .40	2.50 ± .51
3-point belted occupants	2.34 ± .75	2.47 ± .86	2.41 ± .66	1.57 ± .80	2.31 ± .63	1.94 ± .61	1.96 ± .67	2.39 ± .68	2.17 ± .64
Evidently severe or unknown-severity torso injury									
All occupants	3.84 ± .23	4.13 ± .20	3.99 ± .17	3.25 ± .21	3.99 ± .18	3.62 ± .18	3.55 ± .20	4.06 ± .18	3.80 ± .17
Chest injuries	4.07 ± .26	4.34 ± .26	4.20 ± .22	3.54 ± .30	4.28 ± .23	3.91 ± .23	3.80 ± .24	4.31 ± .23	4.06 ± .21
Abdominal injuries	3.65 ± .29	3.70 ± .46	3.67 ± .33	2.86 ± .49	3.51 ± .35	3.19 ± .35	3.26 ± .36	3.61 ± .34	3.43 ± .32
Unbelted occupants	3.65 ± .28	3.76 ± .30	3.70 ± .27	3.09 ± .36	3.70 ± .29	3.40 ± .29	3.37 ± .28	3.73 ± .28	3.55 ± .27
Chest injuries	3.89 ± .33	4.03 ± .33	3.96 ± .27	3.46 ± .40	4.08 ± .31	3.77 ± .28	3.67 ± .33	4.05 ± .31	3.86 ± .26
Abdominal injuries	3.55 ± .48	2.99 ± .72	3.27 ± .52	2.77 ± .75	2.83 ± .59	2.80 ± .61	3.16 ± .54	2.91 ± .62	3.03 ± .53
3-point belted occupants	3.82 ± .34	4.18 ± .36	4.00 ± .31	3.41 ± .29	3.94 ± .31	3.67 ± .26	3.61 ± .27	4.06 ± .30	3.84 ± .27
Chest injuries	4.08 ± .36	4.34 ± .45	4.21 ± .34	3.62 ± .39	4.14 ± .33	3.88 ± .32	3.85 ± .35	4.24 ± .36	4.04 ± .30
Abdominal injuries	3.60 ± .59	3.94 ± .62	3.77 ± .56	3.03 ± .73	3.73 ± .54	3.38 ± .58	3.31 ± .65	3.84 ± .56	3.57 ± .51
Evidently severe torso injury									
All occupants	4.35 ± .31	4.49 ± .31	4.42 ± .28	3.63 ± .38	4.38 ± .28	4.01 ± .29	3.99 ± .29	4.43 ± .27	4.21 ± .26
Chest injuries	4.61 ± .37	4.72 ± .41	4.66 ± .34	4.05 ± .46	4.72 ± .37	4.33 ± .35	4.33 ± .35	4.72 ± .35	4.52 ± .34
Abdominal injuries	3.93 ± .60	3.62 ± .82	3.78 ± .63	2.88 ± .77	3.62 ± .64	3.25 ± .65	3.41 ± .65	3.62 ± .68	3.51 ± .61
Unbelted occupants	4.09 ± .47	4.19 ± .41	4.14 ± .42	3.73 ± .57	4.09 ± .38	3.91 ± .42	3.91 ± .45	4.14 ± .41	4.03 ± .38
3-point belted occupants	4.41 ± .62	4.91 ± .77	4.66 ± .67	3.72 ± .67	4.57 ± .60	4.14 ± .60	4.07 ± .60	4.74 ± .64	4.40 ± .61
Evidently severe or unknown-severity neck injury									
All occupants	3.25 ± .43	3.73 ± .42	3.49 ± .40	2.56 ± .47	3.22 ± .44	2.89 ± .37	2.90 ± .39	3.48 ± .40	3.19 ± .37
Unbelted occupants	3.13 ± .67	3.79 ± .64	3.46 ± .50	2.42 ± .93	3.23 ± .59	2.83 ± .74	2.77 ± .61	3.51 ± .58	3.14 ± .58
3-point belted occupants	3.36 ± .74	3.98 ± .63	3.67 ± .63	2.70 ± .64	3.27 ± .65	2.98 ± .56	3.03 ± .61	3.63 ± .62	3.33 ± .57
Evidently severe neck injury									
All occupants	3.09 ± .53	3.74 ± .50	3.42 ± .43	2.41 ± .53	3.16 ± .39	2.79 ± .39	2.75 ± .42	3.45 ± .36	3.10 ± .38
Unbelted occupants	3.05 ± .78	3.94 ± .58	3.50 ± .58	2.51 ± 1.00	3.30 ± .67	2.91 ± .68	2.78 ± .69	3.62 ± .60	3.20 ± .60
3-point belted occupants	3.03 ± .93	3.79 ± .78	3.41 ± .72	2.13 ± .74	3.00 ± .71	2.57 ± .67	2.58 ± .75	3.40 ± .61	2.99 ± .66

The second row of Table 4-1a shows that head injuries "contributing to death" increased by an average of 2.23 ± .16 percent for drivers, 2.75 ± .20 percent for RF passengers, and 2.49 ± .17 percent overall per year of aging. In other words, risk of head injury increases slower with age than overall fatality risk. All the estimates in this row are close to .60 percentage points lower than the corresponding estimates in the first row, indicating a difference that is consistent for males and females, for drivers and passengers. The next two rows show quite similar aging effects on head injury for unbelted (2.38 ± .22%) and 3-point belted (2.53 ± .25%) occupants. Given the sampling error for these estimates, they cannot be considered significantly different, although the point estimates of the aging effect are slightly higher for the belted occupants, consistent with the findings throughout Chapter 3.

The next three rows limit the analysis to evidently severe head injuries, excluding the many cases where the severity is unknown (typically because the MCOD code says little more than "unspecified head injury"). Much of the data is lost, resulting in confidence bounds that are usually at least twice as wide as the preceding three rows. Nevertheless, the point estimates are quite similar. For example, the overall average effect for head injuries is a 2.49 ± .17 percent fatality increase per year of aging when the unknown-severity head injuries are included and 2.36 ± .35 percent when the analysis is limited to the evidently severe injuries. Results for the identifiably severe injuries are printed in italics as a reminder that they tend to be less reliable due to their wide confidence bounds.

Torso injury risk increases far more with age than head injuries by this analysis, consistent with all the literature (see Section 1.2). The overall average increase per year is 3.80 ± .17 percent, as compared to 2.49 ± .17 percent for head injuries and 3.11 ± .08 percent for overall fatality risk. The other estimates in the torso injuries-all occupants row are all close to .70 percentage points higher than the corresponding estimates for overall fatality risk, indicating a difference that is consistent for males and females, for drivers and passengers. Aging effects are fairly similar for unbelted (3.55 ± .27%) and 3-point belted (3.84 ± .27%) occupants.

When the analysis is limited to specific, evidently severe torso injuries, the observed effects of aging a year are even higher: 4.21 ± .26 percent overall, 4.03 ± .38 percent for unbelted occupants, and 4.40 ± .61 percent for belted occupants. These are the highest point estimates of aging effects in any analysis so far (although the wide confidence bounds indicate a fair amount of uncertainty). It is unknown if the effects intensify because the specific injuries are more severe than the unknown-severity injuries (such as "unspecified injury of thorax") or the body region of an injury is more accurately reported when a specific injury is identified.

The MCOD files have enough data to study two types of torso injury separately: chest and abdominal injuries. Appendices A and B identify which torso injuries are also chest injuries and which are abdominal injuries, as well as other torso injuries that involve the back, shoulder, pelvis, multiple, or unspecified sub-regions. In every case, the effect of aging on chest injuries is slightly larger than its overall effect on torso injuries, the increase in abdominal injuries slightly smaller. For example, the overall average increase per year is 4.06 ± .21 percent for chest injuries, 3.80 ± .17 percent for all torso injuries, and 3.43 ± .32 percent for abdominal injuries. The findings are consistent with the literature, which emphasizes elderly occupants' risk of thoracic injury. The estimated increase in evidently severe chest injuries, 4.52 ± .34 percent, is the largest increase with age in all the MCOD analyses.

The last six rows of Table 4-1a indicate that the effect of aging on neck-injury risk is midway between the effects for head and torso injuries. In fact, the average increase per year in neck injuries, $3.19 \pm .37$ percent is quite similar to the average increase in overall fatality risk, $3.11 \pm .08$ percent.

Table 4-1b compares how the effect of aging changes with age, depending on the body region of the injury, seat position, and belt use. For example, head injury risk of male drivers increases by 2.43 percent a year at ages 21-30 and 2.46 percent a year at 65 to 74. The patterns are actually fairly similar across body regions.

For all three body regions, the tendency of the age effect to intensify at higher ages is stronger for RF passengers than for drivers and stronger for 3-point belted occupants than for unbelted occupants. Thus, for unrestrained drivers, the effects of aging on head and torso injuries are both actually a bit lower at 65 to 74 (2.04% and 3.43%) than they initially were at 21 to 30 (2.36% and 3.96%). Whereas, for 3-point belted RF passengers, the increase in both head and torso injury risk escalates sharply with age (3.78% and 5.60% at 65 to 74 versus 2.26% and 3.39% at 21 to 30). Fundamentally, though, the increase in torso-injury risk is higher than the increase in head-injury risk at all ages: the aging effect on torso-injury risk is high to begin with and it stays high.

In the separate analyses of chest and abdominal injuries, Table 4-1b show chest-injury risk increasing substantially with each year of aging even for young adults 21-30 years old.

Table 4-1c estimates the cumulative effect of all the year-to-year increases in risk. Overall, a 75-year-old male driver is 5.04 times as likely to die as a 21-year-old from similar physical insults; a male RF passenger has 6.70 times the fatality risk at 75 as at 21. The cumulative increases are factors of 3.87 for female drivers and 5.67 for female RF passengers. Head injury risk rises at a substantially slower rate. Thus, a 75-year-old male driver has only 3.63 times the risk of a head injury that "contributes to death" as a 21-year-old. But torso injury risk is 7.53 times as high for 75-year-old male drivers as 21-year-olds, and for male RF passengers, the risk increases by a factor of 9.69 times. In every separate analysis of chest and abdominal injuries, the cumulative increase in chest-injury risk exceeds the increase for the abdominal injuries. When the analyses are limited to evidently severe torso injuries and 3-point belted occupants, the risk for 75-year-old male drivers is 10.12 times as high as at 21, and for RF passengers, 14.58 times as high. Those are the largest cumulative increases in any of the analyses so far. The cumulative increase in neck injury risk is again midway between the results for head and torso injuries.

TABLE 4-1b

INCREASE IN RISK OF "INJURIES THAT CONTRIBUTE TO DEATH" FOR EACH YEAR
THAT AN OCCUPANT GETS OLDER: AT AGE 21-30 VERSUS AGE 65-74
(Given the same crash scenario, FARS MCOD 1987-2007)

BY BODY REGION
CARS AND LTVs (MY 1960-2008), DRIVERS AND RIGHT-FRONT PASSENGERS

	Percent Injury Increase per Year of Aging					
Seat Position →→→→→→→→→→→	Driver			RF Passenger		
Occupant Age Group →→→→→→→	Avg 21-96	21-30	65-74	Avg 21-96	21-30	65-74
OVERALL FATALITY RISK	**3.06 ± .09**	**2.90**	**3.39**	**3.51 ± .09**	**2.90**	**4.58**
Evidently severe or unknown-severity head injury						
All occupants	2.42 ± .19	2.43	2.46	2.92 ± .21	2.50	3.66
Unbelted occupants	2.20 ± .22	2.36	2.04	2.92 ± .28	2.65	3.47
3-point belted occupants	2.51 ± .35	2.43	2.67	2.87 ± .36	2.26	3.78
Evidently severe head injury						
All occupants	*2.42 ± .43*	*2.11*	*2.90*	*2.78 ± .44*	*1.99*	*3.95*
Unbelted occupants	*2.27 ± .51*	*2.11*	*2.54*	*2.98 ± .55*	*2.37*	*3.94*
3-point belted occupants	*2.34 ± .75*	*1.68*	*3.36*	*2.47 ± .86*	*1.45*	*3.90*
Evidently severe or unknown-severity torso injury						
All occupants	3.84 ± .23	3.76	4.13	4.13 ± .20	3.81	4.88
Chest injuries	4.07 ± .26	4.16	4.17	4.34 ± .26	4.20	4.87
Abdominal injuries	3.65 ± .29	3.28	4.29	3.70 ± .46	3.11	4.71
Unbelted occupants	3.65 ± .28	3.96	3.43	3.76 ± .30	3.88	3.82
Chest injuries	3.89 ± .33	4.37	3.48	4.03 ± .33	4.38	3.84
Abdominal injuries	3.55 ± .48	3.73	3.44	2.99 ± .72	3.08	2.95
3-point belted occupants	3.82 ± .34	3.47	4.45	4.18 ± .36	3.39	5.60
Chest injuries	4.08 ± .36	3.84	4.58	4.34 ± .45	3.67	5.57
Abdominal injuries	3.60 ± .59	2.89	4.71	3.94 ± .62	2.93	5.61
Evidently severe torso injury						
All occupants	*4.35 ± .31*	*4.00*	*5.03*	*4.49 ± .31*	*4.00*	*5.45*
Chest injuries	*4.61 ± .37*	*4.53*	*4.98*	*4.72 ± .41*	*4.49*	*5.34*
Abdominal injuries	*3.93 ± .60*	*3.23*	*5.10*	*3.62 ± .82*	*2.82*	*4.83*
Unbelted occupants	*4.09 ± .47*	*4.02*	*4.40*	*4.19 ± .41*	*3.76*	*4.99*
3-point belted occupants	*4.41 ± .62*	*3.94*	*5.26*	*4.91 ± .77*	*4.37*	*6.11*
Evidently severe or unknown-severity neck injury						
All occupants	3.25 ± .43	3.11	3.52	3.74 ± .42	3.12	4.84
Unbelted occupants	3.13 ± .67	2.95	3.43	3.79 ± .64	3.18	4.99
3-point belted occupants	3.36 ± .74	3.28	3.57	3.98 ± .63	3.26	5.32
Evidently severe neck injury						
All occupants	*3.09 ± .53*	*2.80*	*3.59*	*3.74 ± .50*	*2.93*	*5.17*
Unbelted occupants	*3.05 ± .78*	*2.78*	*3.51*	*3.94 ± .58*	*3.25*	*5.31*
3-point belted occupants	*3.03 ± .93*	*2.62*	*3.71*	*3.79 ± .78*	*2.69*	*5.58*

TABLE 4-1c

RISK OF "INJURIES THAT CONTRIBUTE TO DEATH" FOR A 75-YEAR-OLD OCCUPANT
RELATIVE TO A 21-YEAR-OLD OCCUPANT
(Given similar physical insults, FARS MCOD 1987-2007)

CARS AND LTVs (MY 1960-2008), DRIVERS AND RIGHT-FRONT PASSENGERS
BY BODY REGION

| | Risk at Age 75 ÷ Risk at Age 21 | | | |
| | Males | | Females | |
	Drivers	RF Pass	Drivers	RF Pass
OVERALL FATALITY RISK	**5.04**	**6.70**	**3.87**	**5.67**
Evidently severe or unknown-severity head injury				
All occupants	3.63	4.86	2.88	3.79
Unbelted occupants	3.23	4.90	2.61	3.82
3-point belted occupants	3.79	4.67	3.09	3.89
Evidently severe head injury				
All occupants	*3.63*	*4.48*	*2.62*	*3.42*
Unbelted occupants	*3.36*	*5.00*	*3.34*	*3.55*
3-point belted occupants	*3.53*	*3.78*	*2.23*	*3.39*
Evidently severe or unknown-severity torso injury				
All occupants	7.53	9.29	5.34	8.30
Chest injuries	8.41	10.29	6.19	9.81
Abdominal injuries	6.85	7.26	4.35	6.36
Unbelted occupants	6.84	7.48	4.92	7.18
Chest injuries	7.68	8.61	5.97	8.86
Abdominal injuries	6.57	4.86	4.12	4.39
3-point belted occupants	7.49	9.69	5.93	8.07
Chest injuries	8.50	10.40	6.53	9.05
Abdominal injuries	6.70	8.48	4.82	7.27
Evidently severe torso injury				
All occupants	*9.80*	*11.19*	*6.49*	*10.28*
Chest injuries	*11.14*	*12.47*	*8.05*	*12.45*
Abdominal injuries	*7.98*	*6.88*	*4.34*	*6.93*
Unbelted occupants	*8.58*	*9.35*	*6.93*	*8.67*
3-point belted occupants	*10.12*	*14.58*	*6.90*	*11.12*
Evidently severe or unknown-severity neck injury				
All occupants	5.57	7.53	3.76	5.25
Unbelted occupants	5.25	7.94	3.52	5.27
3-point belted occupants	5.90	8.66	4.05	5.27
Evidently severe neck injury				
All occupants	*5.14*	*7.65*	*3.48*	*5.07*
Unbelted occupants	*5.06*	*8.70*	*3.71*	*5.43*
3-point belted occupants	*4.99*	*7.82*	*2.99*	*4.56*

Table 4-1d estimates the increase in fatality risk for a female relative to a male of the same age, given the same crash scenario or physical insult. Females are especially vulnerable to neck injuries. Risk is 39.4 ± 9.4 percent higher than for a male of the same age. For belted occupants, neck-injury risk is 45.0 ± 16.8 percent higher for females than males. The difference between males and females is especially strong for young adults: 60.3 percent risk increase for female relative to male drivers at 21 to 30 and 51.0 percent increase for RF passengers.

By contrast, head- and torso-injury risk is just moderately higher for females than for males: a 14.6 ± 3.1 percent increase of head-injury risk and 13.3 ± 3.6 percent increase for torso injuries. Both of these estimates are slightly below the 17.0 ± 1.5 percent overall increase in fatality risk for a female relative to a male. Moreover, belt use mitigates females' added risk of head and torso injury (to 7.7 ± 4.5% and 9.8 ± 6.5%, respectively), but as noted above, for neck injuries, the added risk for females was especially high (at least as a point estimate) for belted occupants.

Separate analyses of chest and abdominal injuries produce contrasting results. Chest-injury risk is just 8.8 ± 4.6 percent higher for females than males, but abdominal injury, 31.9 ± 8.3 percent higher. Females' increase in abdominal-injury risk is especially high for unrestrained occupants (34.9 ± 18.0%) but also high for belted occupants (26.6 ± 15.1%). The increase reaches 44.6 ± 16.4 for evidently severe abdominal injuries.

In the analyses so far, the effect on risk of being female has often resembled and sometimes differed from the effect of aging. Tables 4-1a and 4-1d show one of the strongest differences: older occupants are especially vulnerable to torso injury (especially thoracic but also abdominal injury), but females to neck and abdominal injury. The deterioration and brittleness of the ribcage and sternum result in multiple fractures that deprive older occupants of protection to their thoracic organs (and the abdominal organs also, to the extent that the lower ribs partially cover them) – whereas skull fracture is less of an issue in head injuries. While it is true that females' ribcage is smaller and of lower mineral bone density than males', the disadvantage is at least partially offset because the trunk that it protects is also smaller and lighter. A male's cervical spine, on the other hand, has deeper facet joints and a wider surface area, resulting in greater spinal-column strength, yet a female's neck is called upon to support and control the motion of a head that is almost as large and heavy as a male's. Thus, it is fairly clear why females are at high risk of neck injury. Their high risk of abdominal injury is harder to explain. It is apparently not primarily an issue of belt fit, for Table 4-1d shows the increase is especially large for unbelted females relative to unbelted males. It is possible that the lowest part of the ribcage surrounding the abdominal organs, which is not well supported by the spine and sternum, may be an especially fragile spot in the female relative to the male anatomy.

TABLE 4-1d: INCREASE IN RISK OF "INJURIES THAT CONTRIBUTE TO DEATH" FOR A FEMALE RELATIVE TO A MALE OF THE SAME AGE **BY BODY REGION** – CARS AND LTVs (MY 1960-2008), DRIVERS AND RIGHT-FRONT PASSENGERS
(Given the same crash scenario, average for occupants 21 to 96, FARS MCOD 1987-2007)

Percent Injury Increase for Females

Seat Position →	Drivers			RF Passengers			Avg of Driver & RF
Occupant Age Group →	21-96	21-30	65-74	21-96	21-30	65-74	21-96
OVERALL FATALITY RISK	**13.4 ± 2.0**	**25.9**	**-1.4**	**20.5 ± 2.2**	**29.2**	**11.4**	**17.0 ± 1.5**
Evidently severe or unknown-severity head injury							
All occupants	13.9 ± 5.0	24.1	1.3	15.2 ± 3.8	27.0	2.2	14.6 ± 3.1
Unbelted occupants	19.3 ± 7.4	29.3	7.0	20.3 ± 5.8	33.5	6.4	19.8 ± 4.3
3-point belted occupants	7.1 ± 6.6	15.1	-3.3	8.3 ± 7.7	15.5	-.1	7.7 ± 4.5
Evidently severe head injury							
All occupants	5.2 ± 10.1	17.6	-10.5	13.6 ± 10.1	25.3	.1	9.4 ± 7.2
Unbelted occupants	14.9 ± 13.9	15.4	14.4	17.9 ± 17.7	34.2	-.1	16.4 ± 10.5
3-point belted occupants	-2.5 ± 15.1	13.9	-22.1	5.9 ± 14.9	9.5	1.3	1.7 ± 10.7
Evidently severe or unknown-severity torso injury							
All occupants	10.1 ± 4.5	25.4	-7.9	16.5 ± 4.9	23.2	9.6	13.3 ± 3.6
Chest injuries	7.1 ± 6.0	20.1	-8.3	10.5 ± 6.3	13.6	7.4	8.8 ± 4.6
Abdominal injuries	25.4 ± 10.5	48.7	-.9	38.3 ± 13.4	47.0	28.9	31.9 ± 8.3
Unbelted occupants	19.4 ± 7.8	36.6	.3	13.0 ± 6.7	16.1	10.4	16.2 ± 5.3
Chest injuries	17.6 ± 9.4	30.8	2.7	5.7 ± 8.7	5.5	6.5	11.7 ± 6.8
Abdominal injuries	34.0 ± 23.0	58.1	6.5	35.7 ± 21.1	40.8	30.3	34.9 ± 18.0
3-point belted occupants	1.6 ± 7.9	9.9	-8.9	17.9 ± 12.0	27.2	7.5	9.8 ± 6.5
Chest injuries	-.7 ± 10.4	8.0	-11.7	14.6 ± 12.4	21.0	7.7	6.9 ± 7.2
Abdominal injuries	20.8 ± 21.9	37.2	1.5	32.3 ± 21.1	42.7	21.2	26.6 ± 15.1
Evidently severe torso injury							
All occupants	10.1 ± 9.6	28.2	-10.7	18.4 ± 9.5	24.1	12.5	14.3 ± 7.4
Chest injuries	5.6 ± 10.7	18.9	-10.1	10.6 ± 10.1	11.6	9.9	8.1 ± 7.4
Abdominal injuries	18.8 ± 15.5	48.4	-13.2	70.3 ± 29.0	73.3	69.2	44.6 ± 16.4
Unbelted occupants	26.7 ± 15.7	39.2	12.7	10.2 ± 11.8	14.7	5.3	18.5 ± 10.8
3-point belted occupants	-.6 ± 14.1	12.5	-16.7	27.2 ± 18.5	42.4	11.2	13.3 ± 12.1
Evidently severe or unknown-severity neck injury							
All occupants	45.9 ± 17.9	60.3	20.3	32.8 ± 12.5	51.0	12.8	39.4 ± 9.4
Unbelted occupants	44.3 ± 23.6	66.8	18.8	24.5 ± 18.5	44.1	2.7	34.4 ± 14.0
3-point belted occupants	42.6 ± 23.2	62.8	19.0	47.4 ± 20.4	76.6	18.0	45.0 ± 16.8
Evidently severe neck injury							
All occupants	52.5 ± 18.3	74.7	27.1	35.8 ± 15.7	57.2	12.7	44.2 ± 11.4
Unbelted occupants	50.3 ± 25.8	68.4	29.4	26.8 ± 21.6	50.2	1.5	38.5 ± 15.8
3-point belted occupants	49.9 ± 29.1	76.7	19.5	58.9 ± 28.0	94.2	24.3	54.4 ± 21.0

4.3   Head, torso, and neck injury by type of impact and occupant protection

Table 4-2a suggests that the effect of aging on head-injury risk is fairly similar across the various types of crashes and occupant protection. The first row of Table 4-2a, copied from the second row of Table 4-1a, indicates that the overall risk of evidently severe or unknown-severity head injuries that "contributed to death" (according to MCOD) increases fairly slowly with age: 2.49 ± .17 percent for each year of aging (average of drivers and RF passengers, males and females). The remaining rows of Table 4-2a show fairly similar numbers for the various impact locations (2.56 ± .24% in frontals, 2.41 ± .35% in nearside impacts, 2.29 ± .37% in far-side impacts, 2.70 ± .60% in first-event rollovers, and 2.25 ± .73% in rear impacts/other crashes). The differences between the point estimates cannot be considered statistically meaningful, given the fairly wide confidence bounds. Further subdivision of the data by belt use or availability of air bags (frontals only) likewise does not produce any estimates significantly different from the overall average effect of 2.49 percent. (Because of the limited number of cases, it would have been futile to obtain separate estimates for belted and unbelted in frontal crashes of vehicles with dual air bags or separate estimates for curtain-plus-torso/combination air bags in side impacts.)

Table 4-2b compares the effects of aging a year for occupants 21 to 30 to occupants 65 to 74. Here, too, there are no large differences by impact location. Table 4-2b shows the customary pattern that the effect of aging levels off rather than intensifies for unbelted occupants, especially drivers (see Tables 3-2b and 3-4b). In vehicles with dual air bags, the aging effect in frontal crashes is quite small for occupants 21 to 30. Table 4-2c shows that the cumulative increase in head-injury risk from 21 to 75 does not differ too much by type of crash or occupant protection.

Table 4-2d suggests females' increased risk of head injury may vary by type of crash and type of occupant protection. However, the variations follow the patterns for overall fatality risk in Section 3.6 and Table 3-4d. The added risk for a female relative to a male appears to be higher in nearside impacts (17.8 ± 9.8%) than in far-side impacts (5.0 ± 10.0%). Dual air bags may nearly eliminate the gender gap in frontal crashes (2.3 ± 12.3%). In first-event rollovers and rear impacts/other crashes, the risk increase for unbelted females relative to unbelted males is exceptionally great, whereas belted females may have equal or even lower risk than belted males. As discussed in Section 3.6, the typically smaller stature and girth of females makes them more vulnerable than males to ejection, when unbelted, but less likely to contact intruding structures, when belted.

TABLE 4-2a: HEAD INJURIES: INCREASE IN RISK FOR EACH YEAR THAT AN OCCUPANT GETS OLDER BY TYPE OF IMPACT AND OCCUPANT PROTECTION – CARS AND LTVs (MY 1960-2008), DRIVERS AND RF PASSENGERS

(Given the same crash scenario, average for occupants age 21-86, FARS MCOD 1987-2007, evidently severe or unknown-severity "injuries that contribute to death")

Percent Head-Injury Increase per Year of Aging

	Males			Females			Average of Males & Females		
	Driver	RF Pass	Average	Driver	RF Pass	Average	Driver	RF Pass	Average
All head injuries	**2.42 ± .19**	**2.93 ± .21**	**2.68 ± .18**	**2.03 ± .18**	**2.57 ± .20**	**2.30 ± .16**	**2.23 ± .16**	**2.75 ± .20**	**2.49 ± .17**
Frontal impacts									
All occupants	2.49 ± .28	2.97 ± .32	2.73 ± .25	2.00 ± .27	2.76 ± .25	2.38 ± .24	2.25 ± .24	2.87 ± .25	2.56 ± .24
No air bag, unbelted occupants	2.40 ± .42	3.20 ± .42	2.80 ± .32	1.91 ± .49	2.97 ± .36	2.44 ± .38	2.16 ± .37	3.09 ± .35	2.62 ± .33
No air bag, 3-pt belted occupants	2.46 ± .60	2.92 ± .75	2.69 ± .59	1.91 ± .82	2.98 ± .53	2.45 ± .61	2.19 ± .65	2.95 ± .65	2.57 ± .60
Dual air bags	2.32 ± .77	2.59 ± .77	2.46 ± .71	1.93 ± .90	2.22 ± .65	2.07 ± .73	2.13 ± .79	2.41 ± .70	2.27 ± .68
Near-side impacts									
All occupants	2.57 ± .48	2.79 ± .58	2.68 ± .38	2.02 ± .55	2.27 ± .58	2.14 ± .40	2.29 ± .43	2.53 ± .55	2.41 ± .35
Unbelted occupants	2.17 ± .66	2.15 ± .84	2.16 ± .54	2.26 ±1.04	1.69 ± .78	1.97 ± .65	2.33 ± .92	1.86 ± .82	2.10 ± .61
3-pt belted occupants	2.61 ± .84	3.25 ±1.11	2.92 ± .70	1.54 ± .73	2.64 ± .84	2.09 ± .56	2.07 ± .69	2.94 ± .94	2.51 ± .58
Far-side impacts									
All occupants	2.01 ± .59	2.85 ± .62	2.43 ± .43	1.68 ± .51	2.60 ± .54	2.14 ± .37	1.85 ± .51	2.72 ± .54	2.29 ± .37
Unbelted occupants	1.98 ± .70	3.01 ± .91	2.50 ± .58	1.50 ± .87	2.58 ± .73	2.04 ± .57	1.74 ± .67	2.80 ± .73	2.27 ± .50
3-pt belted occupants	2.28 ± .99	2.60 ±1.06	2.44 ± .72	1.92 ± .92	2.82 ± .80	2.37 ± .61	2.10 ± .92	2.71 ± .78	2.41 ± .60
First-event rollovers									
All occupants	2.95 ± .71	2.98 ± .69	2.96 ± .63	2.01 ± .79	2.88 ± .60	2.44 ± .66	2.48 ± .66	2.93 ± .58	2.70 ± .60
Unbelted occupants	2.58 ±1.02	2.40 ±1.16	2.49 ± .90	1.77 ±1.19	2.89 ± .80	2.33 ± .85	2.18 ± .93	2.64 ± .93	2.41 ± .84
3-pt belted occupants	3.11 ±1.18	3.19 ± .87	3.15 ± .91	2.50 ±1.14	2.99 ±1.11	2.75 ±1.01	2.80 ±1.06	3.09 ± .82	2.95 ± .84
Rear impacts & other crashes									
All occupants	2.21 ± .75	2.70 ±1.00	2.45 ± .77	2.07 ± .97	2.01 ± .82	2.04 ± .78	2.14 ± .74	2.35 ± .82	2.25 ± .73
Unbelted occupants	1.81 ±1.02	2.35 ±1.50	2.08 ±1.03	1.67 ±1.35	1.56 ± .98	1.62 ±1.04	1.74 ±1.04	1.96 ±1.14	1.85 ± .98
3-pt belted occupants	2.37 ±1.04	2.97 ±1.21	2.67 ±1.07	2.85 ±1.44	2.36 ±1.09	2.61 ±1.10	2.61 ±1.14	2.67 ±1.10	2.64 ±1.01

TABLE 4-2b

HEAD INJURIES: INCREASE IN RISK FOR EACH YEAR
THAT AN OCCUPANT GETS OLDER: AT AGE 21-30 VERSUS AGE 65-74

BY TYPE OF IMPACT AND OCCUPANT PROTECTION
CARS AND LTVs (MY 1960-2008), DRIVERS AND RIGHT-FRONT PASSENGERS
(Given the same crash scenario, average for occupants age 21-86, FARS MCOD 1987-2007, evidently
severe or unknown-severity "injuries that contribute to death")

	Percent Head-Injury Increase per Year of Aging					
Seat Position →→→→→→→→→→→	Driver			RF Passenger		
Occupant Age Group →→→→→→→→	Avg 21-96	21-30	65-74	Avg 21-96	21-30	65-74
All head injuries	**2.42 ± .19**	**2.43**	**2.46**	**2.92 ± .21**	**2.50**	**3.66**
Frontal impacts						
All occupants	2.49 ± .28	2.43	2.64	2.97 ± .32	2.33	4.00
No air bag, unbelted occupants	2.40 ± .42	2.53	2.29	3.20 ± .42	2.71	4.16
No air bag, 3-pt belted occupants	2.46 ± .60	2.43	2.60	2.92 ± .75	2.65	3.46
Dual air bags	2.32 ± .77	1.85	3.00	2.59 ± .77	1.23	4.61
Near-side impacts						
All occupants	2.57 ± .48	2.50	2.76	2.79 ± .58	2.81	3.03
Unbelted occupants	2.17 ± .66	2.62	1.64	2.15 ± .84	2.40	1.92
3-pt belted occupants	2.61 ± .84	2.12	3.46	3.24 ±1.11	3.05	3.82
Far-side impacts						
All occupants	2.01 ± .59	2.03	2.01	2.85 ± .62	2.26	3.58
Unbelted occupants	1.98 ± .71	2.40	1.45	3.01 ± .91	2.90	3.30
3-pt belted occupants	2.28 ± .99	1.55	3.34	2.60 ±1.06	1.61	3.74
First-event rollovers						
All occupants	2.95 ± .71	3.12	2.76	2.98 ± .69	3.24	2.76
Unbelted occupants	2.58 ±1.02	3.15	1.87	2.40 ±1.16	3.27	1.43
3-pt belted occupants	3.11 ±1.18	3.28	2.91	3.19 ± .87	3.25	3.09
Rear impacts & other crashes						
All occupants	2.21 ± .75	1.86	2.72	2.70 ±1.00	1.68	4.30
Unbelted occupants	1.81 ±1.02	1.26	2.57	2.35 ±1.50	2.28	2.60
3-pt belted occupants	2.37 ±1.04	2.66	2.00	2.97 ±1.21	1.53	5.35

TABLE 4-2c

RISK OF **HEAD INJURIES** FOR A 75-YEAR-OLD OCCUPANT
RELATIVE TO A 21-YEAR-OLD OCCUPANT
(Given similar physical insults, FARS MCOD 1987-2007,
evidently severe or unknown-severity "injuries that contribute to death")

BY TYPE OF IMPACT AND OCCUPANT PROTECTION
CARS AND LTVs (MY 1960-2008), DRIVERS AND RF PASSENGERS

	Risk at Age 75 ÷ Risk at Age 21			
	Males		Females	
	Drivers	RF Pass	Drivers	RF Pass
All head injuries	**3.63**	**4.86**	**2.88**	**3.79**
Frontal impacts				
All occupants	3.76	4.99	2.83	4.30
No air bag, unbelted occupants	3.57	5.86	2.72	4.88
No air bag, 3-pt belted occupants	3.65	4.76	2.68	5.06
Dual air bags	3.48	4.10	2.75	3.07
Near-side impacts				
All occupants	3.92	4.57	2.83	3.82
Unbelted occupants	3.19	3.17	3.35	2.33
3-pt belted occupants	4.05	5.94	2.11	3.78
Far-side impacts				
All occupants	2.92	4.47	2.40	3.12
Unbelted occupants	2.88	5.08	2.14	3.81
3-pt belted occupants	3.40	3.85	2.80	4.38
First-event rollovers				
All occupants	4.77	4.99	2.77	4.67
Unbelted occupants	3.92	3.69	2.47	5.07
3-pt belted occupants	5.23	5.64	3.65	4.99
Rear impacts & other crashes				
All occupants	3.25	4.32	3.01	2.64
Unbelted occupants	2.65	3.54	2.47	2.11
3-pt belted occupants	3.51	5.22	4.67	3.24

TABLE 4-2d: HEAD INJURIES: INCREASE IN RISK FOR A FEMALE RELATIVE TO A MALE OF THE SAME AGE
BY TYPE OF IMPACT AND OCCUPANT PROTECTION – CARS AND LTVs (MY 1960-2008), DRIVERS AND RF PASSENGERS

(Given the same crash scenario, average for occupants age 21-86, FARS MCOD 1987-2007,
evidently severe or unknown-severity "injuries that contribute to death")

Percent Head-Injury Increase for Females

Seat Position →→→→→→→→→→	Drivers			RF Passengers			Avg of Driver & RF
Occupant Age Group →→→→→→→→	21-96	21-30	65-74	21-96	21-30	65-74	21-96
All head injuries	**13.9 ± 5.0**	**24.1**	**1.3**	**15.2 ± 3.8**	**27.0**	**2.2**	**14.6 ± 3.1**
Frontal impacts							
All occupants	17.7 ± 6.5	30.9	1.6	11.3 ± 6.0	18.7	3.2	14.5 ± 4.3
No air bag, unbelted occupants	20.0 ± 9.9	32.5	5.1	16.9 ± 9.8	27.1	6.4	18.5 ± 6.6
No air bag, 3-pt belted occupants	18.4 ± 15.1	33.0	1.4	13.0 ± 18.3	10.7	17.9	15.7 ± 10.0
Dual air bags	9.2 ± 22.2	20.1	- 4.8	-4.6 ± 18.5	6.2	- 17.7	2.3 ± 12.3
Near-side impacts							
All occupants	18.0 ± 15.3	34.3	- .7	17.6 ± 12.3	35.0	- 4.0	17.8 ± 9.8
Unbelted occupants	20.3 ± 20.1	19.6	21.9	13.9 ± 18.0	29.6	- 3.5	17.1 ± 13.5
3-pt belted occupants	13.1 ± 24.6	45.3	- 17.8	14.6 ± 17.3	33.0	- 9.5	13.9 ± 15.0
Far-side impacts							
All occupants	9.4 ± 13.7	16.4	- .0	.6 ± 14.5	7.4	- 6.3	5.0 ± 10.0
Unbelted occupants	10.6 ± 19.2	23.3	- 4.9	6.9 ± 19.0	21.9	- 5.7	8.7 ± 13.5
3-pt belted occupants	18.3 ± 22.6	24.4	9.6	- 3.9 ± 21.0	- 9.4	1.8	7.2 ± 15.4
First-event rollovers							
All occupants	- 2.9 ± 14.0	18.9	- 27.2	33.4 ± 19.2	36.4	30.6	15.3 ± 9.4
Unbelted occupants	10.8 ± 21.7	32.7	- 14.1	66.6 ± 38.2	48.3	95.0	38.7 ± 18.4
3-pt belted occupants	- 16.2 ± 17.4	- 5.2	- 30.1	- 3.1 ± 21.1	.0	- 7.9	- 9.7 ± 12.1
Rear impacts & other crashes							
All occupants	24.4 ± 24.5	28.6	19.0	14.9 ± 18.5	39.0	- 10.4	19.7 ± 13.4
Unbelted occupants	55.9 ± 39.0	59.1	51.8	25.2 ± 30.2	53.2	- 2.8	40.5 ± 23.6
3-pt belted occupants	- 3.5 ± 28.1	- 12.1	10.3	6.1 ± 29.1	27.2	- 16.7	1.3 ± 18.4

The next four tables present corresponding results for torso injuries that "contributed to death" (according to MCOD). The effect of aging on torso injuries is much higher than its effect on head injuries, but Table 4-3a shows fairly uniform increases in torso-injury risk across the various types of crashes (3.92 ± .22% in frontals, 3.92 ± .37% in nearside impacts, 3.66 ± .38% in far-side impacts, 3.54 ± .80% in first-event rollovers, and 2.89 ± .88% in rear impacts/other crashes). The effect of aging is also about the same for unbelted and 3-point belted occupants in most of the crash types, and about the same with or without air bags in frontals. Once again, it is important to keep in mind that belts and air bags are effective technologies; the lack of change in the aging effects merely suggests that they are about equally effective for younger and older occupants for torso injuries. Table 4-3b again shows effects of aging starting high and eventually leveling off for unbelted occupants, starting low and escalating with age for belted occupants. Table 4-3c shows many large cumulative increases in torso-injury risk from 21 to 75.

Table 4-3d suggests females' added risk of torso injury relative to males may be mitigated by belt use in all types of crashes and further mitigated by air bags in frontal crashes.

MCOD does not have data for detailed separate analyses of chest and abdominal injury risk for females by type of crash and occupant protection. However, here are some basic point estimates: the risk of abdominal injury is 32 percent higher for a female than a male in frontal impacts, 30 percent higher in nearside impacts, and 37 percent higher in far-side impacts. Within frontal impacts, risk is 29 percent higher for unbelted females than unbelted males in vehicles without air bags; 25 percent higher for belted females than belted males in vehicles without air bags; and 34 percent higher in vehicles with air bags. These point estimates are, statistically speaking, more or less equal (as each estimate has confidence bounds ranging from ± 12 to ± 25 percentage points). In other words, females have high risk of abdominal injury relative to males, apparently regardless of the type of crash and the type of occupant protection, even in vehicles equipped with air bags.

TABLE 4-3a: TORSO INJURIES: INCREASE IN RISK FOR EACH YEAR THAT AN OCCUPANT GETS OLDER BY TYPE OF IMPACT AND OCCUPANT PROTECTION – CARS AND LTVs (MY 1960-2008), DRIVERS AND RF PASSENGERS

(Given the same crash scenario, average for occupants age 21-86, FARS MCOD 1987-2007, evidently severe or unknown-severity "injuries that contribute to death")

Percent Torso-Injury Increase per Year of Aging

	Males			Females			Average of Males & Females		
	Driver	RF Pass	Average	Driver	RF Pass	Average	Driver	RF Pass	Average
All torso injuries	**3.84 ± .23**	**4.13 ± .20**	**3.99 ± .17**	**3.25 ± .21**	**3.99 ± .18**	**3.62 ± .18**	**3.55 ± .20**	**4.06 ± .18**	**3.80 ± .17**
Frontal impacts									
All occupants	3.95 ± .28	4.21 ± .26	4.08 ± .21	3.25 ± .33	4.26 ± .25	3.75 ± .25	3.60 ± .26	4.24 ± .25	3.92 ± .22
No air bag, unbelted occupants	3.87 ± .37	3.78 ± .37	3.83 ± .29	2.98 ± .43	3.98 ± .35	3.48 ± .34	3.24 ± .34	3.88 ± .32	3.65 ± .28
No air bag, 3-pt belted occupants	3.22 ± .68	4.46 ± .58	3.84 ± .59	2.95 ± .67	4.26 ± .60	3.61 ± .53	3.09 ± .74	4.36 ± .57	3.72 ± .56
Dual air bags	4.00 ± .63	3.85 ± .65	3.92 ± .59	3.50 ± .62	3.86 ± .60	3.68 ± .57	3.75 ± .60	3.85 ± .58	3.80 ± .56
Near-side impacts									
All occupants	4.29 ± .55	4.06 ± .64	4.17 ± .42	3.71 ± .60	3.64 ± .60	3.68 ± .42	4.00 ± .48	3.85 ± .57	3.92 ± .37
Unbelted occupants	3.76 ± .74	3.92 ± .75	3.84 ± .53	3.93 ± .76	3.44 ± .58	3.68 ± .48	3.84 ± .64	3.68 ± .57	3.76 ± .43
3-pt belted occupants	4.23 ± .89	3.67 ±1.21	3.95 ± .75	3.40 ± .97	3.35 ±1.22	3.38 ± .78	3.82 ± .94	3.51 ±1.20	3.66 ± .76
Far-side impacts									
All occupants	3.14 ± .59	4.32 ± .63	3.73 ± .43	2.76 ± .60	4.41 ± .58	3.58 ± .42	2.95 ± .52	4.37 ± .56	3.66 ± .38
Unbelted occupants	3.37 ± .71	4.11 ± .72	3.74 ± .50	3.29 ± .73	4.23 ± .68	3.76 ± .50	3.33 ± .66	4.17 ± .64	3.75 ± .46
3-pt belted occupants	2.89 ±1.29	4.43 ±1.03	3.66 ± .83	2.25 ±1.08	4.50 ± .87	3.38 ± .69	2.57 ±1.10	4.47 ± .83	3.52 ± .69
First-event rollovers									
All occupants	3.93 ± .88	3.89 ± .92	3.91 ± .82	2.37 ±1.06	3.97 ± .83	3.17 ± .83	3.15 ± .85	3.93 ± .80	3.54 ± .80
Unbelted occupants	3.71 ±1.04	3.61 ±1.13	3.66 ± .98	1.66 ±1.31	3.86 ± .99	2.76 ± .92	2.68 ± .98	3.74 ± .87	3.21 ± .87
3-pt belted occupants	4.61 ±2.28	4.59 ±2.02	4.60 ±1.89	4.37 ±2.05	4.33 ±1.84	4.35 ±1.95	4.49 ±1.93	4.46 ±1.69	4.47 ±1.83
Rear impacts & other crashes									
All occupants	2.58 ±1.05	2.98 ±1.06	2.78 ± .96	2.99 ±1.08	3.00 ± .93	2.99 ± .91	2.78 ± .99	2.99 ± .95	2.89 ± .88
Unbelted occupants	2.73 ±1.41	2.91 ±1.34	2.82 ±1.24	2.96 ±1.73	2.43 ±1.32	2.70 ±1.38	2.85 ±1.40	2.67 ±1.17	2.76 ±1.27
3-pt belted occupants	1.80 ±1.42	3.34 ±1.41	2.57 ±1.23	3.09 ±1.21	3.38 ±1.51	3.24 ±1.23	2.45 ±1.12	3.36 ±1.36	2.90 ±1.18

TABLE 4-3b

TORSO INJURIES: INCREASE IN RISK FOR EACH YEAR
THAT AN OCCUPANT GETS OLDER: AT AGE 21-30 VERSUS AGE 65-74

BY TYPE OF IMPACT AND OCCUPANT PROTECTION
CARS AND LTVs (MY 1960-2008), DRIVERS AND RIGHT-FRONT PASSENGERS
(Given the same crash scenario, average for occupants age 21-86, FARS MCOD 1987-2007, evidently
severe or unknown-severity "injuries that contribute to death")

	Percent Torso-Injury Increase per Year of Aging					
Seat Position →→→→→→→→→→→	Driver			RF Passenger		
Occupant Age Group →→→→→→→	Avg 21-96	21-30	65-74	Avg 21-96	21-30	65-74
All torso injuries	**3.84 ± .23**	**3.76**	**4.13**	**4.13 ± .20**	**3.81**	**4.88**
Frontal impacts						
All occupants	3.95 ± .28	3.82	4.30	4.21 ± .26	3.82	5.08
No air bag, unbelted occupants	3.87 ± .37	4.29	3.51	3.78 ± .37	4.02	3.70
No air bag, 3-pt belted occupants	3.22 ± .68	2.48	4.42	4.46 ± .58	3.38	6.70
Dual air bags	4.00 ± .63	3.35	5.05	3.85 ± .65	2.59	5.77
Near-side impacts						
All occupants	4.29 ± .55	4.57	4.21	4.06 ± .64	3.73	5.01
Unbelted occupants	3.76 ± .74	4.50	3.05	3.92 ± .75	3.70	4.50
3-pt belted occupants	4.23 ± .89	4.30	4.43	3.67 ±1.21	2.82	5.51
Far-side impacts						
All occupants	3.14 ± .59	2.68	3.85	4.32 ± .63	4.57	4.27
Unbelted occupants	3.37 ± .71	3.32	3.49	4.11 ± .72	5.06	3.42
3-pt belted occupants	2.89 ±1.29	1.28	5.17	4.43 ±1.03	4.42	4.62
First-event rollovers						
All occupants	3.94 ± .88	3.86	4.27	3.89 ± .92	4.51	3.39
Unbelted occupants	3.71 ±1.04	4.00	3.53	3.61 ±1.13	4.53	2.78
3-pt belted occupants	4.61 ±2.28	5.36	3.93	4.59 ±2.02	5.07	4.21
Rear impacts & other crashes						
All occupants	2.58 ±1.05	2.65	2.61	2.98 ±1.06	1.70	4.95
Unbelted occupants	2.73 ±1.41	3.52	1.94	2.91 ±1.34	1.84	4.52
3-pt belted occupants	1.80 ±1.42	1.04	3.09	3.34 ±1.41	2.00	5.65

TABLE 4-3c

RISK OF **TORSO INJURIES** FOR A 75-YEAR-OLD OCCUPANT
RELATIVE TO A 21-YEAR-OLD OCCUPANT
(Given similar physical insults, FARS MCOD 1987-2007,
evidently severe or unknown-severity "injuries that contribute to death")

BY TYPE OF IMPACT AND OCCUPANT PROTECTION
CARS AND LTVs (MY 1960-2008), DRIVERS AND RF PASSENGERS

| | Risk at Age 75 ÷ Risk at Age 21 | | | |
| | Males | | Females | |
	Drivers	RF Pass	Drivers	RF Pass
All torso injuries	**7.53**	**9.29**	**5.34**	**8.30**
Frontal impacts				
All occupants	7.94	9.74	5.33	10.10
No air bag, unbelted occupants	7.66	7.68	4.59	8.69
No air bag, 3-pt belted occupants	5.55	12.35	4.79	10.36
Dual air bags	8.26	7.87	6.18	8.09
Near-side impacts				
All occupants	9.41	9.54	6.75	6.62
Unbelted occupants	7.12	8.40	7.86	5.84
3-pt belted occupants	9.21	8.20	5.76	6.00
Far-side impacts				
All occupants	5.32	10.08	4.31	10.68
Unbelted occupants	5.93	9.31	5.73	10.02
3-pt belted occupants	4.73	10.79	3.47	11.30
First-event rollovers				
All occupants	7.90	8.09	3.22	8.48
Unbelted occupants	6.94	7.04	2.16	8.27
3-pt belted occupants	11.18	11.53	9.60	9.49
Rear impacts & other crashes				
All occupants	3.91	5.06	4.96	5.09
Unbelted occupants	4.26	4.79	4.89	3.40
3-pt belted occupants	2.72	6.48	5.64	6.91

TABLE 4-3d: **TORSO INJURIES:** INCREASE IN RISK FOR A FEMALE RELATIVE TO A MALE OF THE SAME AGE
BY TYPE OF IMPACT AND OCCUPANT PROTECTION – CARS AND LTVs (MY 1960-2008), DRIVERS AND RF PASSENGERS
(Given the same crash scenario, average for occupants age 21-86, FARS MCOD 1987-2007,
evidently severe or unknown-severity "injuries that contribute to death")

Percent Torso-Injury Increase for Females

Seat Position	Drivers			RF Passengers			Avg of Driver & RF
Occupant Age Group	21-96	21-30	65-74	21-96	21-30	65-74	21-96
All torso injuries	**10.1 ± 4.5**	**25.4**	**- 7.9**	**16.5 ± 4.9**	**23.2**	**9.6**	**13.3 ± 3.6**
Frontal impacts							
All occupants	8.0 ± 8.8	26.1	- 12.6	16.2 ± 6.0	16.4	17.5	12.1 ± 4.9
No air bag, unbelted occupants	16.3 ± 13.7	43.1	- 11.8	11.7 ± 9.2	8.7	17.2	14.0 ± 9.2
No air bag, 3-pt belted occupants	2.2 ± 13.6	8.2	- 5.1	21.9 ± 15.7	31.8	11.5	12.1 ± 10.5
Dual air bags	3.9 ± 15.7	15.1	- 9.9	5.4 ± 17.4	3.9	7.9	4.7 ± 10.3
Near-side impacts							
All occupants	16.9 ± 18.8	33.3	- 1.9	16.5 ± 15.2	32.1	- 5.7	16.7 ± 12.1
Unbelted occupants	21.3 ± 29.2	19.5	26.2	2.2 ± 19.3	16.8	- 16.1	11.8 ± 17.5
3-pt belted occupants	7.3 ± 27.9	27.5	- 14.8	11.6 ± 23.9	23.0	- 6.4	9.5 ± 18.4
Far-side impacts							
All occupants	15.8 ± 16.7	22.8	6.1	3.1 ± 12.8	1.2	5.8	9.5 ± 10.5
Unbelted occupants	32.2 ± 22.0	32.4	33.0	12.1 ± 18.8	9.8	16.3	22.1 ± 14.5
3-pt belted occupants	15.9 ± 26.9	25.8	1.5	2.1 ± 24.5	.3	4.5	9.0 ± 18.2
First-event rollovers							
All occupants	7.0 ± 17.8	50.6	- 33.7	46.5 ± 26.7	45.5	49.8	26.7 ± 14.4
Unbelted occupants	9.6 ± 21.8	69.1	- 40.8	57.0 ± 35.2	48.1	72.1	33.3 ± 17.8
3-pt belted occupants	- 20.3 ± 32.8	- 15.4	- 27.1	34.4 ± 52.1	47.5	20.3	7.0 ± 27.3
Rear impacts & other crashes							
All occupants	6.7 ± 20.9	- .7	18.5	13.7 ± 20.6	16.1	11.3	10.2 ± 14.6
Unbelted occupants	29.0 ± 39.9	25.2	36.5	.3 ± 28.9	14.7	- 16.0	14.6 ± 20.9
3-pt belted occupants	- 26.9 ± 26.0	- 43.1	5.2	34.0 ± 41.2	37.0	34.8	3.5 ± 22.2

104

The last four tables present effects of aging and gender on neck injuries. Because there are far fewer neck injuries than head or torso injuries, it is statistically futile to further subdivide any of the crash modes except frontals by the type of occupant protection. In the aggregate, the effect of aging on neck injuries in fatal crashes (3.19 ± .37%) is similar to the effect of aging on overall fatality risk (3.11 ± .08%). The effect of aging on neck injuries is perhaps somewhat higher in frontals (3.45 ± .55%) and it is perhaps lower in side impacts (2.33 ± .70% nearside, 2.39 ± .81% far-side). One point estimate stands out from the others (although it has wide confidence bounds due to the limited data): belted occupants in frontal impacts of vehicles without air bags (4.39 ± 1.24%), especially the RF passengers (4.85 ± 1.36%). It is plausible that such impacts could place exceptional strain on the neck, because the torso is held in place by the belts, while the head has no air bag (and in the case of RF passengers, not even a steering assembly) to arrest its forward motion. This configuration also exhibits in Table 4-4b strong effects of aging even for occupants 21 to 30 and in Table 4-4c exceptionally high cumulative increases from 21 to 75, at least for males.

Table 4-4d shows females have substantially higher risk of neck injury than males in frontals (42.3 ± 13.0%) and in nearside impacts (42.6 ± 27.3%). Here, too, the combination of belt use, frontal impacts, and no air bags appears to be especially noxious for females relative to males (90.9 ± 53.4%). In addition to a risk of strain on the neck due to head excursion, some short-statured females may also find the shoulder belt positioned on or close to the neck. Fortunately, this combination is rapidly becoming less prevalent on the nation's roads, because all passenger cars since MY 1997 and LTVs with GVWR < 8,500 pounds since MY 1999 have been equipped with frontal air bags.[43]

---

[43] Kahane (2004), p. 107.

TABLE 4-4a: NECK INJURIES: INCREASE IN RISK FOR EACH YEAR THAT AN OCCUPANT GETS OLDER
BY TYPE OF IMPACT AND OCCUPANT PROTECTION – CARS AND LTVs (MY 1960-2008), DRIVERS AND RF PASSENGERS

(Given the same crash scenario, average for occupants age 21-86, FARS MCOD 1987-2007, evidently severe or unknown-severity "injuries that contribute to death")

Percent Neck-Injury Increase per Year of Aging

	Males			Females			Average of Males & Females		
	Driver	RF Pass	Average	Driver	RF Pass	Average	Driver	RF Pass	Average
All neck injuries	**3.25 ± .43**	**3.73 ± .42**	**3.49 ± .40**	**2.56 ± .47**	**3.22 ± .44**	**2.89 ± .37**	**2.90 ± .39**	**3.48 ± .40**	**3.19 ± .37**
Frontal impacts									
All occupants	3.45 ± .62	4.09 ± .82	3.77 ± .54	2.74 ± .77	3.54 ± .62	3.14 ± .58	3.09 ± .54	3.81 ± .68	3.45 ± .55
No air bag, unbelted occupants	3.29 ±1.02	3.99 ± .99	3.64 ± .88	2.20 ±1.15	3.50 ±1.06	2.84 ± .92	2.74 ± .93	3.75 ± .99	3.24 ± .86
No air bag, 3-pt belted occupants	4.34 ±1.47	5.74 ±1.41	5.04 ±1.30	3.53 ±1.66	3.96 ±1.33	3.75 ±1.38	3.93 ±1.32	4.85 ±1.36	4.39 ±1.24
Dual air bags	3.62 ±1.47	3.09 ±1.45	3.36 ±1.24	2.55 ±1.77	3.67 ±1.32	3.11 ±1.38	3.08 ±1.49	3.38 ±1.36	3.23 ±1.26
Near-side impacts	2.70 ±1.26	2.30 ±1.06	2.50 ±.82	2.01 ±1.17	2.32 ±1.00	2.17 ± .77	2.36 ±1.07	2.31 ± .90	2.33 ± .70
Far-side impacts	1.99 ±1.17	3.58 ±1.51	2.78 ± .96	1.20 ±1.12	2.79 ±1.04	2.00 ± .77	1.59 ±1.06	3.18 ±1.23	2.39 ± .81
First-event rollovers	3.91 ±1.35	3.87 ±1.03	3.89 ±1.06	2.97 ±1.51	3.24 ±1.14	3.11 ±1.05	3.44 ±1.25	3.56 ± .96	3.50 ±1.03
Rear impacts & other crashes	3.81 ±1.37	4.32 ±1.40	4.06 ±1.17	2.03 ±1.82	3.84 ±1.71	2.93 ±1.49	2.92 ±1.34	4.08 ±1.52	3.50 ±1.26

106

TABLE 4-4b

NECK INJURIES: INCREASE IN RISK FOR EACH YEAR
THAT AN OCCUPANT GETS OLDER: AT AGE 21-30 VERSUS AGE 65-74

BY TYPE OF IMPACT AND OCCUPANT PROTECTION
CARS AND LTVs (MY 1960-2008), DRIVERS AND RIGHT-FRONT PASSENGERS
(Given the same crash scenario, average for occupants age 21-86, FARS MCOD 1987-2007, evidently
severe or unknown-severity "injuries that contribute to death")

	Percent Neck-Injury Increase per Year of Aging					
Seat Position →→→→→→→→→→→→	Driver			RF Passenger		
Occupant Age Group →→→→→→→→	Avg 21-96	21-30	65-74	Avg 21-96	21-30	65-74
All neck injuries	**3.25 ± .43**	**3.11**	**3.52**	**3.74 ± .42**	**3.12**	**4.84**
Frontal impacts						
All occupants	3.45 ± .62	3.10	4.03	4.09 ± .82	3.18	5.77
No air bag, unbelted occupants	3.29 ±1.02	2.90	3.89	3.99 ± .99	3.28	5.41
No air bag, 3-pt belted occupants	4.34 ±1.47	5.23	3.37	5.74 ±1.41	5.53	7.77
Dual air bags	3.62 ±1.47	2.92	4.67	3.09 ±1.45	1.90	4.78
Near-side impacts	2.70 ±1.26	2.16	3.60	2.30 ±1.06	1.91	3.01
Far-side impacts	1.99 ±1.17	1.63	2.55	3.58 ±1.51	4.06	3.25
First-event rollovers	3.91 ±1.35	5.04	2.62	3.87 ±1.03	2.93	5.46
Rear impacts & other crashes	3.81 ±1.37	3.54	4.31	4.32 ±1.40	4.37	4.64

TABLE 4-4c

RISK OF **NECK INJURIES** FOR A 75-YEAR-OLD OCCUPANT
RELATIVE TO A 21-YEAR-OLD OCCUPANT
(Given similar physical insults, FARS MCOD 1987-2007,
evidently severe or unknown-severity "injuries that contribute to death")

BY TYPE OF IMPACT AND OCCUPANT PROTECTION
CARS AND LTVs (MY 1960-2008), DRIVERS AND RF PASSENGERS

	Risk at Age 75 ÷ Risk at Age 21			
	Males		Females	
	Drivers	RF Pass	Drivers	RF Pass
All neck injuries	**5.57**	**7.53**	**3.76**	**5.25**
Frontal impacts				
All occupants	6.19	9.34	4.12	6.22
No air bag, unbelted occupants	5.74	9.09	3.11	6.24
No air bag, 3-pt belted occupants	9.62	26.33	6.24	6.47
Dual air bags	6.81	5.19	3.67	8.17
Near-side impacts	4.27	3.51	2.82	3.66
Far-side impacts	2.93	7.14	1.84	4.26
First-event rollovers	7.88	8.17	4.51	5.15
Rear impacts & other crashes	7.49	10.62	2.77	7.35

TABLE 4-4d: NECK INJURIES: INCREASE IN RISK FOR A FEMALE RELATIVE TO A MALE OF THE SAME AGE BY TYPE OF IMPACT AND OCCUPANT PROTECTION – CARS AND LTVs (MY 1960-2008), DRIVERS AND RF PASSENGERS

(Given the same crash scenario, average for occupants age 21-86, FARS MCOD 1987-2007, evidently severe or unknown-severity "injuries that contribute to death")

Percent Neck-Injury Increase for Females

Seat Position →→→→→→→→→→→	Drivers			RF Passengers			Avg of Driver & RF
Occupant Age Group →→→→→→→	21-96	21-30	65-74	21-96	21-30	65-74	21-96
All neck injuries	**45.9 ± 17.9**	**60.3**	**20.3**	**32.8 ± 12.5**	**51.0**	**12.8**	**39.4 ± 9.4**
Frontal impacts							
All occupants	53.0 ± 27.1	77.7	25.5	31.6 ± 19.4	51.3	9.2	42.3 ± 13.0
No air bag, unbelted occupants	31.7 ± 33.6	63.6	- 1.9	26.7 ± 27.9	42.4	6.7	29.2 ± 20.7
No air bag, 3-pt belted occupants	101. ± 75.6	135.8	66.9	80.5 ± 51.8	205.0	- 4.7	90.9 ± 53.4
Dual air bags	43.9 ± 60.2	84.4	2.3	5.0 ± 27.6	- 11.1	32.0	24.4 ± 28.6
Near-side impacts	42.5 ± 40.0	66.3	16.4	42.8 ± 37.3	41.7	47.5	42.6 ± 27.3
Far-side impacts	30.9 ± 32.4	51.6	2.3	9.1 ± 28.2	32.7	- 11.5	20.0 ± 21.5
First-event rollovers	40.1 ± 36.1	74.3	4.9	11.0 ± 29.8	33.6	- 12.2	25.6 ± 21.6
Rear impacts & other crashes	36.7 ± 44.0	93.0	- 17.2	80.0 ± 66.7	104.2	54.1	58.3 ± 37.3

CHAPTER 5

EFFECT OF OCCUPANT AGE AND GENDER ON NONFATAL INJURIES
ANALYSES OF 1988-2010 NASS-CDS DATA

5.0   Summary

The NASS-CDS file describes all occupants' injuries, including their AIS severity level and body region, for a probability sample of the nation's fatal and nonfatal crashes. Based on analyses of the relative injury risk of two occupants of the same vehicle as a function of their ages and genders, AIS ≥ 2 injury risk is estimated to increase by an average of $1.58 \pm .35$ percent for each year that a front-seat occupant gets older, AIS ≥ 3 injury risk by $2.29 \pm .44$ percent, and AIS ≥ 4 injury risk by $2.65 \pm .61$ percent (whereas, based on FARS data, fatality risk increases by $3.11 \pm .08\%$) – i.e., the more severe the injury, the stronger the effect of age. AIS ≥ 2 injury risk is an average of 28.8 ± 6.0 percent higher for a female than for a male of the same age, in the same seat position, exposed to similar physical insults; the corresponding increases for AIS ≥ 3 is 37.3 ± 10.5 percent; for AIS ≥ 4, 28.9 ± 15.0 percent; but for fatalities (based on FARS), only 17.0 ± 1.5 percent – i.e., no trend to stronger effects for more severe injuries. Older occupants are relatively more vulnerable to torso injuries, especially chest injuries at each AIS level. Females are highly vulnerable to neck injuries and perhaps even more to arm and leg injuries. In fact, the much higher overall injury risk for females than males at the AIS 2 and 3 levels is largely due to their vulnerability to arm and leg injuries, which occur frequently but are rarely life-threatening.

5.1   Preparation of a NASS-CDS analysis file

The Crashworthiness Data System (CDS) of the National Automotive Sampling System (NASS) is a national probability sample of cars and LTVs involved in crashes where at least one vehicle was towed from the scene. CDS has been fairly uniform in terms of the sample design and the definitions of injuries from its inception in 1988 through 2010 (the latest year available as of August 2012).

Injury data: CDS documents occupants' injuries based on data from hospitals, treatment facilities, and autopsies. Injury severity has always been assessed with the abbreviated injury scale (AIS), which rates injuries from 1 (minor) to 6 (maximum).[44] There have been several versions of the AIS over the years. From 1988 to 1992, CDS used the 1985 revision of the AIS; from 1993 to 2009, the 1990/98 revisions; and starting in 2010, the 2005/08 revision. These three versions of the AIS all use a scale from 1 to 6 and have the same names for each level, but some individual types of injury may have changed levels. The analyses of this report consider injuries rated 2 to 6 on the AIS in the CDS file. The survival rates for injuries of each level, based on the records of people with a single reported injury in the National Trauma Data Bank (NTDB), help describe how severity increases from one level to the next. The NTDB is a large aggregation of

---

[44] Gennarelli, T. A., & Wodzin, E. (2006). AIS 2005: A Contemporary Injury Scale. *Injury, 37,* pp. 1083-1091.

injury data from trauma registries in the United States (not limited to injuries in motor-vehicle crashes).[45]

2   Moderate (the survival rate is over 99% in the NTDB data)

3   Serious (survival rate 97% in the NTDB)

4   Severe (survival rate 85% in the NTDB)

5   Critical (survival rate 60% in the NTDB)

6   Maximum (survival rate 21% in the NTDB)

Multiple injuries may be coded for one occupant.

CDS documents the body region of each injury, the lesion, and the system or organ involved. In the analyses that follow, "head injuries" include BODYREG codes H and F (head, face); "torso injuries" include codes B, C, M, P, and S (back, chest, abdomen, pelvis, shoulder); "neck injuries" only include code N (neck); "arm injuries" include A, E, R, W, and X (upper arm, elbow, forearm, wrist/hand, entire arm); and "leg injuries" include K, L, Q, T, and Y (knee, lower leg, ankle/foot, thigh, entire leg). BODYREG codes O (entire body) and U (unknown) are not assigned to any of these groups.

An occupant is defined to have AIS ≥ 2 head injury if any one (or more) of that person's injuries is a head injury with AIS = 2, 3, 4, 5, or 6; similarly for the other body regions, and for AIS ≥ 3 and AIS ≥ 4.

CDS also defines a maximum AIS (MAIS) and a treatment/mortality variable at the occupant level. MAIS is the highest AIS of the occupant's various injuries (excluding injuries with unknown severity). The treatment/mortality variable indicates, among other things, if the occupant died as a result of the crash – regardless of the AIS levels of that person's injuries. It is defined even when there is no information on specific injuries. An occupant is defined to have MAIS ≥ 2 if MAIS = 2, 3, 4, 5, or 6 and/or the occupant died as a result of the crash; similarly for MAIS ≥ 3 and MAIS ≥ 4.

CDS analysis file: The information on a person's individual injuries is used to define six variables at the person level.

- AISGE2 = 1 if MAIS = 2, 3, 4, 5, or 6 and/or the occupant died as a result of the crash; = 0 otherwise; similarly define AISGE3 and AISGE4
- FATAL = 1 it the treatment/mortality variable indicates the occupant died as a result of the crash; = 0 otherwise
- AIS2 = 1 if MAIS = 2 exactly; = 0 otherwise; similarly define AIS3
- HEAD2 = 1 if any one (or more) of that person's injuries is a head injury with AIS = 2, 3, 4, 5, or 6; = 0 otherwise; similarly define HEAD3 and HEAD4
- TORSO2 = 1 if any one (or more) of that person's injuries is a torso injury with AIS = 2, 3, 4, 5, or 6; = 0 otherwise; similarly define TORSO3 and TORSO4

---

[45] *Ibid.*

- NECK2 = 1 if any one (or more) of that person's injuries is a neck injury with AIS = 2, 3, 4, 5, or 6; = 0 otherwise; similarly define NECK3
- ARM2 = 1 if any one (or more) of that person's injuries is an arm injury with AIS = 2, 3, 4, 5, or 6; = 0 otherwise; similarly define ARM3
- LEG2 = 1 if any one (or more) of that person's injuries is a leg injury with AIS = 2, 3, 4, 5, or 6; = 0 otherwise; similarly define LEG3

These person-level records are used to create a database of all CDS cases of cars or LTVs (MY ≥ 1960) involved in fatal crashes and in which the driver and RF passenger seats were occupied; where either the driver, or the RF passenger, or both had at least one reported AIS ≥ 2 injury and/or were fatalities; where both occupants' age and gender must be known, and ages must be in the range of 21 to 96. Furthermore, the driver and RF passenger should have similar belt use and air bag availability (see Section 2.2) – i.e., either both occupants are belted or neither; the vehicle must either have dual air bags or have no air bags at all.

The CDS database contains 6,452 records of driver-RF pairs riding in the same vehicle, at least one or possibly both occupants with MAIS ≥ 2. That is far fewer than the 154,467 records of the FARS database of Chapters 2 and 3, because CDS is a sample, not a census like FARS. It is also far fewer than the effective size of the FARS-MCOD database of Chapter 4, which furnished 33,998 cases for the basic analysis of head injuries and 30,378 cases for the basic analysis of torso injuries. It means the analyses of CDS cannot delve into as many subgroups as the FARS and FARS-MCOD analyses. There are simply not enough cases to subdivide the data into model-year groups, or by impact location. Subdivisions that will be considered are those uniquely possible with CDS, namely by AIS level and AIS/body region. The objective is to see how the effects of aging and gender vary with the severity of the injuries. Double-pair comparison analysis has the great advantage of allowing unbiased measurement of how different groups of people respond to similar physical insults, but its price is that the data must be limited to driver-RF pairs; the many vehicle cases of unaccompanied drivers cannot be included. Appendix F lists the N of cases for each of the individual regression analyses.

Another potential complication is that CDS is not merely a collection of individual cases, but a probability sample of the nation's towaway crashes. Each case has a ratio weight factor (RATWGT) equal to the inverse of its probability of selection.[46] For unbiased national estimates of totals or simple injury rates, each case needs to be weighted by RATWGT. But computing statistics based on weighted data tends to increase the statistical uncertainty (which is already high due to the limited number of cases), because a few high-RATWGT cases in one cell or another can distort the results. Unweighted data can yield a more statistically precise estimate, at the expense of unknown bias. Use of unweighted data is inadvisable, as stated above, in estimating national totals or population injury rates. It might be less of an issue with double-pair comparison analysis, which is limited to vehicles in which at least one person has an AIS ≥ 2 injury, and which compares the RF passenger to the driver of the same vehicle (and therefore with the same RATWGT, a variable that has the same value for all the occupants and vehicles in a given crash).

---

[46] In CY 1988-1989, the case-weight variable was called NATWGT.

Ideally (when there is sufficient data), the weighted and unweighted CDS analyses will have fairly consistent, mutually reinforcing results – as, for example, in NHTSA's evaluation of head impact protection, which, however, was not a double-pair comparison analysis.[47] In another situation, with fewer data points – NHTSA's double-pair comparison analysis of booster seats – the results from the weighted CDS data were inconsistent with the unweighted data and non-significant, with large confidence intervals.[48] Here, too, the double-pair comparison analysis with weighted CDS data has so much sampling error that not a single estimate of the effects of aging or gender in the basic analysis of AIS ≥ 2 injuries is statistically significant (as will be discussed in the next section). That leaves a choice of accepting the analysis of unweighted cases or ending up without any CDS analysis and without any insight on how the effects of aging and gender vary by AIS level.

5.2   Effects of aging and gender on injury risk by MAIS level

Techniques combining the logistic regression coefficients with a double-pair comparison analysis generate estimates of the increase in injury risk, given similar physical insults, as a male or female occupant ages one year and the injury risk for a female relative to a male of the same age. Table 5-1a compares the effects of aging on injury risk by MAIS level – overall and separately for unbelted and 3-point belted occupants – for people involved in towaway crashes, including mostly nonfatal but also fatal crashes, based on unweighted 1988-2010 CDS data. As discussed in Section 2.11, the confidence bounds take into account that CDS is a cluster sample and has a design effect (sampling error that is larger than what it would have been for a simple random sample of the same size).

The first row of Table 5-1a is the estimated overall increase in fatality risk, based on the limited number of cases in CDS that were fatal to the driver and/or the RF passenger. The average effect for male and female drivers and RF passengers is a risk increase of 3.32 ± .77 percent for each year that a person ages. The large sampling error limits the utility of this estimate. The second row of Table 5-1a copies the first row from Table 3-1a: the corresponding estimate from the much larger FARS database: the overall increase is 3.11 ± .08 percent. The FARS estimate is quite close to the CDS estimate; it is easily inside the confidence bounds of the CDS estimate. Furthermore, all the individual FARS estimates in the second row of Table 5-1a are well within the confidence bounds of the CDS estimates directly above them, and in many cases quite close to CDS. The FARS and CDS estimates can be considered equivalent and the CDS estimates for the various levels of nonfatal injury can be compared to the precise FARS estimate for fatalities.

[47] Kahane, C. J. (2011). *Evaluation of the 1999-2003 Head Impact Upgrade of FMVSS No. 201 – Upper-Interior Components: Effectiveness of Energy-Absorbing Materials Without Head-Protection Air Bags.* (Report No. DOT HS 811 538, chapter 2). Washington, DC: National Highway Traffic Safety Administration. Available at www-nrd.nhtsa.dot.gov/Pubs/811538.PDF

[48] Sivinski, R. (2010). *Booster Seat Effectiveness Estimates Based on CDS and State Data.* (Report No. DOT HS 811 338). Washington, DC: National Highway Traffic Safety Administration. Available at www-nrd.nhtsa.dot.gov/Pubs/811338.pdf

TABLE 5-1a: INCREASE IN INJURY RISK FOR EACH YEAR THAT AN OCCUPANT GETS OLDER
BY MAIS SEVERITY LEVEL AND BELT USE – CARS AND LTVs, DRIVERS AND RIGHT-FRONT PASSENGERS
(Given the same crash scenario, average for occupants 21 to 96, CDS unweighted data, 1988-2010)

Percent Injury Increase per Year of Aging

	Males			Females			Average of Males & Females		
	Driver	RF Pass	Average	Driver	RF Pass	Average	Driver	RF Pass	Average
Fatal Injuries (CDS data)	3.38 ± .86	3.73 ±1.09	3.55 ± .98	2.54 ±1.00	3.63 ± .88	3.09 ± .77	2.96 ± .77	3.68 ± .94	3.32 ± .77
Fatal Injuries (FARS data)	3.06 ± .09	3.51 ± .09	3.28 ± .08	2.60 ± .10	3.27 ± .08	2.93 ± .08	2.83 ± .08	3.39 ± .07	3.11 ± .08
Unbelted occupants	2.92 ±.11	3.38 ±.10	3.15 ±.09	2.52 ±.14	3.13 ±.10	2.82 ±.11	2.72 ±.11	3.25 ±.09	2.99 ±.10
3-pt belted occupants	3.11 ±.17	3.53 ±.17	3.32 ±.15	2.64 ±.18	3.35 ±.14	3.00 ±.15	2.87 ±.17	3.44 ±.15	3.16 ±.15
MAIS ≥ 4 All occupants	2.39 ± .75	3.08 ± .80	2.74 ± .71	2.25 ± .95	2.89 ± .80	2.57 ± .78	2.32 ± .73	2.98 ± .69	2.65 ± .73
Unbelted occupants	1.84 ±1.13	2.51 ±1.12	2.17 ±1.03	1.46 ±1.38	2.20 ±1.05	1.83 ± .90	1.65 ±1.01	2.35 ±1.07	2.00 ± .94
3-pt belted occupants	2.84 ±1.22	3.53 ± .81	3.19 ± .92	2.37 ±1.08	3.43 ±1.25	2.90 ±1.12	2.61 ±1.01	3.48 ± .88	3.04 ± .90
MAIS ≥ 3 All occupants	2.04 ± .46	2.62 ± .64	2.33 ± .50	1.98 ± .59	2.51 ± .56	2.24 ± .55	2.01 ± .50	2.56 ± .53	2.29 ± .50
Unbelted occupants	1.41 ± .75	2.23 ± .78	1.82 ± .71	1.12 ± .84	1.69 ± .82	1.41 ± .77	1.27 ± .70	1.96 ± .76	1.61 ± .67
3-pt belted occupants	2.75 ± .83	2.92 ± .92	2.84 ± .85	1.97 ± .91	3.12 ± .90	2.54 ± .79	2.36 ± .80	3.02 ± .87	2.69 ± .80
MAIS ≥ 2 All occupants	1.40 ± .43	1.66 ± .44	1.53 ± .37	1.57 ± .38	1.70 ± .33	1.64 ± .33	1.48 ± .33	1.68 ± .35	1.58 ± .33
Unbelted occupants	.84 ± .64	1.28 ± .74	1.06 ± .66	.71 ± .61	1.08 ± .49	.90 ± .52	.78 ±.55	1.18 ±.62	.98 ± .57
3-pt belted occupants	1.79 ± .59	1.96 ± .68	1.87 ± .50	1.87 ± .56	2.10 ± .49	1.98 ± .52	1.83 ± .52	2.03 ± .56	1.93 ± .49
MAIS 3 exactly All occupants	.75 ± .78	2.09 ±1.14	1.92 ± .94	1.76 ± .82	2.18 ± .71	1.97 ± .71	1.76 ± .74	2.13 ± .88	1.94 ± .81
Unbelted occupants	.81 ±1.41	1.79 ±1.96	1.30 ±1.55	.42 ±2.17	.84 ±1.98	.63 ±1.95	.62 ±1.66	1.32 ±1.86	.97 ±1.73
3-pt belted occupants	2.70 ±1.13	2.39 ±1.95	2.54 ±1.37	1.98 ±1.06	2.97 ± .92	2.47 ± .82	2.34 ±1.01	2.68 ±1.36	2.51 ±1.13
MAIS 2 exactly All occupants	.84 ± .96	.52 ± .76	.68 ± .69	1.05 ± .85	.84 ± .61	.94 ± .61	.94 ± .80	.68 ± .60	.81 ± .61
Unbelted occupants	.11 ±1.52	-.29 ±1.23	-.09 ±1.27	.07 ±1.82	.32 ±1.32	.20 ±1.45	.09 ±1.54	.02 ±1.26	.05 ±1.39
3-pt belted occupants	.13 ± .98	1.10 ±1.03	1.11 ± .78	1.61 ±1.05	1.22 ± .77	1.42 ± .81	1.37 ± .96	1.16 ± .74	1.27 ± .74

114

The salient feature of Table 5-1a is that the effect of aging decreases for lower MAIS levels. The risk increase per year of aging is $3.11 \pm .08$ percent for fatalities, $2.65 \pm .73$ percent for MAIS $\geq$ 4 (severe) injuries, $2.29 \pm .50$ percent for MAIS $\geq$ 3 injuries ($1.94 \pm .81\%$ for MAIS = 3 exactly), and only $1.58 \pm .33$ percent for MAIS $\geq$ 2 injuries ($0.81 \pm .61\%$ for MAIS = 2 exactly). Drivers and RF passengers, males and females all have the same trend to lower effects at lower MAIS. Intuitively, fatality risk increases with age because fragility (risk of injury given similar physical insults) and frailty (risk of death given the same injury) both increase with age. For the nonfatal injuries, frailty is not an issue, only fragility. An additional factor may be that the distribution of injuries changes with AIS; for example, a large proportion of the AIS 2 injuries, but hardly any of the AIS $\geq$ 4 injuries are to the arms or legs.

Another feature of Table 5-1a is that the effect of aging is consistently higher for belted than for unrestrained occupants. Numerous analyses of fatality risk in Chapters 3 and 4 have already shown that pattern. But it appears to intensify at the nonfatal levels. The increase in fatality risk per year of aging is 3.16 percent for belted and a still fairly similar 2.99 percent for unrestrained occupants. At MAIS $\geq$ 4, these estimates are 3.04 and 2.00 percent, respectively; at MAIS $\geq$ 3, 2.69 and 1.61 percent; and at MAIS $\geq$ 2, 1.93 and 0.98 percent. Belts are relatively more effective for young occupants than for the elderly, but whereas the difference in effectiveness is relatively small for fatality reduction, it is apparently stronger at the nonfatal levels. The results are also consistent with the findings of Zhou, Rouhana, and Melvin that tolerance to belt load decreases faster with age than tolerance to blunt impact.[49]

Table 5-1b compares how the effect of aging changes with age, depending on the MAIS level of the injury, seat position, and belt use. For example, FARS-based fatality risk of male drivers increases by 2.90 percent a year at 21 to 30 and 3.39 percent a year at 65 to 74. Given the limited size of the CDS database, the individual point estimates for the effects at 21 to 30 or 65 to 74 cannot be considered precise. Nevertheless, CDS at the MAIS $\geq$ 4, MAIS $\geq$ 3, and MAIS $\geq$ 2 levels shares one pattern with the FARS results: the effect of aging is almost always higher at 65 to 74 than at 21 to 30, especially for RF passengers.

Table 5-1c estimates the cumulative effect of all the year-to-year increases in risk. Based on FARS, a 75-year-old male driver is 5.04 times as likely to die as a 21-year-old from similar physical insults; a male RF passenger has 6.70 times the fatality risk at 75 as at 21. The lower the MAIS level, the smaller the cumulative increases: for male drivers, the MAIS $\geq$ 4 injury risk is 3.71 time as high at 75 as at 21; MAIS $\geq$ 3, 3.07 times; and MAIS $\geq$ 2, just 2.16 times as high.

Table 5-1d estimates the increase in fatality risk for a female relative to a male of the same age, given the same crash scenario or physical insult. In Chapters 3 and 4, fatality-risk increases for females often paralleled the effects of aging. This is not true for nonfatal injuries. Table 5-1a showed that the effect of aging decreased steadily as injury severity decreased, but Table 5-1d shows females may be even more vulnerable at the lower injury levels. Whereas, according to FARS, the fatality risk of a female is 17.0 ± 1.5 percent higher than for a male of the same age (average of driver and RF passenger), females have 28.9 ± 15.6 percent higher risk of MAIS $\geq$ 4 injury than males, 37.3 ± 9.9 percent higher MAIS $\geq$ 3, and 28.8 ± 5.4 percent higher MAIS $\geq$ 2.

---

[49] Zhou, Rouhana, & Melvin (1996).

TABLE 5-1b: INCREASE IN INJURY RISK FOR EACH YEAR THAT AN OCCUPANT GETS OLDER, AT AGE 21-30 VERSUS AGE 65-74
(Given the same crash scenario, CDS unweighted data, 1988-2010)

BY MAIS SEVERITY LEVEL AND BELT USE
CARS AND LTVs, DRIVERS AND RIGHT-FRONT PASSENGERS

		Percent Injury Increase per Year of Aging					
Seat Position →→→→→→→→→→→		Driver			RF Passenger		
Occupant Age Group →→→→→→→→		Avg 21-96	21-30	65-74	Avg 21-96	21-30	65-74
Fatal Injuries (CDS data)		3.38 ± .86	2.06	5.29	3.73 ±1.09	1.98	6.58
Fatal Injuries (FARS data)		*3.06 ± .09*	*2.90*	*3.39*	*3.51 ± .09*	*2.90*	*4.58*
	Unbelted occupants	*2.92 ± .11*	*2.99*	*2.95*	*3.38 ± .10*	*2.98*	*4.16*
	3-pt belted occupants	*3.11 ± .17*	*2.69*	*3.78*	*3.53 ± .17*	*2.60*	*5.03*
MAIS ≥ 4	All occupants	2.39 ± .75	.87	4.63	3.08 ± .80	1.60	5.47
	Unbelted occupants	1.84 ±1.13	.92	3.13	2.51 ±1.12	.80	5.37
	3-pt belted occupants	2.84 ±1.22	.84	5.89	3.53 ± .81	2.75	5.02
MAIS ≥ 3	All occupants	2.04 ± .46	1.25	3.22	2.62 ± .64	1.65	4.05
	Unbelted occupants	1.41 ± .75	1.26	1.65	2.23 ± .78	1.59	3.31
	3-pt belted occupants	2.75 ± .83	1.55	4.46	2.92 ± .92	2.11	4.09
MAIS ≥ 2	All occupants	1.40 ± .43	.98	1.99	1.66 ± .44	1.09	2.41
	Unbelted occupants	.84 ± .64	.64	1.13	1.28 ± .74	.90	1.79
	3-pt belted occupants	1.79 ± .59	1.33	2.43	1.96 ± .68	1.47	2.59
MAIS 3 exactly	All occupants	1.75 ± .78	1.81	1.71	2.09 ±1.14	1.77	2.55
	Unbelted occupants	.81 ±1.41	1.70	- .48	1.79 ±1.96	2.22	1.37
	3-pt belted occupants	2.70 ±1.13	2.49	3.07	2.39 ±1.95	2.11	2.77
MAIS 2 exactly	All occupants	.84 ± .96	.84	.83	.52 ± .76	.91	.02
	Unbelted occupants	.11 ±1.52	-.22	.59	- .29 ±1.23	.42	-1.16
	3-pt belted occupants	1.13 ± .98	1.47	.63	1.10 ±1.03	1.48	.62

TABLE 5-1c

INJURY RISK FOR A 75-YEAR-OLD OCCUPANT
RELATIVE TO A 21-YEAR-OLD OCCUPANT
(Given similar physical insults, CDS unweighted data, 1988-2010)

BY MAIS SEVERITY LEVEL AND BELT USE
CARS AND LTVs, DRIVERS AND RIGHT-FRONT PASSENGERS

		Risk at 75 ÷ Risk at Age 21			
		Males		Females	
		Drivers	RF Pass	Drivers	RF Pass
Fatal Injuries (CDS data)		6.08	7.82	3.81	7.64
Fatal Injuries (FARS data)		*5.04*	*6.70*	*3.87*	*5.67*
	Unbelted occupants	*4.69*	*6.25*	*3.71*	*5.23*
	3-pt belted occupants	*5.23*	*6.91*	*3.98*	*6.07*
MAIS ≥ 4	All occupants	3.71	5.45	3.40	4.77
	Unbelted occupants	2.72	4.09	2.21	3.21
	3-pt belted occupants	4.77	6.73	3.51	6.67
MAIS > 3	All occupants	3.07	4.18	2.94	3.92
	Unbelted occupants	2.15	3.47	1.83	2.33
	3-pt belted occupants	4.44	4.85	2.81	5.82
MAIS ≥ 2	All occupants	2.16	2.47	2.38	2.56
	Unbelted occupants	1.59	2.00	1.47	1.78
	3-pt belted occupants	2.65	2.86	2.75	3.23
MAIS 3 exactly	All occupants	2.55	3.04	2.56	3.26
	Unbelted occupants	1.51	2.67	1.20	1.45
	3-pt belted occupants	4.20	3.54	2.77	5.36
MAIS 2 exactly	All occupants	1.57	1.33	1.78	1.65
	Unbelted occupants	1.07	.86	1.04	1.30
	3-pt belted occupants	1.81	1.82	2.44	1.99

TABLE 5-1d: INCREASE IN INJURY RISK FOR A FEMALE RELATIVE TO A MALE OF THE SAME AGE BY MAIS SEVERITY LEVEL AND BELT USE – CARS AND LTVs, DRIVERS AND RIGHT-FRONT PASSENGERS

(Given the same crash scenario, average for occupants 21 to 96, CDS unweighted data, 1988-2010)

Percent Injury Increase for Females

Seat Position →	Drivers			RF Passengers			Avg of Driver & RF
Occupant Age Group →	21-96	21-30	65-74	21-96	21-30	65-74	21-96
Fatal Injuries (CDS data)	40.7 ± 29.6	66.1	11.4	32.1 ± 32.2	31.6	34.1	36.4 ± 23.2
Fatal Injuries (FARS data)							
	13.4 ± 2.0	*25.9*	*-1.4*	*20.5 ± 2.2*	*29.2*	*11.4*	*17.0 ± 1.5*
Unbelted occupants	*15.7 ± 2.9*	*27.7*	*1.9*	*21.2 ± 3.1*	*30.3*	*11.9*	*18.4 ± 2.1*
3-pt belted occupants	*10.2 ± 3.7*	*20.9*	*-3.3*	*16.1 ± 4.6*	*22.3*	*9.1*	*13.2 ± 2.7*
MAIS ≥ 4							
All occupants	29.8 ± 19.8	34.8	22.8	28.0 ± 27.6	34.6	20.2	28.9 ± 15.6
Unbelted occupants	19.4 ± 35.4	27.9	7.4	18.8 ± 28.7	27.0	7.3	19.1 ± 17.6
3-pt belted occupants	58.6 ± 42.6	85.1	28.4	20.7 ± 40.7	20.6	24.3	39.7 ± 27.1
MAIS ≥ 3							
All occupants	47.7 ± 15.7	50.7	43.8	26.9 ± 15.7	30.0	23.7	37.3 ± 9.9
Unbelted occupants	25.5 ± 20.9	34.4	12.5	25.5 ± 15.7	48.2	3.2	25.5 ± 10.5
3-pt belted occupants	77.6 ± 35.0	112.0	42.2	15.9 ± 28.2	5.7	31.1	46.8 ± 15.6
MAIS ≥ 2							
All occupants	28.0 ± 8.8	25.5	31.2	29.6 ± 8.0	26.5	32.9	28.8 ± 5.4
Unbelted occupants	13.2 ± 11.3	15.9	9.3	23.5 ± 9.6	28.3	17.4	18.4 ± 6.9
3-pt belted occupants	36.3 ± 17.7	38.5	33.7	34.4 ± 16.0	26.6	44.4	35.3 ± 11.7
MAIS 3 exactly							
All occupants	68.1 ± 37.6	69.0	68.6	30.7 ± 25.1	27.7	36.1	49.4 ± 23.2
Unbelted occupants	30.9 ± 49.2	43.4	15.1	39.1 ± 53.0	74.6	5.0	35.0 ± 36.3
3-pt belted occupants	101.0 ± 61.8	135.8	65.0	14.3 ± 45.3	-2.1	41.3	57.6 ± 25.8
MAIS 2 exactly							
All occupants	10.2 ± 17.7	5.9	16.7	34.2 ± 24.0	22.8	50.6	22.2 ± 11.6
Unbelted occupants	-4.0 ± 25.5	-2.9	-5.4	24.6 ± 30.3	5.9	54.1	10.3 ± 17.1
3-pt belted occupants	9.7 ± 22.8	-1.3	27.4	55.7 ± 34.7	48.5	65.6	32.7 ± 17.7

The analyses of fatal crashes usually showed females at substantially higher risk than males up to age 35, but after that the "gender gap" steadily diminishes and, in the case of drivers, eventually changes to lower risk for females than males. Table 5-1d, based on FARS shows 25.9 percent higher fatality risk for female drivers 21 to 30 than for males of the same age, but 1.4 percent lower risk for female drivers 65 to 74. The paradigm shifts for nonfatal injuries. At the MAIS ≥ 4 and MAIS ≥ 3 levels, the increase for females is still fairly strong at 65 to 74. At MAIS ≥ 2, the risk increase <u>intensifies</u> with age, (from 25.5% at 21-30 to 31.2% at 65 to 74 for drivers, and from 26.5% to 32.9% for RF passengers). Intuitively, the gender gap closes for fatalities because females, although initially more fragile than males, become frail at a substantially slower pace than males. For nonfatal injuries, frailty is not an issue; older females do not have an advantage that offsets their higher fragility. Furthermore, a large proportion of the AIS 2 injuries are arm and leg fractures, which may be especially prevalent in older females due to osteoporosis and related conditions.

Fatality-risk increase is slightly lower for belted females relative to belted males of the same age (13.2 ± 2.7%) than for unrestrained females relative to unrestrained males (18.4 ± 2.1%). But the opposite appears to be the case at the nonfatal injury levels. The gender gap is clearly higher at MAIS ≥ 2 for belted occupants (35.3 ± 11.7% higher risk for females) than unrestrained occupants (18.4 ± 6.9% higher risk for females). It may also be higher at MAIS ≥ 3 (46.8 ± 15.6% versus 25.5 ± 10.5%) and, perhaps, at MAIS ≥ 4 (39.7 ± 27.1% versus 19.1 ± 17.6%). Belts may be somewhat less effective for females than males at these lower injury levels because they are of limited effectiveness against leg injuries, which account for a large share of females' AIS 2 and AIS 3 injuries.

The basic analyses of MAIS ≥ 2 injuries were repeated with weighted CDS cases (using RATWGT, the inverse of the sampling fraction). The overall average effect of aging one year is a non-significant risk increase of 1.15 ± 2.41 percent (as compared to the statistically significant 1.58 ± .33% with the unweighted data). The average risk increase for a female relative to a male of the same age is a non-significant 24.1 + 58.9 percent (as compared to the significant 28.8 ⊥ 5.4% with the unweighted data). None of the individual estimates by seat position (or by gender and seat position for the effect of aging) are statistically significant with the weighted data, but the unweighted point estimate is always within the confidence bounds of the weighted estimate. In other words, the estimates based on weighted data do not offer statistically meaningful information, but at least they are not inconsistent with the results based on unweighted data.

5.3   Effects of aging and gender on injury risk by body region and AIS

Table 5-2a estimates the effect of aging one year on injury risk, by body region and AIS level. Head-injury risk increases gradually with age at all AIS levels: by 1.54 ± 1.10 percent a year for AIS ≥ 4, by 2.07 ± 1.12 percent for AIS ≥ 3, and by 1.35 ± .68 percent for AIS ≥ 2. These rates of increase are lower than the corresponding estimates for MAIS (2.65%, 2.29%, and 1.58%, respectively, in Table 5-1a). In other words, risk of head injury increases slower with age than overall injury risk. The rates of increase are fairly similar for males and females. They are slightly higher for RF passengers than drivers, at least at the AIS ≥ 2 and AIS ≥ 3 levels.

Torso-injury risk increases strongly with age. The higher the AIS, the greater the rate of increase: by 4.17 ± 1.05 percent a year for AIS ≥ 4, by 2.86 ± .92 percent for AIS ≥ 3, and by 2.39 ± .49

percent for AIS $\geq$ 2. All of these are higher than the corresponding rates for MAIS. Here, too, the rates of increase are fairly similar for males and females and they are slightly higher for RF passengers than drivers. The results are quite consistent with the MCOD findings that the effect of aging in fatal crashes is strongest for torso injuries, as well as with the literature, which shows high rates of torso injury for older occupants.

CDS has just enough data for a separate look at two types of torso injury: chest (BODYREG = C) and abdominal injuries (BODYREG = M). At each AIS level, the effect of aging on chest injuries is slightly larger than its overall effect on torso injuries: at the AIS $\geq$ 4 level, the overall average increase per year is 4.43 ± 1.14 percent for chest injuries, 4.17 ± 1.05 percent for all torso injuries; at AIS $\geq$ 3 the corresponding estimates are 3.30 ± 1.03 percent for chest injuries, 2.86 ± .92 percent for all torso injuries; at AIS $\geq$ 2, 3.34 ± .68 percent for chest, 2.39 ± .49 percent for all torso injuries. Statistically meaningful estimates for abdominal injuries are only possible at the AIS $\geq$ 2 level; the estimate, 2.57 ± .88 percent is about the same as the overall increase in torso injury risk (2.39 ± .49%) and lower than the estimated increase in chest injury risk (3.34 ± .68%). The findings are consistent with the literature, which emphasizes elderly occupants' risk of thoracic injury, as well as with the MCOD analyses in Table 4-1a.

Neck injuries are less common than head or torso injuries and it is statistically futile to perform detailed analyses at the higher AIS levels. At the AIS $\geq$ 2 level however, risk increases by a relatively strong 3.15 ± 1.24 percent per year of aging. (At AIS $\geq$ 3, the estimate is a rather imprecise 3.68 ± 1.91%.)

Arm- and leg-injury risk also increase moderately with age. At the AIS $\geq$ 2 level, arm injuries increase by 2.44 ± .65 percent for each year that an occupant gets older (2.55 ± 1.58% at the AIS $\geq$ 3 level). AIS $\geq$ 2 leg-injury risk escalates by 1.46 ± .53 percent per year of aging (2.06 ± .82% for AIS $\geq$ 3).

Table 5-2b compares the effects of aging a year for occupants 21 to 30 to occupants 65 to 74. For head- and torso-injury risk, the rates of increase intensify at the higher ages for all three AIS levels and both seat positions. Arm and leg injuries, however, have a more nearly constant rate of increase across all ages (although it is difficult to say for sure, given the sampling error).

Table 5-2c shows considerable divergence in the cumulative increase of injury risk from 21 to 75 depending on the body region and AIS level. Head injury risk is typically 2 to 3 times as high at 75 as at 21, given similar physical insults; leg-injury risk increases by similar multiples. But torso-injury risk increases by a factor of 3.14 to 4.06 at the AIS $\geq$ 2 level, 3.65 to 5.88 at AIS $\geq$ 3, and 6.01 to 13.70 at the AIS $\geq$ 4 level. Again, this is consistent with the literature – e.g., Ridella, Rupp, and Poland's graphs of AIS $\geq$ 3 injury risk.[50]

---

[50] Ridella, Rupp, & Poland (2012).

TABLE 5-2a: INCREASE IN INJURY RISK FOR EACH YEAR THAT AN OCCUPANT GETS OLDER BY BODY REGION AND AIS SEVERITY LEVEL – CARS AND LTVs, DRIVERS AND RIGHT-FRONT PASSENGERS

(Given the same crash scenario, average for occupants 21 to 96, CDS unweighted data, 1988-2010)

Percent Injury Increase per Year of Aging

		Males			Females			Average of Males & Females		
		Driver	RF Pass	Average	Driver	RF Pass	Average	Driver	RF Pass	Average
HEAD	AIS ≥ 4	1.50 ± .97	1.55 ±1.55	1.53 ±1.18	1.44 ±1.32	1.65 ±1.28	1.55 ±1.19	1.47 ± .90	1.60 ±1.39	1.54 ±1.10
	AIS ≥ 3	1.84 ± .77	2.69 ±1.52	2.27 ±1.07	1.63 ±1.28	2.13 ±1.37	1.88 ±1.26	1.73 ± .99	2.41 ±1.35	2.07 ±1.12
	AIS ≥ 2	1.16 ± .58	1.67 ± .82	1.42 ± .64	.89 ± .97	1.66 ± .72	1.28 ± .78	1.03 ± .72	1.67 ± .75	1.35 ± .68
TORSO	AIS ≥ 4	3.38 ±1.33	4.72 ± .91	4.05 ±1.00	4.13 ±1.17	4.43 ±1.09	4.28 ±1.06	3.76 ±1.15	4.58 ± .97	4.17 ±1.05
	AIS ≥ 3	2.62 ± .75	3.29 ± .16	2.95 ± .91	2.45 ± .96	3.10 ± .81	2.77 ± .88	2.53 ± .91	3.19 ± .93	2.86 ± .92
	AIS ≥ 2	2.11 ± .46	2.61 ± .79	2.36 ± .56	2.33 ± .53	2.51 ± .55	2.42 ± .52	2.22 ± .46	2.56 ± .61	2.39 ± .49
Chest	AIS ≥ 4	3.57 ±1.36	4.91 ± .15	4.24 ±1.13	4.41 ±1.30	4.83 ±1.21	4.62 ±1.20	3.99 ±1.29	4.87 ±1.09	4.43 ±1.14
	AIS ≥ 3	3.01 ± .90	3.67 ±1.30	3.34 ±1.05	3.00 ±1.07	3.51 ± .91	3.25 ±1.00	3.00 ± .97	3.59 ±1.07	3.30 ±1.03
	AIS ≥ 2	3.09 ± .83	3.69 ± .95	3.39 ± .85	3.26 ± .63	3.30 ± .81	3.28 ± .65	3.18 ± .66	3.49 ± .79	3.34 ± .68
Abdomen	AIS ≥ 2	2.04 ±1.06	3.21 ±1.42	2.63 ±1.02	2.30 ±1.07	2.73 ±1.00	2.52 ± .93	2.17 ± .93	2.97 ±1.13	2.57 ± .88
NECK	AIS ≥ 2	3.41 ±1.50	3.54 ±1.63	3.47 ±1.33	2.43 ±1.56	3.21 ±1.26	2.82 ±1.28	2.92 ±1.35	3.38 ±1.27	3.15 ±1.24
ARM	AIS ≥ 2	2.38 ± .79	2.17 ± .99	2.27 ± .81	2.66 ± .85	2.56 ± .87	2.61 ± .67	2.52 ± .66	2.37 ± .87	2.44 ± .65
LEG	AIS ≥ 2	1.17 ± .71	1.09 ± .96	1.13 ± .57	1.86 ± .78	1.71 ± .50	1.78 ± .56	1.51 ± .54	1.40 ± .65	1.46 ± .53

TABLE 5-2b

INCREASE IN INJURY RISK FOR EACH YEAR THAT AN OCCUPANT GETS OLDER AT AGE 21-30 VERSUS AGE 65-74
(Given the same crash scenario, CDS unweighted data, 1988-2010)

BY BODY REGION AND AIS SEVERITY LEVEL
CARS AND LTVs, DRIVERS AND RIGHT-FRONT PASSENGERS

Seat Position →→→→→→→→→→		Percent Injury Increase per Year of Aging					
		Driver			RF Passenger		
Occupant Age Group →→→→→→→		Avg 21-96	21-30	65-74	Avg 21-96	21-30	65-74
HEAD	AIS ≥ 4	1.50 ± .97	-.60	4.70	1.55 ±1.55	-.85	5.09
	AIS ≥ 3	1.84 ± .77	1.27	2.71	2.69 ±1.52	1.00	5.52
	AIS ≥ 2	1.16 ± .58	.40	2.37	1.67 ± .82	.96	2.69
TORSO	AIS ≥ 4	3.38 ±1.33	2.73	4.48	4.72 ± .91	4.01	6.52
	AIS ≥ 3	2.62 ± .76	1.66	4.03	3.29 ±1.16	2.48	4.58
	AIS ≥ 2	2.11 ± .46	1.58	2.89	2.61 ± .79	2.08	3.36
Chest	AIS ≥ 4	3.57 ±1.36	3.08	4.45	4.91 ±1.15	4.68	5.98
	AIS ≥ 3	3.01 ± .90	2.08	4.40	3.67 ±1.30	2.98	4.85
	AIS ≥ 2	3.09 ± .83	3.01	3.27	3.69 ± .95	3.35	4.30
Abdomen	AIS ≥ 2	2.04 ±1.06	1.03	3.60	3.21 +1.42	2.04	5.28
NECK	AIS ≥ 2	3.41 ±1.50	3.51	3.45	3.54 ±1.63	1.91	6.18
ARM	AIS ≥ 2	2.38 ± .79	2.56	2.18	2.17 ± .99	2.01	2.41
LEG	AIS ≥ 2	1.17 ± .71	1.54	.67	1.09 ± .96	.70	1.59

122

TABLE 5-2c

INJURY RISK FOR A 75-YEAR-OLD OCCUPANT
RELATIVE TO A 21-YEAR-OLD OCCUPANT
(Given similar physical insults, CDS unweighted data, 1988-2010)

BY BODY REGION AND AIS SEVERITY LEVEL
CARS AND LTVs, DRIVERS AND RIGHT-FRONT PASSENGERS

		Risk at Age 75 ÷ Risk at Age 21			
		Males		Females	
		Drivers	RF Pass	Drivers	RF Pass
HEAD	AIS ≥ 4	2.38	2.36	2.28	2.57
	AIS ≥ 3	2.76	4.67	2.42	2.99
	AIS ≥ 2	1.92	2.46	1.61	2.45
TORSO	AIS ≥ 4	6.01	13.70	9.24	10.41
	AIS ≥ 3	4.06	5.88	3.65	5.25
	AIS ≥ 2	3.14	4.06	3.58	3.84
Chest	AIS ≥ 4	6.58	14.71	10.82	13.42
	AIS ≥ 3	4.99	7.21	4.94	6.66
	AIS ≥ 2	5.17	7.36	5.73	5.75
Abdomen	AIS ≥ 2	3.07	5.97	3.56	4.03
NECK	AIS ≥ 2	6.06	7.00	3.45	5.65
ARM	AIS ≥ 2	3.55	3.19	4.16	4.17
LEG	AIS ≥ 2	1.86	1.79	2.82	2.70

123

Table 5-2d indicates that females have substantially higher injury risk than males of the same age, given similar physical insults: especially for neck, arm, and leg injuries and for head and torso injuries at the lower severity levels. A female's added risk of head injury is 15.4 ± 21.8 percent at the AIS ≥ 4 level, but 31.2 ± 21.6 percent at the AIS ≥ 3 level and 22.1 ± 16.0 percent at AIS ≥ 2. The corresponding increases in torso injury are 26.6 ± 17.0 percent (AIS ≥ 4), 34.6 ± 16.7 percent (AIS ≥ 3), and 31.1 ± 10.7 percent (AIS ≥ 2). The increases in chest injury are, at each AIS level, slightly less than the increases in overall torso injury, whereas the 38.5 ± 28.4 percent increase in abdominal injury, estimated only at the AIS ≥ 2 level, is somewhat higher than the increase in overall torso injury ($31.1 \pm 10.7\%$). That is consistent with the MCOD results, although the observed difference between the chest- and abdominal-injury estimates is not as intense as in MCOD.

Females are especially vulnerable to neck injuries. Risk of AIS ≥ 2 neck injury is 44.7 ± 34.0 percent higher than for a male of the same age. The estimate is quite similar to the neck-injury increase in MCOD: 39.4 ± 9.4 percent (Table 4-1d). The difference between males and females is especially strong for young adults: identical 67.0 percent risk increases for females relative to males, for both drivers and RF passengers at 21 to 30. That is also consistent with MCOD. A male's cervical spine has deeper facet joints and a wider surface area, resulting in greater spinal-column strength, yet a female's neck is called upon to support and control the motion of a head that is almost as large and heavy as a male's.

Arm and leg injuries, however, are where females have the highest risk relative to males. Arm-injury risk at the AIS ≥ 2 level is 58.2 ± 20.6 percent higher for a female than a male. AIS ≥ 2 leg-injury risk is 79.7 ± 16.3 percent higher. Evidently, females' bones have lower bone-mineral density and lower strength than males', but are exposed to the same contacts with vehicle structures. Furthermore, the added risk for females increases rather than decreases with age, perhaps because osteoporosis and related conditions are even more common in older females than in older males, whereas the fact that females are less frail than males of the same age is of little advantage in these injuries, which are rarely life-threatening. Female drivers have an especially high risk of leg injury ($98.5 \pm 30.8\%$ higher than males). Several factors have been suggested. Short stature may cause a driver to move the seat closer to the instrument panel to have pedal contact; the close seat position changes the angle of the lower leg in relationship with the instrument panel. The length of the lower leg could affect what part of the leg contacts the knee bolster. The size of the foot (smaller for female) could affect motion of the leg toward the toe pan.[51]

Thus, Tables 5-1d and 5-2d are consistent with the findings of Bose, Segui-Gomez, and Crandall that females' risk of MAIS ≥ 3 and MAIS ≥ 2 injuries is considerably higher than males', but they also suggest that arm and leg injuries – usually not life-threatening – account for much of the increase.[52]

---

[51] Dischinger, P. C., Kerns, T. J., & Kufera, J. A. (1995). Lower Extremity Fractures in Motor Vehicle Collisions: The Role of Driver Gender and Height. *Accident Analysis and Prevention, 27*, pp. 601-606.
[52] Bose, Segui-Gomez, & Crandall (2011).

TABLE 5-2d: INCREASE IN INJURY RISK FOR A FEMALE RELATIVE TO A MALE OF THE SAME AGE
BY BODY REGION AND AIS SEVERITY LEVEL – CARS AND LTVs, DRIVERS AND RIGHT-FRONT PASSENGERS
(Given the same crash scenario, average for occupants 21 to 96, CDS unweighted data, 1988-2010)

Percent Injury Increase for Females

Seat Position → → → → → → → → → → →		Drivers			RF Passengers			Avg of Driver & RF
Occupant Age Group → → → → → → → →		21-96	21-30	65-74	21-96	21-30	65-74	21-96
HEAD	AIS ≥ 4	12.6 ± 29.8	12.9	11.7	18.3 ± 41.0	13.1	26.2	15.4 ± 21.8
	AIS ≥ 3	42.4 ± 32.7	47.4	35.0	20.0 ± 22.8	39.6	-3.1	31.2 ± 21.6
	AIS ≥ 2	28.2 ± 22.3	36.7	16.9	16.0 ± 15.5	16.1	16.1	22.1 ± 16.0
TORSO	AIS ≥ 4	40.9 ± 29.8	21.6	75.9	12.4 ± 31.0	26.7	-3.2	26.6 ± 17.0
	AIS ≥ 3	39.0 ± 22.0	45.6	32.1	30.1 ± 19.4	34.7	25.5	34.6 ± 16.7
	AIS ≥ 2	27.5 ± 13.9	22.7	34.7	34.7 ± 16.3	37.4	31.4	31.1 ± 10.7
Chest	AIS ≥ 4	35.0 ± 28.9	13.2	75.3	2.0 ± 24.9	7.4	-3.8	18.5 ± 18.4
	AIS ≥ 3	35.1 ± 19.4	35.6	35.5	24.7 ± 18.5	26.8	23.3	29.9 ± 17.2
	AIS ≥ 2	17.7 ± 19.6	13.5	24.3	35.2 ± 20.8	45.7	24.1	26.4 ± 13.6
Abdomen	AIS ≥ 2	48.1 ± 38.2	39.5	62.0	28.9 ± 46.7	48.3	5.3	38.5 ± 28.4
NECK	AIS ≥ 2	36.8 ± 37.9	67.0	5.3	52.6 ± 60.0	67.0	37.7	44.7 ± 34.0
ARM	AIS ≥ 2	63.0 ± 43.6	57.6	72.1	53.4 ± 42.6	38.8	76.5	58.2 ± 20.6
LEG	AIS ≥ 2	98.5 ± 30.8	74.7	141.0	61.0 ± 34.3	39.1	95.2	79.7 ± 16.3

CHAPTER 6

SPECIFIC INJURIES PREVALENT AMONG OLDER FATALITIES AND WOMEN – ANALYSES OF 1999-2007 FARS-MCOD ICD-10 DATA

6.0 Summary

The FARS-MCOD file lists a fatally injured occupant's injuries that "contributed to death." For any specific injury, a high average age of the victims indicates that older occupants are vulnerable; a high percentage of female victims indicates that females are vulnerable. In general, thoracic injuries are prevalent among older occupants, while abdominal injuries and fractures of the neck, arms, or legs are prevalent among females. Specifically, hip fractures are the injuries with the highest average age of the victims, followed by multiple rib fractures, sternum fractures, and associated thoracic injuries. The average age of people with subdural hemorrhage is considerably higher than the ages for other head injuries. Fractures of the first or second cervical vertebrae and injuries of the spleen, liver, or gallbladder are injuries with exceptionally high percentages of female victims – the neck fractures even more so for belted occupants.

6.1 Ranking the specific injuries by victims' average age or percent females

The FARS-Multiple Cause of Death file, as discussed in Section 4.1, lists each fatally injured occupant's injuries that "contributed to the fatality" according to the death certificate. For any specific injury – e.g., traumatic pneumothorax – it is possible to compute the average age of the victims and the percent that are female. The various specific injuries can be ranked by average age from oldest to youngest or by percent female victims from highest to lowest. Intuitively, there are two factors that might contribute to associating a specific injury with considerably higher average age (or percent of female victims) than the typical injury:

- Older occupants (or women) are anatomically or physiologically more vulnerable to this type of injury in the current automotive-crash environment.
- Older occupants (or women) are relatively more likely to be involved in the types of impacts that result in physical insults conducive to that type of injury.

The second factor is probably of negligible importance for most injuries. The archetypal "older driver's crash" is perhaps the left turn across traffic, but it is doubtful that any specific injury is typically associated with this type of crash and not with other crashes. There does not appear to be an archetypal "women's crash." There are, on the other hand, some crashes especially associated with younger and/or male drivers, namely those involving much alcohol, high speeds, and extreme risk-taking. Thus, if burn victims have a low average age and are mostly males, the most likely explanation is that young males have many of the exceedingly severe crashes resulting in fire – and not that young male flesh, being exceptionally tender, burns more easily than other flesh. But aside from burns and other non-impact harm such as alcohol poisoning, drowning, or asphyxiation, it is probable that the ranking of the injuries from impact trauma by average age or percent female primarily reflects the first factor. A high average age means older people are especially vulnerable (relative to young people) to this type of injury, at least in vehicles of the past 50 model years. High percentages of female victims mean women are especially vulnerable.

The purpose of the analysis is exploration, not estimation. We are searching for patterns: individual injuries and groups of injuries that are especially common among older occupants and/or women, indicating vulnerability. The numbers themselves – the average age or percent female for a given injury – are not particularly important; no statistical procedures are used to test for statistical significance or estimate confidence bounds.

Average age and percent female are tabulated for front-seat-outboard occupants (drivers and RF passengers combined) and rear-seat-outboard occupants; then separately for unrestrained and belted front-outboard and rear-outboard occupants. Data is limited to passenger cars with MY ≥ 1960 (i.e., excluding LTVs), for a more homogeneous crash environment. All adults, age 18 to 96, are included. Unlike Chapters 2-5, there is no need to start at 21, where the full trend of risk increasing with age begins, because this analysis does not attempt to quantify trends.

Injury data: MCOD uses the International Classification of Diseases to classify injuries. As explained in Section 4.1, MCOD used the 9th revision of that system (ICD-9) from 1987 to 1998 and the 10th revision of that system (ICD-10) from 1999 onwards (2007 is the latest year of FARS-MCOD data as of October 2012).[53] The two versions differ substantially. There is no simple one-to-one correspondence between specific injury codes. For analyses of specific injury (unlike the analysis of Chapter 4 that only considered the body region), there is no easy way to combine the codes. We must choose one or the other. We selected the ICD-10 codes, because the ICD-9 analysis would have been limited to pre-2000 cars. The analysis is based on nine years of data, 1999 to 2007.

An ICD-10 code consists of a letter, a two-digit number, and, possibly, a decimal point and another digit. Relevant to this study are the codes starting with S and T, comprising all types of injuries as well as poisonings and consequences of trauma (see Appendix B for a listing of the codes). FARS-MCOD has space to list up to 15 ICD-10 codes per fatally injured person, but usually lists only one or two, only occasionally four or more. Unlike Chapter 4, this will be an injury-level database, with multiple records for occupants who have multiple injuries.

MCOD does not gauge the severity of injuries with a scale such as the AIS; it only claims listed injuries "contributed to death." MCOD does not reveal how the injury data is obtained – e.g., from autopsies, hospital reports, or just external observation. Only about ¼ of the MCOD injuries are ICD-10 codes that identify specific injuries, such as S02.0: fracture of the vault of the skull. About ½ of the listings specify only a body region – e.g., S09.9: unspecified injury of head – and perhaps a hint of the lesion or organ – e.g., S02.9: fracture of skull and facial bones, part unspecified (which is definitely a fracture, but could be anything from a broken nose to a severe skull fracture). The remaining ¼ of the listings do not even specify a body region or much other useful information – e.g., T07 (unspecified multiple injuries), T14.9 (injury, unspecified), or T14.8 (other injuries of unspecified body region). Notwithstanding these caveats, FARS-MCOD is a unique resource that offers insight about the distribution of injuries in a large number of fatal car crashes.

---

[53] CDC (a); WHO (2005).

6.2 Injuries ranked by victims' average age – front seat versus back seat

Table 6-1 ranks the injuries of fatally injured drivers and RF passengers of cars by the average age of the victim, from oldest to youngest. The key to identifying patterns is the color coding: bold blue print for head injuries, bold red for torso, and bold green for neck. Head injuries include brain, skull, and face. Torso injuries include chest, abdomen, thoracic or lumbar spine, pelvis, clavicle, and scapula. Neck injuries include the cervical spine and throat. Plain black print indicates the arms or legs, plus harm such as burns or drowning, which is not associated with impact to a specific body region. Injuries to multiple or unspecified body regions are generally excluded from the table, except one group, "injury, unspecified" (which also includes "unspecified multiple injuries" and "other injuries of unspecified body region") is shown in bold black type as a benchmark, indicating the overall average of the MCOD injuries for this group of occupants. Two other benchmarks, also printed in bold black letters, are the average age of driver/RF passenger fatalities in passenger cars in 1999-2007 FARS (i.e., the average at the occupant level, not the injury level) and the weighted average age of drivers and RF passengers of cars involved in crashes (fatal or nonfatal) in 1999-2007 CDS. Table 6-1 displays the average age of the victims (left column), the percent female (second column), and the number of times the injury is reported in FARS-MCOD (right column). Table 6-1 is limited to injuries that occurred at least 125 times; that allows printing the results on a single page, to make it easier to look for patterns.

The salient feature of Table 6-1 is the sea of red print at the top: torso injuries. It is consistent with the literature and all the findings of Chapters 2 to 5: older occupants have high risk of torso injury. Head injuries, the blue print, are concentrated at the end of the table. Neck injuries (green) are scattered, but mostly in the middle. The only group as low as or lower than the head injuries are the non-impact events, such as burns, drowning, or alcohol poisoning, typically associated with "young people's crashes."

It is interesting to compare the specific injuries to the three benchmark lines in the bold black letters. The average victim age of the unspecified injuries (44.15) and the average age of the fatally injured occupants (43.76) are both near the middle of the table. (They tend to be close to one another because they measure almost the same thing.) The average age of people involved in nonfatal or fatal crashes (36.65, based on weighted CDS data) is near the low end of the table. Risk increases with age for almost every type of life-threatening injury; it just increases more for some types of injury than others. The few injuries that will have average age as low as 36.65 or lower are those whose risk does not increase at all with age plus those unusual injuries associated with crashes that primarily involve young people: burns, alcohol poisoning, and asphyxiation. The other two benchmarks (43.76 and 44.15) indicate how much the "typical" injury's risk – or overall fatality risk – escalate with age. Thus, the risks of the injuries listed in the table above these two benchmarks escalate with age even more than the "typical" injury: older occupants are exceptionally vulnerable to them. The risks of the injuries listed below those two benchmarks but above the third benchmark (36.65) also escalate with age to some extent, but at a slower rate than the "typical" injury. Most of the torso injuries are above the 43.76/44.15 benchmarks and most of the head injuries are below it.

AVERAGE AGE	PERCENT FEMALE	INJURY	N
57.00	44.79	Fracture of hip and/or fracture of neck of femur	163
57.30	45.36	Multiple fractures of ribs	1083
55.54	38.34	Fracture of sternum	396
55.22	40.69	Flail chest	457
53.38	45.21	Traumatic pneumothorax	492
53.36	40.51	Fracture of rib	1512
53.20	41.49	Traumatic hemopneumothorax	166
53.01	51.28	Injury of unspecified intra-abdominal organ	271
52.34	44.22	Fracture of thoracic vertebra	189
51.87	46.41	Fracture of second cervical vertebra (axis)	157
51.77	47.08	Fracture of other and unspecified parts of lumbar spine and pelvis	839
51.49	37.62	Injury of heart, unspecified	726
50.89	35.46	Other specified injuries of thorax	182
50.78	38.60	Other injuries of heart	1264
50.66	37.82	Traumatic subdural haemorrhage	1083
50.38	43.44	Injury of other specified intrathoracic organs	209
49.66	34.50	Injury of heart with hemopericardium	171
49.32	44.05	Fracture of shaft of tibia	168
49.12	38.32	Crushed chest	335
48.79	41.28	Traumatic hemothorax	1314
48.68	38.46	Fracture of lower limb, level unspecified	273
48.66	44.31	Other specified injuries of abdomen, lower back and pelvis	571
48.25	34.22	Unspecified injury of thorax	18525
48.18	45.95	Traumatic shock	1075
47.90	38.34	Injury of thoracic aorta	3521
47.86	38.04	Injury of unspecified intrathoracic organ	602
47.35	43.58	Fracture of clavicle	142
47.17	46.05	Unspecified injury of lower limb, level unspecified	367
46.95	49.68	Fracture of upper limb, level unspecified	155
46.80	38.27	Other and unspecified injuries of cervical spinal cord	358
46.30	37.65	Other specified injuries of trunk, level unspecified	1004
45.53	42.65	Multiple fractures of cervical spine	188
46.17	37.58	Other injuries of lung	1708
45.25	43.10	Unspecified injury of abdomen, lower back and pelvis	4664
45.42	39.18	Injury of other intra-abdominal organs	1231
45.13	40.84	Multiple injuries of thorax	1132
44.50	34.78	Other specified injuries of neck	154
44.55	35.83	Other intracranial injuries	988
44.36	40.15	Unspecified injury of trunk, level unspecified	2753
44.15	40.16	Injury, unspecified	69431
44.07	40.64	Fracture of first cervical vertebra (atlas)	257
43.76	38.80	average for driver and RF passenger fatalities, FARS 1999-2007	
43.73	45.25	Fracture of neck, part unspecified	1156
43.67	43.73	Injuries of intrathoracic, intra-abdominal & pelvic organs	1515
43.62	34.84	Fracture of femur, part unspecified	577
43.47	38.72	Traumatic subarachnoid haemorrhage	719
43.25	43.60	Injury of spleen	1490
42.94	46.88	Fracture of shaft of humerus	160
42.24	43.72	Other multiple injuries of abdomen, lower back and pelvis	231
41.96	31.97	Foreign body in respiratory tract, part unspecified	147
41.77	40.38	Unspecified injury of neck	6422
41.36	45.48	Injury of liver or gallbladder	3081
40.99	45.57	Dislocation of cervical vertebra	1038
40.73	40.17	Multiple injuries of neck	381
39.80	38.53	Multiple injuries of head	1986
39.64	38.73	Injury of multiple intra-abdominal organs	173
39.44	35.09	Diffuse brain injury	2405
39.27	37.13	Other specified injuries of head	800
39.02	42.01	Other and unspecified injuries of neck	169
39.02	38.28	Unspecified injury of head	37211
38.87	37.81	Injury of kidney	254
38.47	38.68	Drowning	1466
37.27	30.97	Other specified gases, fumes and vapors	494
37.25	34.65	Open wound of head, part unspecified	202
37.20	34.39	Fracture of skull and facial bones, part unspecified	3582
36.94	33.48	Crushing injury of head, part unspecified	260
36.65	50.29	weighted average for crash-involved drivers and RF passengers, CDS 1999-2007	
36.26	30.25	Asphyxiation	1223
36.06	32.29	Multiple fractures involving skull and facial bones	350
36.03	36.95	Fracture of base of skull	1453
35.95	34.60	Intracranial injury, unspecified	5774
35.65	30.61	Burn of unspecified body region, unspecified degree	1212
34.63	28.93	Burns involving 90% or more of body surface	197
33.85	20.25	Toxic effect of alcohol	326
31.40	34.52	Traumatic cerebral edema	252

Hip fractures (i.e. fractures of the hip, of the neck of the femur, or of both) are at the top of the list, by a margin of almost four years. That is not surprising, because hip fractures are a well-known impact trauma of the elderly, by no means limited to car crashes. The next-oldest group of injuries includes rib fractures, sternum fractures, and flail chest, plus resultant soft-tissue injuries such as pneumothorax, hemothorax, and heart injuries. These are the injuries to older occupants that are repeatedly mentioned in the literature. They are followed by other thoracic injuries.

Table 6-1 also includes some outliers: blues and greens in the top of the chart or reds much lower down. One neck injury – fracture of the second cervical vertebra (axis) – and one head injury – subdural hemorrhage – are near the top of the chart and appear to be special vulnerabilities of older occupants. Consistent with the statistical findings of Table 4-1a, injuries to abdominal organs such as the kidney, liver, gallbladder, or spleen have substantially lower average age than the thoracic injuries (although some abdominal injuries such as "unspecified intra-abdominal organ" and "other specified injuries of abdomen, lower back, or pelvis" are fairly high on the list). One head injury, traumatic cerebral edema, is at the end of the list by a margin of several years and well below the typical head injuries. This might actually be an injury to which young people are physiologically susceptible (as opposed to burns and alcohol poisoning, which are low on the list because young people have more high-speed crashes). Skull fractures are also prevalent among fairly young occupants, close to the benchmark average age of all crash-involved drivers and RF passengers. That might suggest the skull deteriorates slower with age than other bones, or it could to some extent reflect higher rates of ejection among the younger occupants (because of more rollovers and less belt use).

Table 6-2 presents the corresponding injury ranking for rear-outboard occupants. There are, of course, far fewer adult fatalities in the back seat than at the driver and RF positions. Even though Table 6-2 includes any injuries that occurred at least 25 times (as compared to a minimum of 125 times in Table 6-1), it is still a shorter list than Table 6-1. Many injuries from Table 6-1 – e.g., hip fracture, the first injury on the earlier list – are missing here. Nevertheless, the overall pattern is quite similar to the front-seat occupants, with torso injuries mostly at the top, then head injuries, and finally the non-impact phenomena such as drowning and burns. Rib fractures are at the top of the list – the more fractures, the higher on the list: flail chest, then multiple rib fractures, then fracture of rib. Heart and lung injuries are next. As in Table 6-1, victims of subdural hemorrhage are substantially older than victims of the typical head injury, but here, subarachnoid hemorrhage is also exceptionally high on the list. Neck and abdominal injuries are in similar ranks as for front-seat occupants and skull fractures are again near the young end of the list. There does not appear to be any common injury or group of injury whose rank for back-seat occupants greatly differs from its rank for front-seat occupants.

6.3   Injuries ranked by percentage of the victims that are females

Table 6-3 lists exactly the same injuries as Table 6-1 (car drivers and RF passengers), but ranked by percent female rather than average age: the higher the percent female, the higher on the list. One similarity of Tables 6-1 and 6-3 is that the torso injuries (red) are mostly in the upper half of the chart and the head injuries (blue) are almost all in the lower half.

TABLE 6-2: MCOD INJURIES OF REAR-OUTBOARD PASSENGER FATALITIES IN CARS
RANKED BY AVERAGE AGE OF THE VICTIM
(1999-2007 FARS, OCCUPANTS AGE 18-96)

AVERAGE AGE	PERCENT FEMALE	INJURY	N
53.75	67.86	Flail chest	28
50.89	52.83	Multiple fractures of ribs	53
48.03	52.05	Fracture of rib	73
46.13	33.33	Other injuries of heart	39
45.48	43.75	Traumatic subarachnoid haemorrhage	48
44.89	54.62	Unspecified injury of abdomen, lower back and pelvis	249
44.83	37.29	Traumatic subdural haemorrhage	59
43.62	46.88	Unspecified injury of thorax	800
43.07	57.14	Other specified injuries of abdomen, lower back and pelvis	28
42.47	50.91	Traumatic shock	55
42.08	52.07	Fracture of neck, part unspecified	217
41.58	38.71	Fracture of femur, part unspecified	31
41.12	39.22	Other injuries of lung	102
41.00	40.00	Multiple injuries of thorax	60
40.53	47.26	Injury of thoracic aorta	146
39.53	56.60	Other specified injuries of trunk, level unspecified	53
39.00	46.90	Injury, unspecified	3738
38.52	37.50	Injury of other intra-abdominal organs	64
38.48	35.94	Fracture of other and unspecified parts of lumbar spine and pelvis	64
38.38	44.26	average for back-seat outboard passenger fatalities, FARS 1999-2007	
38.34	34.04	Other specified injuries of head	47
38.32	45.43	Unspecified injury of neck	328
37.96	44.23	Traumatic hemothorax	52
37.80	43.16	Unspecified injury of trunk, level unspecified	360
37.24	42.31	Injury of spleen	78
37.17	48.08	Dislocation of cervical vertebra	52
36.45	42.99	Injury of liver or gallbladder	107
35.33	39.64	Multiple injuries of head	111
34.83	44.19	Injuries of intrathoracic, intra-abdominal & pelvic organs	86
34.27	39.88	Unspecified injury of head	2199
33.50	35.51	Diffuse brain injury	138
33.28	49.15	weighted average for back-seat outboard passengers, CDS 1999-2007	
32.83	39.66	Other intracranial injuries	58
32.78	36.00	Drowning	50
32.45	38.28	Intracranial injury, unspecified	384
31.83	36.68	Fracture of skull and facial bones, part unspecified	229
31.03	37.18	Burn of unspecified body region, unspecified degree	78
30.93	28.33	Asphyxiation	60
29.27	40.74	Fracture of base of skull	81
24.59	31.25	Other specified gases, fumes and vapors	32

TABLE 6-3: MCOD INJURIES OF CAR DRIVER AND RF-PASSENGER FATALITIES
RANKED BY PERCENTAGE OF VICTIMS THAT ARE FEMALES
(1999-2007 FARS, OCCUPANTS AGE 18-96)

PERCENT FEMALE	AVERAGE AGE	INJURY	N
51.23	53.01	Injury of unspecified intra-abdominal organ	271
50.23	36.65	weighted average for crash-involved drivers and RF passengers, CDS 1999-2007	
49.68	46.95	Fracture of upper limb, level unspecified	155
49.44	50.28	Injury of other specified intrathoracic organs	263
48.64	44.07	Fracture of first cervical vertebra (atlas)	257
48.41	51.67	Fracture of second cervical vertebra (axis)	157
47.04	51.77	Fracture of other and unspecified parts of lumbar spine and pelvis	868
46.88	42.94	Fracture of shaft of humerus	160
46.05	47.17	Unspecified injury of lower limb, level unspecified	367
45.95	48.18	Traumatic shock	1075
45.90	43.25	Injury of spleen	1499
45.67	40.99	Dislocation of cervical vertebra	1036
45.48	41.38	Injury of liver or gallbladder	2082
45.34	57.30	Multiple fractures of ribs	1083
45.25	43.73	Fracture of neck, part unspecified	4155
45.24	53.86	Traumatic pneumothorax	462
44.78	81.00	Fracture of hip and/or fracture of neck of femur	183
44.31	43.58	Other specified injuries of abdomen, lower back and pelvis	671
44.22	52.34	Fracture of thoracic vertebra	198
44.05	49.32	Fracture of shaft of tibia	168
43.75	43.57	Injuries of intrathoracic, intra-abdominal & pelvic organs	1525
43.72	42.24	Other multiple injuries of abdomen, lower back and pelvis	231
43.66	47.36	Fracture of clavicle	142
43.10	45.69	Unspecified injury of abdomen, lower back and pelvis	4824
42.85	48.53	Multiple fractures of cervical spine	136
42.01	38.02	Other and unspecified injuries of neck	195
41.49	53.29	Traumatic hemopneumothorax	188
40.09	56.22	Flail chest	437
40.04	45.13	Multiple injuries of thorax	1132
40.51	52.88	Fracture of rib	1811
40.38	41.77	Unspecified injury of neck	5422
40.28	48.78	Traumatic hemothorax	1314
40.17	40.73	Multiple injuries of neck	361
40.16	44.15	Injury, unspecified	89431
40.15	44.33	Unspecified injury of trunk, level unspecified	5753
39.22	43.25	Unspecified injury of thorax	13526
39.16	45.42	Injury of other intra-abdominal organs	1291
38.95	38.03	Fracture of base of skull	1453
38.92	48.12	Crushed chest	335
38.84	47.39	Injury of unspecified intrathoracic organ	582
38.80	43.78	all driver and RF passenger fatalities, FARS 1999-2007	
38.73	39.64	Injury of multiple intra-abdominal organs	173
38.68	38.47	Drowning	1466
38.53	39.80	Multiple injuries of head	1988
38.46	48.68	Fracture of lower limb, level unspecified	273
38.34	47.80	Injury of thoracic aorta	3621
38.27	40.80	Other and unspecified injuries of cervical spinal cord	853
37.82	50.56	Traumatic subdural haemorrhage	1083
37.65	45.60	Other specified injuries of trunk, level unspecified	1094
37.61	38.67	Injury of kidney	294
37.55	48.17	Other injuries of lung	1788
37.51	47.57	Injury of heart, unspecified	740
37.13	39.27	Other specified injuries of head	300
36.72	43.47	Traumatic subarachnoid haemorrhage	719
36.50	50.73	Other injuries of heart	1054
36.46	50.68	Other specified injuries of thorax	492
36.34	55.64	Fracture of sternum	386
36.26	39.02	Unspecified injury of head	37211
35.88	44.55	Other intracranial injuries	988
35.09	38.44	Diffuse brain injury	2405
34.84	43.62	Fracture of femur, part unspecified	577
34.78	44.59	Other specified injuries of neck	194
34.65	37.25	Open wound of head, part unspecified	202
34.60	35.95	Intracranial injury, unspecified	5774
34.52	31.40	Traumatic cerebral edema	252
34.50	42.59	Injury of heart with haemopericardium	171
34.39	37.20	Fracture of skull and facial bones, part unspecified	3582
33.46	33.84	Crushing injury of head, part unspecified	260
32.29	36.08	Multiple fractures involving skull and facial bones	350
31.97	41.96	Foreign body in respiratory tract, part unspecified	147
30.97	37.27	Other specified gases, fumes and vapors	494
30.61	35.65	Burn of unspecified body region, unspecified degree	1212
30.25	36.26	Asphyxiation	1223
28.93	34.63	Burns involving 90% or more of body surface	197
20.25	33.85	Toxic effect of alcohol	326

132

Neck injuries have high percentages of female victims, consistent with the findings of Chapter 4. But not all neck injuries: mainly the fractures, especially fractures of the first and second vertebrae (atlas and axis). Other and unspecified types of neck injury are widely dispersed in the chart. The high risk of fractures appears related to the anatomy of a typical female: thin vertebrae called upon to support a head almost as heavy as a male's. Arm and leg fractures (upper limb, shaft of humerus, and shaft of tibia) are also high on the list, consistent with the findings in Section 5.3. So is traumatic shock, a consequence of injury.

Another resemblance is that the non-impact phenomena such as burns and asphyxiation are at the end of the list – but here, except for drowning, they are even lower on the list. Alcohol poisoning is more than eight percentage points below the next lowest, burns. These phenomena are even more characteristic of "men's crashes" than of "young people's crashes."

Whereas Tables 6-1 and 6-3 both have a lot of red print near the top, the order of the injuries is reversed. Females appear to be somewhat more prone to abdominal injuries such as "unspecified intra-abdominal organ" (the first injury on the list), spleen, liver, and gallbladder. Heart injuries are relatively low on the list. That is consistent with the statistical findings of Table 4-1d, which showed a high risk of abdominal injury for females relative to males.

As in Table 6-1, the benchmark lines for the overall percentage of fatalities and of unspecified injuries that are females are near the middle of the chart. That is understandable, because these are close to the "typical" injury. On the other hand, the weighted average for all crash-involved occupants in CDS is almost at the top of the chart: 50.23 percent females. In fact, for most of the injuries on the chart, less than 50 percent of the victims are female, in some cases, substantially less. This is not so because females are intrinsically less vulnerable to injury given similar physical insults – because all the double-pair comparison analyses of Chapters 2 to 5 demonstrate they are, on the average, more vulnerable. It is because, even though females account for about half of the VMT and half of the routine, nonfatal crashes, they are substantially underrepresented in fatal crashes, since they are less likely to drink and drive or engage in other high-risk driving behaviors.

Table 6-4 ranks injuries to back-seat occupants by the percentage of female victims. Here (unlike the comparison of Tables 6-1 and 6-2), there are some noteworthy differences from the front-seat occupants. In the back seat, the pattern of female vulnerability in Table 6-4 is quite similar to the pattern for elderly occupants in Table 6-2. Flail chest is highest on the list by a margin of over 10 percentage points; "multiple fractures of ribs" and "fracture of rib" are also near the top of the list. These three injuries also had the highest average age in Table 6-2. Unlike the front seat, injuries to abdominal organs such as the liver, gallbladder, or spleen are not particularly high on the list, while some thoracic injuries have a high percent of female victims. Subarachnoid and subdural hemorrhages were high on the list for front-seat occupants, but not here. It is unknown to what extent these contrasting results are due to real differences in the design of the front and rear seats and their surrounding structures, or just an artifact of the relatively limited data on back-seat occupants. However, one similarity in the results for front- and back-seat occupants is that a high percentage of the neck fractures are to female occupants.

TABLE 6-4: MCOD INJURIES OF REAR-OUTBOARD PASSENGER FATALITIES IN CARS
RANKED BY PERCENTAGE OF VICTIMS THAT ARE FEMALES
(1999-2007 FARS, OCCUPANTS AGE 18-96)

PERCENT FEMALE	AVERAGE AGE	INJURY	N
67.86	53.75	Flail chest	28
57.14	43.07	Other specified injuries of abdomen, lower back and pelvis	28
56.60	39.53	Other specified injuries of trunk, level unspecified	53
54.62	44.89	Unspecified injury of abdomen, lower back and pelvis	249
52.83	50.89	Multiple fractures of ribs	53
52.07	42.08	Fracture of neck, part unspecified	217
52.05	48.03	Fracture of rib	73
50.91	42.47	Traumatic shock	55
49.15	33.28	weighted average for back-seat outboard passengers, CDS 1999-2007	
48.08	37.17	Dislocation of cervical vertebra	52
47.26	40.53	Injury of thoracic aorta	146
46.90	39.00	Injury, unspecified	3738
46.88	43.82	Unspecified injury of thorax	800
45.43	38.32	Unspecified injury of neck	328
44.26	38.38	all back-seat outboard passenger fatalities, FARS 1999-2007	
44.23	37.96	Traumatic hemothorax	52
44.19	34.83	Injuries of intrathoracic, intra-abdominal & pelvic organs	86
43.75	45.48	Traumatic subarachnoid haemorrhage	48
43.16	37.80	Unspecified injury of trunk, level unspecified	380
42.99	36.45	Injury of liver or gallbladder	107
42.31	37.24	Injury of spleen	78
40.74	29.27	Fracture of base of skull	81
40.00	41.00	Multiple injuries of thorax	80
39.88	34.27	Unspecified injury of head	2199
39.66	32.83	Other intracranial injuries	58
39.64	35.33	Multiple injuries of head	111
39.22	41.12	Other injuries of lung	102
38.71	41.58	Fracture of femur, part unspecified	31
38.28	32.45	Intracranial injury, unspecified	384
37.50	38.52	Injury of other intra-abdominal organs	64
37.29	44.83	Traumatic subdural haemorrhage	59
37.18	31.03	Burn of unspecified body region, unspecified degree	78
36.68	31.83	Fracture of skull and facial bones, part unspecified	229
36.00	32.78	Drowning	50
35.01	19.68	Fracture of other and unspecified parts of lumbar spine and pelvis	64
35.51	33.80	Diffuse brain injury	100
34.04	38.34	Other specified injuries of head	47
33.33	46.13	Other injuries of heart	39
31.25	24.59	Other specified gases, fumes and vapors	32
28.33	30.93	Asphyxiation	60

134

6.4   Injuries ranked by victims' average age – unrestrained versus belted

Table 6-5 ranks the injuries of fatally injured <u>unrestrained</u> drivers and RF passengers by the average age of the victim, from oldest to youngest. It is limited to injuries that occurred at least 50 times, a threshold that generates just enough injury codes to fill up for the table to fill a page. Table 6-6 presents the corresponding rankings for <u>belted</u> drivers and RF passengers; it is limited to injuries that occurred at least 35 times. (In the future, when more MCOD data is available, it would be worthwhile to further limit the analysis of Table 6-6 to belted occupants of cars equipped with air bags.) The most remarkable feature of Tables 6-5 and 6-6 is how similar the rankings are. Hip fractures and multiple rib fractures are number one and two in average occupant age on both tables. Most of the other injuries are also in similar locations. Not a single injury is near the top of one table and the bottom of the other. Here are the few injuries that have moderately different ranks for unrestrained and belted occupants:

- Fractures of the first or second cervical vertebrae and fractures of the vault of the skull are somewhat higher in the unrestrained than in the belted table.
- Traumatic subdural hemorrhage and kidney injuries are somewhat higher in the belted table
- Fracture of the shaft of the humerus is listed only in the unbelted table; it was not frequent enough with belted occupants to be included in the belted table
- Concussions are listed only in the belted table

The great similarity of the relative rankings for unrestrained and belted occupants implies that these rankings are largely intrinsic to the aging process and only secondarily affected by safety technologies. Older people are highly susceptible to hip fractures and multiple rib fractures, regardless of whether they are unrestrained or belted – or, for that matter, whether they were in a car crash or fell down in their homes.

Even though the relative rankings are quite similar, the absolute ages of the belted casualties are approximately 10 years older than the unrestrained. For example, the average age of hip fractures is 53.54 for unrestrained and 64.09 for belted occupants. Two factors contribute almost equally to the difference, as may be discerned from the benchmark entries on the tables. Older people are historically more likely to buckle up, as evidenced by a 32.82 average age for unrestrained (and not necessarily injured, let alone killed) drivers and RF passengers on CDS, versus 37.82 for belted. In addition, because belts are somewhat more effective for younger occupants, the average unrestrained <u>fatality</u> is 38.83 years old, but the average belted fatality, 48.81.

It is not the primary objective of this chapter, given the limited data on specific injuries, to examine how the pattern of injuries may have changed over time (as, for example, in the analyses of Section 3.3). Nevertheless it might be worthwhile to look only at belted front-seat occupants in relatively late-model vehicles equipped with frontal air bags and compare the injury distribution to Table 6-6 (belted occupants of vehicles of all model years). Table 6-6a limits Table 6-6 to MY 1999 and later vehicles, all equipped with dual air bags. It was necessary to go back as far as MY 1999 to garner adequate numbers of injury cases, especially because the MCOD data are only available through 2007. Table 6-6a includes injuries that occurred at least 25 times, but it still yields a shorter list than in Table 6-6 (which required 35 cases).

AVERAGE AGE	PERCENT FEMALE	INJURY	N
50.64	37.50	Fracture of hip and/or fracture of neck of femur	56
50.40	38.25	Multiple fractures of ribs	434
49.31	31.89	Flail chest	186
49.04	24.14	Fracture of sternum	145
47.25	28.33	Fracture of second cervical vertebra (axis)	60
45.78	32.66	Injury of heart, unspecified	343
45.50	43.33	Fracture of thoracic vertebra	60
45.83	24.00	Other injuries of heart	535
45.84	29.79	Fracture of rib	875
44.74	35.58	Traumatic pneumothorax	199
44.35	35.40	Fracture of other and unspecified parts of lumbar spine and pelvis	338
44.35	36.11	Traumatic hemopneumothorax	72
44.19	32.76	Fracture of shaft of tibia	58
43.85	30.92	Injury of unspecified intrathoracic organ	482
43.64	32.74	Crushed chest	334
43.80	33.33	Injury of heart with hemopericardium	84
43.33	34.46	Unspecified injury of lower limb, level unspecified	148
43.18	30.11	Injury of unspecified intra-abdominal organ	98
43.19	28.52	Injury of thoracic aorta	1650
43.06	29.13	Fracture of lower limb, level unspecified	127
42.91	33.23	Traumatic hemothorax	526
42.80	38.80	Injury of other specified intrathoracic organs	83
42.69	30.67	Other specified injuries of thorax	75
42.41	34.92	Multiple fractures of cervical spine	63
42.22	31.41	Unspecified injury of thorax	5514
42.19	29.17	Traumatic subdural haemorrhage	432
41.53	30.95	Other specified injuries of trunk, level unspecified	462
41.52	36.94	Fracture of first cervical vertebra (atlas)	101
40.81	30.19	Other specified injuries of abdomen, lower back and pelvis	232
40.77	23.37	Other and unspecified injuries of cervical spinal cord	143
40.48	38.13	Traumatic shock	417
40.17	37.88	Fracture of upper limb, level unspecified	66
40.08	32.30	Injury of other intra-abdominal organs	583
39.80	26.11	Other intracranial injuries	434
39.81	43.59	Fracture of shaft of humerus	78
39.78	35.23	Multiple injuries of thorax	568
39.53	21.44	Other injuries of lung	757
39.33	32.40	Injury, unspecified	30048
39.24	25.00	Other specified injuries of neck	60
39.22	30.47	Injuries of intrathoracic, intra-abdominal & pelvic organs	721
39.18	34.78	Unspecified injury of abdomen, lower back and pelvis	1346
39.01	32.65	Unspecified injury of trunk, level unspecified	3028
38.95	36.25	Fracture of neck, part unspecified	1878
38.89	30.15	Fracture of femur, part unspecified	262
38.83	31.38	average for unrestrained driver and RF passenger fatalities, FARS 1999-2007	
38.41	34.36	Open wound of other parts of head	84
38.14	28.90	Traumatic subarachnoid haemorrhage	916
37.88	33.91	Other multiple injuries of abdomen, lower back and pelvis	116
37.67	33.33	Foreign body in respiratory tract, part unspecified	60
37.43	28.88	Injury of spleen	813
37.42	32.58	Injury of multiple intra-abdominal organs	78
36.93	31.90	Unspecified injury of neck	2958
36.79	33.62	Multiple injuries of head	1059
36.71	30.89	Injury of liver or gallbladder	935
36.51	24.01	Crushing of trunk/unspecified parts of abdomen, lower back and pelvis	7
36.43	29.41	Fracture of vault of skull	51
36.41	36.22	Dislocation of cervical vertebra	464
36.35	33.33	Cocaine poisoning	72
36.08	34.29	Multiple injuries of neck	175
35.67	33.64	Drowning	645
35.65	29.11	Diffuse brain injury	1216
35.61	29.83	Unspecified injury of head	15670
35.58	28.04	Other specified injuries of head	428
35.54	33.33	Fracture of mandible	54
35.31	25.33	Other and unspecified injuries of neck	75
34.48	30.00	Crushing injury of head, part unspecified	150
34.47	23.43	Fracture of skull and facial bones, part unspecified	1857
33.89	33.53	Fracture of base of skull	862
33.37	25.64	Burn of unspecified body region, unspecified degree	312
32.99	18.32	Toxic effect of alcohol	191
32.97	26.52	Multiple fractures involving skull and facial bones	181
32.95	21.43	Other specified gases, fumes and vapors	140
32.95	27.82	Intracranial injury, unspecified	2788
32.82	43.49	weighted average for unrestrained crash-involved drivers and RF passengers, CDS 1999-2007	
32.53	27.84	Injury of kidney	97
32.51	28.85	Open wound of head, part unspecified	108
32.39	26.23	Asphyxiation	671
30.53	22.81	Burns involving 90% or more of body surface	57
28.80	29.48	Traumatic cerebral edema	128

136

AVERAGE AGE	PERCENT FEMALE	INJURY	N
64.09	51.65	Fracture of hip and/or fracture of neck of femur	91
62.61	51.35	Multiple fractures of ribs	582
61.74	51.85	Traumatic pneumothorax	235
60.77	46.33	Fracture of sternum	218
60.24	36.00	Multiple fractures of lumbar spine and pelvis	50
60.19	45.70	Flail chest	248
59.93	55.00	Injury of unspecified intra-abdominal organ	160
59.19	44.00	Traumatic hemopneumothorax	100
58.58	45.83	Fracture of thoracic vertebra	96
58.63	45.85	Fracture of rib	899
58.22	74.00	Fractures of other parts of lower leg	50
57.22	45.75	Traumatic subdural haemorrhage	553
56.61	43.16	Other specified injuries of thorax	95
55.58	55.66	Injury of other specified intrathoracic organs	153
55.17	54.12	Fracture of other and unspecified parts of lumbar spine and pelvis	571
55.90	43.69	Injury of heart, unspecified	366
55.80	45.34	Other injuries of heart	545
55.04	45.54	Traumatic hemothorax	518
54.53	36.62	Injury of heart with hemopericardium	73
54.50	59.26	Fractures involving multiple regions of both lower limbs	54
54.25	46.70	Crushed chest	584
54.23	51.38	Fracture of second cervical vertebra (axis)	68
54.14	47.35	Unspecified injury of thorax	5705
54.12	54.34	Other specified injuries of abdomen, lower back and pelvis	279
53.82	53.73	Traumatic shock	577
53.80	45.08	Fracture of lower limb, level unspecified	122
52.36	49.75	Injury of unspecified intrathoracic organ	201
52.92	49.76	Injury of thoracic aorta	1604
52.48	51.06	Fracture of shaft of tibia	94
52.29	44.65	Other injuries of lung	871
52.33	45.93	Fracture of clavicle	87
52.20	46.12	Other specified injuries of trunk, level unspecified	451
52.19	51.56	Multiple fractures of cervical spine	84
51.76	60.82	Concussion	49
51.74	56.41	Fracture of upper limb, level unspecified	78
51.60	45.85	Other and unspecified injuries of cervical spinal cord	195
51.43	56.15	Unspecified injury of lower limb, level unspecified	187
51.07	51.23	Unspecified injury of abdomen, lower back and pelvis	2324
50.87	47.09	Injury of other intra-abdominal organs	587
50.83	48.41	Multiple injuries of thorax	502
49.79	41.79	Other specified injuries of neck	67
49.76	48.10	Unspecified injury of trunk, level unspecified	3185
49.58	52.24	Fracture of fibula alone	67
48.94	43.57	Other intracranial injuries	482
48.87	48.00	Injury, unspecified	33546
48.81	48.80	average for belted driver and RF passenger fatalities, FARS 1999-2007	
48.63	53.48	Injury of spleen	774
48.51	54.33	Fracture of neck, part unspecified	1988
48.30	46.70	Traumatic subarachnoid haemorrhage	349
48.29	46.40	Injuries of intrathoracic, intra-abdominal & pelvic organs	887
47.82	57.00	Other multiple injuries of abdomen, lower back and pelvis	100
47.64	38.87	Fracture of femur, part unspecified	283
46.80	52.49	Injury of liver or gallbladder	960
46.43	49.11	Unspecified injury of neck	3073
46.19	47.79	Injury of kidney	113
46.13	59.15	Fracture of first cervical vertebra (atlas)	142
45.72	46.30	Unspecified injury of upper limb, level unspecified	54
45.72	52.03	Other specified injuries of head	298
45.63	46.39	Multiple injuries of neck	108
45.37	52.75	Dislocation of cervical vertebra	510
45.28	31.58	Foreign body in respiratory tract, part unspecified	76
44.86	51.35	Fracture of shaft of humerus	74
44.19	43.22	Diffuse brain injury	1055
44.11	48.24	Open wound of head, part unspecified	85
43.45	44.73	Multiple injuries of head	816
43.12	44.36	Unspecified injury of head	19164
43.11	58.82	Other and unspecified injuries of neck	85
42.53	43.68	Injury of multiple intra-abdominal organs	87
42.12	42.76	Drowning	587
41.29	42.74	Fracture of skull and facial bones, part unspecified	1495
41.06	36.96	Asphyxiation	487
40.92	37.70	Burns involving 90% or more of body surface	61
40.61	42.55	Multiple fractures involving skull and facial bones	141
40.49	39.66	Other specified gases, fumes and vapors	174
39.79	39.60	Crushing injury of head, part unspecified	101
39.43	44.90	Fracture of vault of skull	49
38.17	42.67	Intracranial injury, unspecified	2587
38.93	35.97	Burn of unspecified body region, unspecified degree	417
38.20	44.60	Fracture of base of skull	722
37.82	52.06	weighted average for belted crash-involved drivers and RF passengers, CDS 1999-2007	
36.50	36.70	Traumatic cerebral edema	108
34.72	24.47	Toxic effect of alcohol	94

137

TABLE 6-6a: MY 1999-2008 CARS WITH DUAL AIR BAGS; MCOD INJURIES OF BELTED CAR DRIVER AND RF-PASSENGER FATALITIES RANKED BY AVERAGE AGE OF THE VICTIM (1999-2007 FARS, OCCUPANTS AGE 18-96)

AVERAGE AGE	PERCENT FEMALE	REC_CD	N
67.32	48.00	Fracture of hip and/or fracture of neck of femur	25
66.05	72.97	Injury of unspecified intra-abdominal organ	37
63.52	40.74	Traumatic hemopneumothorax	27
61.45	48.25	Multiple fractures of ribs	143
61.03	42.21	Traumatic subdural haemorrhage	154
59.64	48.72	Fracture of sternum	39
58.44	55.56	Fracture of second cervical vertebra (axis)	27
58.03	50.89	Fracture of other and unspecified parts of lumbar spine and pelvis	169
57.95	50.82	Flail chest	61
57.67	54.90	Traumatic pneumothorax	51
56.36	38.54	Injury of heart, unspecified	96
56.19	51.82	Fracture of rib	274
55.92	64.00	Fracture of upper limb, level unspecified	25
55.60	53.73	Unspecified injury of lower limb, level unspecified	67
54.61	48.97	Other injuries of heart	145
54.41	44.81	Traumatic hemothorax	154
54.24	60.00	Injury of other specified intrathoracic organs	50
54.17	58.02	Traumatic shock	162
54.08	47.84	Unspecified injury of thorax	2335
53.61	64.29	Fracture of shaft of tibia	28
53.17	42.96	Multiple injuries of thorax	135
52.93	63.33	Other specified injuries of abdomen, lower back and pelvis	30
52.86	47.75	Injury of thoracic aorta	400
52.67	48.39	Crushed chest	93
51.88	41.77	Other injuries of lung	237
51.41	48.83	Injury of other intra-abdominal organs	179
50.98	38.64	Other intracranial injuries	132
50.93	55.67	Traumatic subarachnoid haemorrhage	97
50.92	51.20	Unspecified injury of trunk, level unspecified	1123
50.63	57.87	Fracture of neck, part unspecified	572
50.51	53.48	Unspecified injury of abdomen, lower back and pelvis	647
49.85	48.78	Injury of unspecified intrathoracic organ	41
49.82	58.82	Other specified injuries of trunk, level unspecified	51
49.79	50.00	Other and unspecified injuries of cervical spinal cord	48
48.99	52.35	Injuries of intrathoracic, intra-abdominal & pelvic organs	170
48.98	**48.12**	**average for belted driver and RF fatalities, FARS 1999-2007, MY 1999-2008**	
48.97	49.49	Injury, unspecified	10231
48.61	51.02	Unspecified injury of neck	929
48.43	31.88	Fracture of femur, part unspecified	69
48.12	58.37	Injury of spleen	204
47.21	48.84	Multiple injuries of neck	47
46.97	56.86	Dislocation of cervical vertebra	153
46.40	65.12	Fracture of first cervical vertebra (atlas)	43
45.12	43.48	Multiple injuries of head	230
44.37	56.40	Injury of liver or gallbladder	211
43.97	43.23	Diffuse brain injury	310
43.96	48.33	Drowning	180
43.80	46.20	Unspecified injury of head	4907
42.75	35.44	Asphyxiation	158
41.64	41.91	Fracture of skull and facial bones, part unspecified	451
40.54	34.09	Burn of unspecified body region, unspecified degree	132
40.52	40.00	Other specified gases, fumes and vapors	60
39.67	43.63	Fracture of base of skull	204
39.58	41.96	Intracranial injury, unspecified	827
39.47	46.67	Other and unspecified injuries of neck	30
38.24	**54.32**	**wtd avg for belted crash-involved drivers & RF, CDS 1999-2007, MY 1999-2008**	
37.40	36.00	Traumatic cerebral edema	25
31.23	30.00	Toxic effect of alcohol	30

A comparison of Tables 6-6 and 6-6a shows some injuries missing from Table 6-6a because there were fewer than the 25 cases required to get listed in the table. That might be a consequence of frontal air bags and other recent safety technologies especially mitigating those injuries or it could just reflect the limited database available for Table 6-6a. These missing injuries include: crushing of the head, concussion, vault fracture of the skull, multiple fractures of the cervical spine, fracture of the clavicle, heart injury with hemopericardium, fracture of a thoracic vertebra, and kidney injury. But the remaining injuries are listed in quite similar order in Tables 6-6 and 6-6a. It suggests that frontal air bags and other safety improvements in late-model vehicles reduced injury risk in absolute terms for occupants of all ages, but did not substantially change how the injuries were distributed by occupant age.

Tables 6-7 and 6-8 rank the injuries by average age for unrestrained and belted <u>back-seat</u> occupants. "Belted" occupants include people at seats equipped with a 3-point belt or just a lap belt; however, in 1999-2007 FARS vehicles with decodable VINs, 93 percent of the belted adults used 3-point belts. Because of limited data, in order to fill even half a page it is necessary to list unbelted injuries if they occurred as few as 18 times and belted injuries, 12 times. Such small numbers reduce the precision of the averages.

Here, too, most of the injuries have similar rankings for unrestrained and belted occupants, but the differences are somewhat stronger than in the front seat. Tables 6-7 and 6-8 both tend to show more torso injuries (red) for the older occupants and more head injuries (blue) for the younger occupants, but the tendency is visibly stronger for the belted occupants. A plausible explanation is that back-seat belts are quite effective for head injuries at all ages, but for torso injuries, they are not so effective for older occupants.

Injuries of the liver or the gallbladder, fractures of the lumbar spine or pelvis, and "other injuries of lung" have relatively higher average age in the belted table – consistent with the hypothesis that belts in the back seat are not so effective for some types of torso injury in older occupants. Traumatic subdural hemorrhage is at the top of the list for belted back-seat occupants. It was also high on the list for belted front-seat occupants (Table 6-6). There are several hypotheses why subdural hemorrhage is especially prevalent in belted occupants and why, unlike the typical head injury, risk increases rapidly with age: "Older adults are at increased risk for subdural hematomas due to fragility of bridging cerebral veins. As cerebral atrophy develops, the brain shrinks away from the dura and bridging veins are predisposed to tearing due to increased stress…Anticoagulant and antiplatelet therapy are associated with an increased risk for spontaneous subdural hematomas."[54]

---

[54] Karnath, B. (2004). Subdural Hematoma, Presentation and Management in Older Adults. *Geriatrics, 58*, pp. 18-23. https://ssl-w03dnn0374.websiteseguro.com/sbn-neurocirurgia/site/download/artigos/article.pdf

TABLE 6-7: MCOD INJURIES OF UNRESTRAINED REAR-OUTBOARD PASSENGER FATALITIES IN CARS
RANKED BY AVERAGE AGE OF THE VICTIM
(1999-2007 FARS, OCCUPANTS AGE 18-96)

AVERAGE AGE	PERCENT FEMALE	INJURY	N
47.50	43.33	Multiple fractures of ribs	30
43.17	39.13	Fracture of femur, part unspecified	23
42.45	27.59	Other injuries of heart	29
41.09	44.68	Fracture of rib	47
40.92	34.21	Traumatic subarachnoid haemorrhage	38
39.92	50.35	Unspecified injury of abdomen, lower back and pelvis	143
39.81	50.00	Other specified injuries of trunk, level unspecified	28
39.59	28.13	Other specified injuries of head	32
39.18	42.62	Unspecified injury of thorax	542
39.00	47.37	Injury of heart, unspecified	19
38.54	35.90	Multiple injuries of thorax	39
38.23	47.37	Fracture of neck, part unspecified	152
37.43	29.73	Traumatic subdural haemorrhage	37
37.09	43.78	Unspecified injury of neck	217
36.56	30.23	Injury of other intra-abdominal organs	43
35.86	42.14	Injury, unspecified	2468
35.30	35.42	Injury of thoracic aorta	96
35.24	39.70	average for unrestrained back-seat outboard fatalities, FARS 1999-2007	
34.85	45.45	Traumatic shock	33
34.68	40.54	Dislocation of cervical vertebra	37
33.79	37.99	Unspecified injury of trunk, level unspecified	229
33.03	37.77	Unspecified injury of head	1607
32.99	37.50	Multiple injuries of head	80
32.85	33.82	Other injuries of lung	68
32.64	36.36	Fracture of other and unspecified parts of lumbar spine and pelvis	44
32.42	40.35	Injuries of intrathoracic, intra-abdominal & pelvic organs	57
32.25	33.33	Injury of spleen	48
31.97	33.33	Traumatic hemothorax	33
31.92	36.00	Burn of unspecified body region, unspecified degree	25
31.76	43.10	weighted average, unrestrained back-seat outboard passengers, CDS 1999-2007	
31.67	33.81	Intracranial injury, unspecified	281
31.57	30.39	Diffuse brain injury	102
31.55	29.79	Asphyxiation	47
31.11	34.27	Fracture of skull and facial bones, part unspecified	178
30.97	28.17	Injury of liver or gallbladder	71
30.91	32.56	Other intracranial injuries	43
30.61	33.33	Drowning	33
26.91	37.93	Fracture of base of skull	58

TABLE 6-8: MCOD INJURIES OF BELTED REAR-OUTBOARD PASSENGER FATALITIES IN CARS
RANKED BY AVERAGE AGE OF THE VICTIM
(1999-2007 FARS, OCCUPANTS AGE 18-96)

AVERAGE AGE	PERCENT FEMALE	INJURY	N
62.11	55.56	Traumatic subdural haemorrhage	18
61.54	53.57	Other injuries of lung	28
60.92	41.67	Fracture of other and unspecified parts of lumbar spine and pelvis	12
60.74	69.57	Fracture of rib	23
60.15	76.92	Crushed chest	13
58.53	73.68	Multiple fractures of ribs	19
56.93	64.29	Traumatic shock	14
56.54	59.70	Unspecified injury of thorax	201
55.36	72.00	Fracture of neck, part unspecified	50
55.32	65.88	Unspecified injury of abdomen, lower back and pelvis	85
52.79	73.68	Injury of thoracic aorta	38
49.50	62.08	Injury, unspecified	873
48.97	73.33	Injury of liver or gallbladder	30
48.80	66.67	Traumatic hemothorax	15
48.73	59.04	average for belted back-seat outboard passenger fatalities, FARS 1999-2007	
48.35	59.05	Unspecified injury of trunk, level unspecified	105
47.59	59.09	Multiple injuries of head	22
47.13	68.75	Injury of other intra-abdominal organs	16
47.08	46.15	Multiple injuries of thorax	13
46.16	68.42	Injury of spleen	19
43.84	57.89	Injuries of intrathoracic, intra-abdominal & pelvic organs	19
43.39	56.96	Unspecified injury of neck	79
41.61	62.50	Other specified injuries of trunk, level unspecified	16
40.54	53.85	Drowning	13
39.87	51.04	Unspecified injury of head	431
39.84	56.00	Diffuse brain injury	25
38.32	50.00	Fracture of skull and facial bones, part unspecified	34
37.12	55.07	Intracranial injury, unspecified	69
36.35	50.00	Fracture of base of skull	20
36.15	30.77	Burn of unspecified body region, unspecified degree	13
34.70	53.95	weighted average for belted back-seat outboard passengers, CDS 1999-2007	

6.5   Injuries ranked by percent female victims – unrestrained versus belted

Table 6-9 lists exactly the same injuries as Table 6-5 (unrestrained car drivers and RF passengers), but ranked by percent female rather than average age. Table 6-10 presents the corresponding rankings for belted front-seat occupants. The great majority of the injuries have similar ranks in the unrestrained and belted tables (just as the unrestrained and belted ranks by average age were similar for most injuries in Tables 6-5 and 6-6). But there are some important exceptions:

- Neck injuries, especially fractures of the first or second cervical vertebrae are near the top of the list for belted occupants, but not nearly so much for the unrestrained (where fracture of the second vertebra is actually quite low on the list). This is consistent with the findings in Table 4-1d and 4-4d that belted females are highly susceptible to neck injury, especially in frontal crashes in vehicles without air bags.
- Lower-leg injuries are near the top of the list for belted occupants (especially "fractures of other parts of lower leg" which has the highest percentage of females by nine percentage points), but are lower on the list or not on it all for the unrestrained. This may reflect belts' effectiveness against injuries to body regions other than the lower legs.
- By contrast, several thoracic injuries – fracture of a thoracic vertebra, traumatic hemopneumothorax, and heart injury with hemopericardium – are high on the unrestrained list and low on the belted list. The likely explanation is that belts protect females well from some types of thoracic injuries.

Table 6-10a (like Table 6-6a) is limited to belted front-seat occupants of MY 1999 and later cars, all equipped with dual frontal air bags. Most of the injuries occupy similar relative positions on Table 6-10 (all model years, including no vehicles without air bags) and Table 6-10a. Three injuries, however, dropped to a lower position on Table 6-10a, possibly indicating that air bags and other recent safety technologies have mitigated the excess risk for females: fracture of the second cervical vertebra, hip fractures, and multiple rib fractures. Three types of harm moved up in relative rank: fracture of the shaft of the tibia, traumatic subarachnoid hemorrhage, and drowning. That may reflect the effectiveness of late-model safety technologies in mitigating females' risk of harm <u>other than</u> these three types.[55]

---

[55] "Other injuries of heart" also moved up on Table 6-10a, but that is more or less offset by the quite similar "injury of heart, unspecified" moving down.

TABLE 6-8: MCOD INJURIES OF BELTED REAR-OUTBOARD PASSENGER FATALITIES IN CARS
RANKED BY AVERAGE AGE OF THE VICTIM
(1999-2007 FARS, OCCUPANTS AGE 18-96)

AVERAGE AGE	PERCENT FEMALE	INJURY	N
62.11	55.56	Traumatic subdural haemorrhage	18
61.54	53.57	Other injuries of lung	28
60.92	41.67	Fracture of other and unspecified parts of lumbar spine and pelvis	12
60.74	69.57	Fracture of rib	23
60.15	76.92	Crushed chest	13
58.53	73.68	Multiple fractures of ribs	19
56.93	64.29	Traumatic shock	14
56.54	59.70	Unspecified injury of thorax	201
55.36	72.00	Fracture of neck, part unspecified	50
55.32	65.88	Unspecified injury of abdomen, lower back and pelvis	85
52.79	73.68	Injury of thoracic aorta	38
49.50	60.09	Injury, unspecified	873
48.97	73.33	Injury of liver or gallbladder	30
48.80	66.67	Traumatic hemothorax	15
48.73	59.04	average for belted back-seat outboard passenger fatalities, FARS 1999-2007	
48.35	59.05	Unspecified injury of trunk, level unspecified	105
47.59	59.09	Multiple injuries of head	22
47.13	68.75	Injury of other intra-abdominal organs	16
47.08	46.15	Multiple injuries of thorax	13
46.16	68.42	Injury of spleen	19
43.84	57.89	Injuries of intrathoracic, intra-abdominal & pelvic organs	19
43.39	56.96	Unspecified injury of neck	79
41.81	62.50	Other specified injuries of trunk, level unspecified	16
40.54	53.85	Drowning	13
39.87	51.04	Unspecified injury of head	431
39.64	56.00	Diffuse brain injury	25
38.32	50.00	Fracture of skull and facial bones, part unspecified	34
37.12	55.07	Intracranial injury, unspecified	69
36.35	50.00	Fracture of base of skull	20
36.15	30.77	Burn of unspecified body region, unspecified degree	13
34.70	53.95	weighted average for belted back-seat outboard passengers, CDS 1999-2007	

PERCENT FEMALE	AVERAGE AGE	INJURY	N
74.00	58.22	Fractures of other parts of lower leg	50
65.00	39.89	Injury of unspecified intra-abdominal organ	160
61.36	54.23	Fracture of second cervical vertebra (axis)	88
59.26	54.50	Fractures involving multiple regions of both lower limbs	54
59.15	48.13	Fracture of first cervical vertebra (atlas)	142
58.82	43.11	Other and unspecified injuries of neck	85
57.00	47.26	Other multiple injuries of abdomen, lower back and pelvis	100
56.41	51.74	Fracture of upper limb, level unspecified	78
56.15	51.43	Unspecified injury of lower limb, level unspecified	187
55.58	56.56	Injury of other specified intrathoracic organs	113
55.64	54.12	Other specified injuries of abdomen, lower back and pelvis	270
54.33	48.51	Fracture of neck, part unspecified	1988
54.12	55.17	Fracture of other and unspecified parts of lumbar spine and pelvis	571
53.73	53.82	Traumatic shock	577
53.48	48.30	Injury of spleen	774
52.75	45.37	Dislocation of cervical vertebra	510
52.24	49.58	Fracture of fibula alone	67
52.18	48.50	Injury of liver or gallbladder	960
52.06	37.82	weighted average for belted crash-involved drivers and RF passengers, CDS 1999-2007	
52.03	45.72	Other specified injuries of head	296
51.89	51.71	Traumatic pneumothorax	296
51.85	54.09	Fracture of hip and/or fracture of neck of femur	81
51.58	52.19	Multiple fractures of cervical spine	64
51.55	52.61	Multiple fractures of ribs	562
51.35	44.86	Fracture of shaft of humerus	74
51.29	51.07	Unspecified injury of abdomen, lower back and pelvis	2324
51.06	52.48	Fracture of shaft of tibia	94
49.76	53.38	Injury of unspecified intrathoracic organ	201
49.11	46.43	Unspecified injury of neck	3073
48.99	58.83	Fracture of rib	383
48.78	60.18	Flail chest	343
48.78	52.32	Injury of thoracic aorta	1604
48.64	55.04	Traumatic haemothorax	619
48.48	48.23	Injuries of intrathoracic, intra-abdominal & pelvic organs	397
48.24	44.11	Open wound of head, part unspecified	85
48.12	49.73	Unspecified injury of trunk, level unspecified	3185
48.00	48.87	Injury, unspecified	33546
47.78	48.19	Injury of kidney	419
47.39	54.14	Unspecified injury of thorax	8708
47.09	50.07	Injury of other intra-abdominal organs	597
46.80	48.81	all belted driver and RF passenger fatalities, FARS 1999-2007	
46.70	48.30	Traumatic subarachnoid haemorrhage	349
46.70	54.25	Crushed chest	604
49.41	50.09	Multiple injuries of thorax	502
46.38	45.83	Multiple injuries of neck	168
46.33	50.77	Fracture of sternum	219
46.30	45.72	Unspecified injury of upper limb, level unspecified	54
46.12	52.20	Other specified injuries of trunk, level unspecified	451
45.99	51.96	Fracture of clavicle	87
45.85	51.50	Other and unspecified injuries of cervical spinal cord	189
45.83	48.88	Fracture of thoracic vertebra	
45.75	57.22	Traumatic subdural haemorrhage	553
45.08	53.80	Fracture of lower limb, level unspecified	122
44.90	39.43	Fracture of vault of skull	49
44.70	10.10	Multiple injuries of head	512
44.00	08.20	Fracture of base of skull	702
44.55	52.35	Other injuries of lung	371
44.33	43.12	Unspecified injury of head	18164
44.00	52.78	Traumatic haemopneumothorax	100
43.88	66.08	Injury of heart, unspecified	325
43.84	55.80	Other injuries of heart	554
43.83	42.23	Injury of multiple intra-abdominal organs	67
43.57	48.84	Other intracranial injuries	482
43.22	44.19	Diffuse brain injury	1055
43.16	55.83	Other specified injuries of thorax	26
42.76	42.12	Drowning	587
42.74	41.29	Fracture of skull and facial bones, part unspecified	1485
42.87	39.17	Intracranial injury, unspecified	2597
42.55	40.81	Multiple fractures involving skull and facial bones	141
41.79	46.79	Other specified injuries of neck	97
40.82	51.76	Concussion	49
39.66	40.49	Other specified gases, fumes and vapors	174
39.60	39.79	Crushing injury of head, part unspecified	161
38.87	47.64	Fracture of femur, part unspecified	283
38.00	50.24	Multiple fractures of lumbar spine and pelvis	50
37.70	40.92	Burns involving 90% or more of body surface	61
36.96	41.06	Asphyxiation	487
36.70	38.50	Traumatic cerebral edema	108
35.97	38.93	Burn of unspecified body region, unspecified degree	417
34.62	54.53	Injury of heart with hemopericardium	73
01.50	46.28	Foreign body in respiratory tract, part unspecified	76
24.47	34.72	Toxic effect of alcohol	94

TABLE 6-10a: MY 1999-2008 CARS WITH DUAL AIR BAGS; MCOD INJURIES OF BELTED CAR DRIVER AND RF-
PASSENGER FATALITIES RANKED BY PERCENTAGE OF VICTIMS THAT ARE FEMALES
(1999-2007 FARS, OCCUPANTS AGE 18-96)

PERCENT FEMALE	AVERAGE AGE	REC_CD	N
72.97	66.05	Injury of unspecified intra-abdominal organ	37
65.12	46.40	Fracture of first cervical vertebra (atlas)	43
64.29	53.61	Fracture of shaft of tibia	28
64.00	55.92	Fracture of upper limb, level unspecified	25
63.33	52.93	Other specified injuries of abdomen, lower back and pelvis	30
60.00	54.24	Injury of other specified intrathoracic organs	50
58.82	49.82	Other specified injuries of trunk, level unspecified	51
58.02	54.17	Traumatic shock	162
57.87	50.63	Fracture of neck, part unspecified	572
56.86	46.97	Dislocation of cervical vertebra	153
56.40	44.37	Injury of liver or gallbladder	211
56.37	48.12	Injury of spleen	204
55.67	50.93	Traumatic subarachnoid haemorrhage	97
55.56	58.44	Fracture of second cervical vertebra (axis)	27
54.90	57.67	Traumatic pneumothorax	51
54.32	38.24	wtd avg for belted crash-involved drivers & RF, CDS 1999-2007, MY 1999-2008	
53.73	55.60	Unspecified injury of lower limb, level unspecified	67
53.48	50.51	Unspecified injury of abdomen, lower back and pelvis	647
52.35	48.99	Injuries of intrathoracic, intra-abdominal & pelvic organs	170
51.82	58.19	Fracture of rib	274
51.20	50.92	Unspecified injury of trunk, level unspecified	1123
51.02	48.61	Unspecified injury of neck	929
50.89	58.03	Fracture of other and unspecified parts of lumbar spine and pelvis	169
50.62	57.95	Flail chest	81
50.00	49.79	Other and unspecified injuries of cervical spinal cord	48
49.49	48.97	Injury, unspecified	10231
48.97	54.61	Other injuries of heart	145
48.94	47.21	Multiple injuries of neck	47
48.78	49.85	Injury of unspecified intrathoracic organ	41
48.72	59.84	Fracture of sternum	39
48.39	52.67	Crushed chest	93
48.33	43.96	Drowning	180
48.25	61.45	Multiple fractures of ribs	143
48.12	48.98	average for belted driver and RF fatalities, FARS 1999-2007, MY 1999-2008	
48.00	67.32	Fracture of hip and/or fracture of neck of femur	25
47.84	54.08	Unspecified injury of thorax	2335
47.75	52.68	Injury of thoracic aorta	400
46.93	51.41	Injury of other intra-abdominal organs	179
46.67	39.47	Other and unspecified injuries of neck	30
46.20	43.80	Unspecified injury of head	4907
44.81	54.41	Traumatic hemothorax	154
43.63	39.67	Fracture of base of skull	204
43.48	45.12	Multiple injuries of head	230
43.23	43.97	Diffuse brain injury	310
42.96	53.17	Multiple injuries of thorax	135
42.21	61.03	Traumatic subdural haemorrhage	154
41.96	39.58	Intracranial injury, unspecified	827
41.91	41.64	Fracture of skull and facial bones, part unspecified	451
41.77	51.88	Other injuries of lung	237
40.74	63.52	Traumatic hemopneumothorax	27
40.00	40.52	Other specified gases, fumes and vapors	60
38.64	50.98	Other intracranial injuries	132
38.54	56.36	Injury of heart, unspecified	96
36.00	37.40	Traumatic cerebral edema	25
35.44	42.75	Asphyxiation	158
34.09	40.54	Burn of unspecified body region, unspecified degree	132
31.88	48.43	Fracture of femur, part unspecified	69
30.00	31.23	Toxic effect of alcohol	30

Tables 6-11 and 6-12 rank the injuries for unrestrained and belted <u>back-seat</u> occupants by the percentage of victims that are females. Although many injuries have similar rankings for unrestrained and belted occupants, there are two distinct groups of injuries that do not:

- Abdominal injuries – liver, gallbladder, spleen, and "other intra-abdominal organs" – as well as two thoracic injuries (aorta and hemothorax) are near the top of the list for belted occupants (Table 6-12), but even lower on the list for the unrestrained (Table 6-11) than on the combined list for belted and unrestrained back-seat occupants (Table 6-4). Furthermore, "crushed chest" has the highest percentage of females on the belted list but does not even appear on the unrestrained list.
- Conversely, two head injuries – fracture of base of skull and "unspecified injury of head" – are in the middle of the unrestrained list but near the end of the belted list. . .

Both contrasts would appear to reflect the performance of belts for back-seat occupants. These belts have not been fully successful in protecting occupants from thoracic and abdominal injuries – but especially not for females, who are more fragile than males in those areas.[56] They have been quite successful in protecting occupants from head injuries, but probably even more so for females, whose shorter stature would tend to limit head excursion and contact with interior surfaces.

---

[56] Morgan (1999), pp. 108-110.

TABLE 6-11: MCOD INJURIES OF UNRESTRAINED REAR-OUTBOARD PASSENGER FATALITIES IN CARS
RANKED BY PERCENTAGE OF VICTIMS THAT ARE FEMALES
(1999-2007 FARS, OCCUPANTS AGE 18-96)

PERCENT FEMALE	AVERAGE AGE	INJURY	N
50.35	39.92	Unspecified injury of abdomen, lower back and pelvis	143
50.00	39.61	Other specified injuries of trunk, level unspecified	28
47.37	39.00	Injury of heart, unspecified	19
47.37	38.23	Fracture of neck, part unspecified	152
45.45	34.85	Traumatic shock	33
44.68	41.09	Fracture of rib	47
43.78	37.09	Unspecified injury of neck	217
43.33	47.50	Multiple fractures of ribs	30
43.10	31.76	weighted average, unrestrained back-seat outboard passengers, CDS 1999-2007	
42.82	39.18	Unspecified injury of thorax	542
42.14	35.86	Injury, unspecified	2468
40.54	34.68	Dislocation of cervical vertebra	37
40.35	32.42	Injuries of intrathoracic, intra-abdominal & pelvic organs	57
39.70	35.24	all unrestrained back-seat outboard passenger fatalities, FARS 1999-2007	
39.13	43.17	Fracture of femur, part unspecified	23
37.99	33.79	Unspecified injury of trunk, level unspecified	229
37.93	26.91	Fracture of base of skull	58
37.77	33.03	Unspecified injury of head	1607
37.50	32.99	Multiple injuries of head	80
36.36	32.64	Fracture of other and unspecified parts of lumbar spine and pelvis	44
36.00	31.92	Burn of unspecified body region, unspecified degree	25
35.90	38.54	Multiple injuries of thorax	39
35.42	35.30	Injury of thoracic aorta	96
34.27	31.11	Fracture of skull and facial bones, part unspecified	178
34.21	40.92	Traumatic subarachnoid haemorrhage	38
33.82	32.85	Other injuries of lung	68
33.81	31.67	Intracranial injury, unspecified	281
33.33	32.25	Injury of spleen	48
33.33	31.97	Traumatic hemothorax	33
33.33	30.61	Drowning	33
32.56	30.91	Other intracranial injuries	43
30.39	31.57	Diffuse brain injury	102
30.23	36.58	Injury of other intra-abdominal organs	43
29.79	31.55	Asphyxiation	47
29.73	37.43	Traumatic subdural haemorrhage	37
28.17	30.97	Injury of liver or gallbladder	71
28.13	39.59	Other specified injuries of head	32
27.59	42.45	Other injuries of heart	29

147

TABLE 6-12: MCOD INJURIES OF BELTED REAR-OUTBOARD PASSENGER FATALITIES IN CARS
RANKED BY PERCENTAGE OF VICTIMS THAT ARE FEMALES
(1999-2007 FARS, OCCUPANTS AGE 18-96)

PERCENT FEMALE	AVERAGE AGE	INJURY	N
76.92	60.15	Crushed chest	13
73.68	58.53	Multiple fractures of ribs	19
73.68	52.79	Injury of thoracic aorta	38
73.33	48.97	Injury of liver or gallbladder	30
72.00	55.36	Fracture of neck, part unspecified	50
69.57	60.74	Fracture of rib	23
68.75	47.13	Injury of other intra-abdominal organs	16
68.42	46.16	Injury of spleen	19
66.67	48.80	Traumatic hemothorax	15
65.88	55.32	Unspecified injury of abdomen, lower back and pelvis	85
64.29	56.93	Traumatic shock	14
62.50	41.81	Other specified injuries of trunk, level unspecified	16
62.08	49.50	Injury, unspecified	873
59.70	58.54	Unspecified injury of thorax	201
59.09	47.59	Multiple injuries of head	22
59.05	48.35	Unspecified injury of trunk, level unspecified	105
59.04	48.73	all belted back-seat outboard passenger fatalities, FARS 1999-2007	
57.89	43.84	Injuries of intrathoracic, intra-abdominal & pelvic organs	19
56.96	43.39	Unspecified injury of neck	79
56.00	39.64	Diffuse brain injury	25
55.56	62.11	Traumatic subdural haemorrhage	18
55.07	37.12	Intracranial injury, unspecified	69
53.95	34.70	weighted average for belted back-seat outboard passengers, CDS 1999-2007	
53.85	40.54	Drowning	13
53.57	61.54	Other injuries of lung	28
51.04	39.87	Unspecified injury of head	431
50.00	38.32	Fracture of skull and facial bones, part unspecified	34
50.00	36.35	Fracture of base of skull	20
46.15	47.08	Multiple injuries of thorax	13
41.67	60.92	Fracture of other and unspecified parts of lumbar spine and pelvis	12
30.77	36.15	Burn of unspecified body region, unspecified degree	13

148

CHAPTER 7

SPECIFIC AIS ≥ 2 INJURIES PREVALENT AMONG OLDER OCCUPANTS AND WOMEN – ANALYSES OF 1988-2010 NASS-CDS DATA

7.0   Summary

NASS-CDS lists all occupants' individual fatal or nonfatal injuries, permitting analyses of injury risk for crash survivors as well as fatalities. For any specific injury, a high average age of the victims indicates that older occupants are vulnerable; high percentages of female victims indicate that females are vulnerable. At the AIS ≥ 2 injury level, thoracic and neck injuries are prevalent among older occupants of passenger cars, while injuries of the legs and arms are highly prevalent among females. Specifically, chest fractures and associated injuries to thoracic organs are the injuries with the highest average age of the victims. Neck fractures and certain types of abdominal injury have a relatively higher percentage of elderly and female victims among belted occupants than among the unrestrained.

7.1   Ranking the specific injuries by victims' average age or percent females

The NASS-CDS file, as discussed in Section 5.1, describes the individual AIS ≥ 2 injuries of each occupant of a vehicle involved in a towaway crash: nonfatal as well as fatal crashes. For any specific injury it is possible to compute the average age of the victims and the percent that are female. The injuries can be ranked by average age from oldest to youngest or by percent female victims from highest to lowest, creating a set of tables just like the FARS-MCOD analyses of Chapter 6. A high average age means older people are especially vulnerable (relative to young people) to this type of injury, at least in vehicles of the past 50 model years. A high percentage female victims means women are especially vulnerable. The principal differences from the FARS-MCOD analysis are:

- The injuries are not limited to those that "contributed to death" but include any injury with AIS ≥ 2 to any occupant – fatality or survivor. Among other things, that will add many arm and leg injuries.
- Injuries are determined from documents provided by treatment facilities, hospitals, or medical examiners.
- Injuries are specified in detail; categories like "unspecified injury of head" or "multiple injuries of thorax" were prevalent in MCOD, but not here. . .

The purpose of the analysis is to search for patterns: individual injuries and groups of injuries that are especially common among older occupants and/or women, indicating vulnerability. The specific numbers – the average age or percent female for a given injury – are not particularly important; no statistical procedures are used to test for statistical significance or estimate confidence bounds. As in Chapter 5, the analyses are based on unweighted CDS data to avoid the misleading results if a few especially old or young occupants with high sampling weights have a particular type of injury.

149

Average age and percent female are tabulated for front-seat-outboard occupants (drivers and RF passengers combined) and rear-seat-outboard occupants; then separately for unrestrained and belted front-outboard and rear-outboard occupants. Data is limited to passenger cars with MY ≥ 1960 (i.e., excluding LTVs), for a more homogeneous crash environment. All adults, age 18 to 96, are included. Unlike Chapters 2-5, there is no need to start at 21, where the full trend of risk increasing with age begins, because this analysis does not attempt to quantify trends.

Injury data: CDS documents occupants' injuries based on data from hospitals, treatment facilities, and autopsies. Injury severity has always been assessed with the abbreviated injury scale (AIS), which rates injuries from 1 (minor) to 6 (maximum).[57] There have been several versions of the AIS over the years. From 1988 to 1992, CDS used the 1985 revision of the AIS; from 1993 to 2009, the 1990/98 revisions; and starting in 2010, the 2005/08 revision. These three versions of the AIS all use a scale from 1 to 6 and have the same names for each level, but some individual types of injury may have changed levels. The analyses of this report consider injuries rated 2 to 6 on the AIS in the CDS file (survival rates for the various levels are discussed in Section 5.1):

2   Moderate

3   Serious

4   Severe

5   Critical

6   Maximum

Multiple injuries may be coded for one occupant. One consistent coding on CDS since its inception in 1988 is a 3-letter code indicating the body region (BODYREG), the LESION, and the system or organ (SYSORG). The tables in this chapter simply list the various possible 3-letter combinations. For example, CLA would be chest-laceration-artery. The listings in the tables are sometimes abbreviated if it would be redundant to specify the body region and the organ or the lesion and the organ. For example, if the spleen is injured, it would be redundant to say that the body region is the abdomen; if somebody has a concussion, it would be redundant to say the body region is the head and the organ is the brain. Furthermore, as documented in Appendix C, infrequent but fairly similar codes are grouped together to obtain a larger cluster with enough data to obtain a meaningful average age and percent female. For example, abrasions, contusions, lacerations, avulsions, and ruptures of the ear are grouped as "ear external injury."

7.2   Injuries ranked by victims' average age – front seat versus back seat

Table 7-1 ranks the AIS ≥ 2 injuries of drivers and RF passengers of cars by the average age of the victim, from oldest to youngest. The key to identifying patterns is the color coding, based on the CDS body region codes: bold blue print for head injuries (CDS codes H and F: head and face); bold red for torso (B, C, M, P, and S: back, chest, abdomen, pelvis, and shoulder); and bold green for neck (N). Plain black print indicates the arms (A, E, R, W, and X: upper arm, elbow, forearm, wrist/hand, entire arm) or legs (K, L, Q, T, and Y: knee, lower leg, ankle/foot,

---

[57] Gennarelli & Wodzin (2006).

thigh, entire leg), plus all burns. Two benchmarks, printed in bold black letters, are the weighted average age of drivers and RF passengers with MAIS ≥ 2 in passenger cars in 1988-2010 CDS (i.e., the average at the occupant level, not the injury level) and the weighted average age of drivers and RF passengers of cars involved in crashes (fatal or nonfatal) in 1988-2010 CDS. Table 7-1 displays the average age of the victims (left column), the percent female (second column), the number of times the injury is reported in CDS (next-to-last column), and the average AIS of these injuries. Table 7-1 is limited to injuries that occurred at least 20 times if the average AIS < 3.5 or at least 10 times if the average AIS ≥ 3.5; that allows printing the results on a single page, to make it easier to look for patterns. The information on the average AIS is furnished to help identify the more severe injuries.

Table 7-1 resembles Table 6-1, the corresponding MCOD analysis of injuries that "contributed to death" in many respects but differs in some. Like Table 6-1, the great majority of the injuries at the top of the list are torso injuries (red), while head injuries (blue) are concentrated at the end of the table. Furthermore, in both tables, chest fractures (ribs, sternum) and related soft-tissue injuries (heart, lungs, or blood vessels) are at or near the top of the list. It is consistent with the literature and all the findings of Table 5-2a: older occupants have high risk of torso injury, especially thoracic injuries, even at the nonfatal level.

One difference from Table 6-1 is that quite a few neck injuries (green) cluster near the top of the list. That is consistent with Tables 4-1a and 5-2a, which show about the same effects of aging on neck-injury risk at the fatal and nonfatal levels, whereas for other body regions, aging has less effect at the nonfatal levels. Another difference is that torso injuries are not limited to the top half of the list: there are quite a few in the middle and some even at the end. The wider dispersion may partly reflect that Table 7-1 is based on less data; thus the observed estimates of average age are less precise. Another factor is that the increase of torso-injury risk with age is simply not so large at the nonfatal levels. However, a closer examination of Table 7-1 indicates that the torso injuries near the top of the list are largely (but not always) chest injuries, while those further down on the list are often (but not always) abdominal injuries. That is directionally similar to Table 6-1 and consistent with the statistics in Table 5-2a. Hip fractures were at the very top of Table 6-1, but pelvis fracture is not so high here (40.45). Perhaps it would be accurate to say that hip fracture per se is not so much an old person's injury, but that it is far more likely to have life-threatening consequences in older people.

TABLE 7-1: AIS 2-6 INJURIES OF CAR DRIVERS AND RF PASSENGERS RANKED BY AVERAGE AGE OF THE VICTIM
(1988-2010 CDS, OCCUPANTS AGE 18-96)

AVG AGE	PERCENT FEMALE	INJURY	N	AVG AIS
48.04	44.29	CHEST FRACTURES	9196	2.90
47.56	34.78	HEART LACERATION	1081	3.37
45.90	42.77	HEART CONTUSION	491	3.23
45.54	39.62	THORACOLUMBAR CORD LACERATION	156	4.88
45.41	47.06	NECK AMPUTATION	17	6.00
45.02	31.25	CERVICAL CORD CONTUSION	320	4.06
44.36	38.70	CHEST LACERATED ARTERY	1516	4.52
44.30	42.68	THORACOLUMBAR DISLOCATION	82	2.29
44.24	36.05	HEART RUPTURE	96	5.76
43.56	36.02	ABDOMEN ARTERY LACERATION	260	3.50
42.95	39.55	TORSO ARTERY TRANSECTION/RUPTURE	177	5.47
42.94	36.76	DIAPHRAGM CONTUSION	49	2.00
41.52	52.76	ARM FRACTURE	7656	2.38
41.60	37.70	LUNG LACERATION	1318	3.25
41.47	43.73	CONTUSION OF GENITALS	107	2.00
41.22	39.27	NECK FRACTURE	2773	2.31
41.11	34.63	DIAPHRAGM LACERATION	588	3.00
40.73	43.32	NECK DISLOCATED VERTEBRA	584	2.21
40.72	46.09	SHOULDER FRACTURE	3036	2.00
40.61	37.65	NECK RUPTURED VERTEBRA	85	2.13
40.59	54.09	LOWER LEG FRACTURE	10996	2.18
40.51	46.91	TORSO SUPERFICIAL INJURY	418	2.04
40.47	34.55	BRAIN STEM TRANSECTION	55	6.00
40.46	63.95	PELVIS FRACTURE	5743	2.32
40.28	37.50	THORACOLUMBAR CORD CONTUSION	98	3.95
40.21	52.14	weighted average of drivers and RF passengers with MAIS ≥ 2, CDS 1988-2010		
40.17	57.14	TRACHEA LACERATION	35	3.23
40.09	37.94	INTESTINAL CONTUSION	798	2.62
40.05	42.49	ARM SUPERFICIAL INJURY	506	2.08
39.78	42.25	CERVICAL CORD LACERATION	213	5.72
39.66	56.96	LEG DISLOCATED JOINT	869	2.16
39.56	36.40	DIAPHRAGM RUPTURE	101	3.75
39.41	39.19	LEG AMPUTATION	74	3.22
39.29	47.01	LIVER CONTUSION	488	2.10
39.27	47.12	ARM DISLOCATED JOINT	278	2.28
39.24	31.37	CHEST CRUSHED	51	5.34
39.19	41.21	SHOULDER DISLOCATED	563	2.00
39.13	37.69	INTESTINAL LACERATION	504	2.98
38.93	55.41	NECK ARTERY LACERATION	74	3.03
38.99	46.72	SPLEEN CONTUSION	351	2.06
38.56	42.18	THORACOLUMBAR FRACTURE	3793	2.09
38.82	37.89	ARM AMPUTATION	95	2.45
38.80	18.75	INTESTINAL RUPTURE	16	4.00
38.62	39.77	LUNG CONTUSION	3884	3.32
38.39	36.69	BRAIN UNK INJURY	8056	3.65
38.15	46.90	PELVIS CRUSHED	81	4.23
38.04	44.76	LIVER LACERATION	2528	2.71
37.88	55.81	ARM CRUSHED	43	2.53
37.85	41.53	SPLEEN LACERATION	1982	2.56
37.60	40.00	KIDNEY RUPTURE	15	4.60
37.56	40.43	FEMUR FRACTURE	3651	3.00
37.27	28.33	ARM JOINT LACERATION	60	2.02
37.22	42.40	HEAD SUPERFICIAL INJURY	533	2.05
37.06	30.93	HIP JOINT SEPARATION	451	2.99
36.97	32.10	EYE AVULSION/DETACHMENT	82	3.17
36.83	42.11	SPLEEN RUPTURE	190	2.65
36.78	35.47	BRAIN CONTUSION	2120	3.41
36.52	42.61	FACIAL LACERATION	1036	2.02
36.45	42.68	KIDNEY CONTUSION	558	2.08
36.40	50.15	weighted average of crash-involved drivers and RF passengers, CDS 1988-2010		
36.24	34.02	HIP DISLOCATION	650	2.35
36.10	52.94	LEG SUPERFICIAL INJURY	1071	2.07
35.81	48.19	CONCUSSION	9150	2.00
35.77	41.80	LACERATION OF GENITALS	255	2.79
35.66	35.79	FACIAL FRACTURE	4193	2.18
35.52	36.90	HEAD ARTERY LACERATION	84	3.09
35.43	48.15	FACIAL DISLOCATION	91	2.01
34.90	40.43	NECK SUPERFICIAL INJURY	47	2.00
34.87	30.43	EAR EXTERNAL INJURY	46	2.00
34.83	55.65	KNEE OR ANKLE SPRAIN	611	2.00
34.67	33.48	BRAIN LACERATION	923	4.89
34.50	36.36	ARM ARTERY LACERATION	22	2.86
34.48	25.81	LEG ARTERY LACERATION	31	2.84
34.27	31.95	SKULL FRACTURE	3593	2.90
34.15	36.16	KIDNEY LACERATION	525	2.80
33.94	31.79	HEAD CRUSHED	151	6.00
33.83	53.70	RUPTURE OF GENITALS	62	3.89
33.30	48.31	LEG JOINT LACERATION	296	2.18
32.90	40.00	LIVER RUPTURE	30	4.57
32.91	26.13	BURNS	173	5.03
31.13	79.17	LEG CRUSHED	24	2.79

152

It is interesting to compare the specific injuries to the two benchmark lines in the bold black letters. The average age of the occupants with MAIS ≥ 2 (40.21) is approximately ⅓ of the way down the table. The average age of people involved in nonfatal or fatal crashes (36.40) is almost ¾ of the way down. Principally chest and neck injuries, plus some abdominal injuries have average age above 40.21, ranging all the way up to 48.04 for chest fractures. A fair number of not-too-severe head injuries (superficial head injury, facial laceration, concussion, facial fracture, facial dislocation, and ear external injury) have average age quite close to 36.40, suggesting that risk barely increases with age, if at all, for these injuries. The injuries with average age several years below 36.40 include some associated with severe crashes that primarily involve young people, including crashes that result in ejection: burns, "head crushed," and skull fractures. For other injuries near the end of the list – kidney laceration, laceration/rupture of genitals – it is not so clear if they are associated with "young people's crashes" or if younger people might actually be physiologically more susceptible to them.

Table 7-1 includes various arm and leg injuries. Both are dispersed throughout the table, but the leg injuries are somewhat more often near the end – consistent with Table 5-2a, which showed a possibly greater increase with age for arm injuries than for leg injuries.

Table 7-2 presents the corresponding injury ranking for rear-outboard occupants. There are, of course, fewer adults in the back seat than at the driver and RF positions. Table 7-2 includes injuries that occurred at least 10 times if the average AIS < 3.5 or at least 5 times if the average AIS ≥ 3.5 (as compared to minima of 20 and 10 Table 7-1), but it is still a shorter list than Table 7-1. The overall pattern is quite similar to the front-seat occupants and also to the MCOD injuries of rear-seat occupants (Table 6-2). Torso injuries constitute almost all the injuries at the top of the list, along with neck and leg fractures. Head injuries are mostly near the end of the list. Possible differences from Table 7-1 are that injuries to torso arteries account for three of the top four listings and three types of hip injuries – pelvis crushed, hip joint separation, and hip dislocation – are quite high on the list. These could indicate special vulnerabilities for older occupants in the back seat, especially belted occupants. Tables 7-5 and 7-6 will provide additional information on the role of belts in these injuries.

7.3   Injuries ranked by percentage of the victims that are females

Table 7-3 lists exactly the same injuries as Table 7-1 (car drivers and RF passengers), but ranked by percent female rather than average age: the higher the percent female, the higher on the list. The salient feature of Tables 7-3 is the extraordinary concentration at the top of leg injuries and then arm injuries, especially fractures and dislocations. Almost all of these injuries are AIS 2 or 3. Most are AIS 2. The results are consistent with Table 5-2d showing high risk of arm and, especially, leg injuries for female occupants. They are also consistent with the literature showing females to have a substantially higher risk of <u>nonfatal</u> injuries than males.[58] At the top of the list: leg crushed, 79.17 percent of the victims are women.

---

[58] Bose, Segui-Gomez, & Crandall (2011).

TABLE 7-2: AIS 2-6 INJURIES OF REAR-OUTBOARD PASSENGERS IN CARS
RANKED BY AVERAGE AGE OF THE VICTIM
(1988-2010 CDS, OCCUPANTS AGE 18-96)

AVG AGE	PERCENT FEMALE	INJURY	N	AVG AIS
46.17	66.67	TORSO ARTERY TRANSECTION/RUPTURE	6	5.67
45.70	50.00	ABDOMEN ARTERY LACERATION	20	3.35
44.35	54.02	CHEST FRACTURES	435	2.96
41.06	42.86	CHEST LACERATED ARTERY	84	4.68
40.10	56.32	NECK FRACTURE	190	2.34
38.89	33.33	HEART LACERATION	57	4.05
38.60	52.00	LOWER LEG FRACTURE	250	2.15
38.50	40.00	TORSO SUPERFICIAL INJURY	10	2.10
38.20	40.00	PELVIS CRUSHED	5	4.20
38.09	34.38	HIP JOINT SEPARATION	32	3.00
37.85	42.50	HIP DISLOCATION	40	2.30
37.64	45.45	THORACOLUMBAR CORD LACERATION	11	5.00
37.39	65.79	FACIAL LACERATION	38	2.05
37.37	46.34	INTESTINAL CONTUSION	41	2.00
37.12	45.10	INTESTINAL LACERATION	51	2.49
37.11	55.42	FEMUR FRACTURE	249	3.00
36.63	53.51	SHOULDER FRACTURE	185	2.00
36.61	55.56	HEART CONTUSION	18	3.11
36.44	50.60	PELVIS FRACTURE	334	2.27
36.04	39.13	LUNG LACERATION	92	3.26
35.84	42.11	ARM SUPERFICIAL INJURY	19	2.11
35.81	52.38	SPLEEN CONTUSION	21	2.05
35.38	50.00	HEAD SUPERFICIAL INJURY	42	2.02
35.28	42.16	weighted average of backseat outboard passengers with MAIS ≥ 2, CDS 1988-2010		
35.02	52.23	ARM FRACTURE	358	2.39
34.96	46.43	CERVICAL CORD CONTUSION	28	3.86
34.50	55.56	DIAPHRAGM LACERATION	18	3.00
34.17	45.24	NECK DISLOCATED VERTEBRA	42	2.21
33.70	59.46	KNEE OR ANKLE SPRAIN	37	2.00
33.39	39.67	BRAIN UNK INJURY	547	3.60
33.38	40.00	LUNG CONTUSION	210	3.38
33.13	51.61	weighted average of crash-involved backseat outboard passengers, CDS 1988-2010		
32.77	43.59	SPLEEN LACERATION	117	2.81
32.40	45.32	LIVER LACERATION	139	2.70
31.91	38.30	SHOULDER DISLOCATED	47	2.00
31.83	22.22	LEG DISLOCATED JOINT	18	2.33
31.70	43.70	THORACOLUMBAR FRACTURE	801	2.10
31.64	21.43	LACERATION OF GENITALS	14	2.57
31.57	42.86	ARM DISLOCATED JOINT	14	2.07
31.04	38.19	BRAIN CONTUSION	144	3.42
30.91	46.07	CONCUSSION	471	2.31
30.87	41.57	FACIAL FRACTURE	255	2.22
30.85	36.36	LEG SUPERFICIAL INJURY	33	2.21
30.32	47.37	KIDNEY LACERATION	38	2.61
30.00	30.00	DIAPHRAGM RUPTURE	10	3.60
29.79	53.57	LIVER CONTUSION	28	2.18
29.46	53.85	THORACOLUMBAR CORD CONTUSION	13	3.92
29.26	31.56	SKULL FRACTURE	244	2.93
29.11	22.22	HEAD CRUSHED	9	6.00
28.88	34.15	KIDNEY CONTUSION	41	2.00
28.50	12.50	CERVICAL CORD LACERATION	8	5.63
28.35	30.65	BRAIN LACERATION	62	4.82
26.21	28.57	BURNS	14	5.21

TABLE 7-3: AIS 2-6 INJURIES OF CAR DRIVERS AND RF PASSENGERS RANKED BY PERCENTAGE OF VICTIMS THAT ARE FEMALES
(1988-2010 CDS, OCCUPANTS AGE 18-96)

PERCENT FEMALE	AVG AGE	INJURY	N	AVG AIS
79.17	31.13	LEG CRUSHED	24	2.79
58.70	38.83	RUPTURE OF GENITALS	62	3.36
57.14	40.17	TRACHEA LACERATION	35	3.23
56.96	39.66	LEG DISLOCATED JOINT	869	2.16
55.81	37.88	ARM CRUSHED	43	2.53
55.65	34.83	KNEE OR ANKLE SPRAIN	611	2.00
55.41	38.33	NECK ARTERY LACERATION	74	3.03
54.09	40.59	LOWER LEG FRACTURE	10996	2.18
53.35	40.45	PELVIS FRACTURE	5743	2.32
52.94	36.10	LEG SUPERFICIAL INJURY	1071	2.07
52.76	41.52	ARM FRACTURE	7656	2.38
52.14	40.21	weighted average of drivers and RF passengers with MAIS ≥ 2, CDS 1988-2010		
50.15	38.40	weighted average of crash-involved drivers and RF passengers, CDS 1988-2010		
48.31	33.30	LEG JOINT LACERATION	296	2.18
47.12	39.27	ARM DISLOCATED JOINT	278	2.28
47.06	45.41	NECK AMPUTATION	17	2.00
47.01	39.29	LIVER CONTUSION	483	2.19
46.73	41.47	CONTUSION OF GENITALS	107	2.00
46.72	38.26	SPLEEN CONTUSION	351	2.05
46.15	35.43	FACIAL DISLOCATION	81	2.01
46.05	40.72	SHOULDER FRACTURE	3033	2.00
45.81	40.61	TORSO SUPERFICIAL INJURY	418	2.04
45.20	38.15	PELVIS CRUSHED	31	4.23
44.73	38.04	LIVER LACERATION	2823	2.71
44.25	48.04	CHEST FRACTURES	6136	2.90
43.32	40.73	NECK DISLOCATED VERTEBRA	564	2.21
43.18	35.81	CONCUSSION	9150	2.30
42.77	45.80	HEART CONTUSION	491	3.26
42.63	38.45	KIDNEY CONTUSION	588	2.05
42.58	44.50	THORACOLUMBAR DISLOCATION	62	2.20
42.61	36.52	FACIAL LACERATION	1035	2.02
42.49	40.05	ARM SUPERFICIAL INJURY	506	2.08
42.40	37.22	HEAD SUPERFICIAL INJURY	533	2.05
42.25	39.78	CERVICAL CORD LACERATION	219	5.72
42.16	38.86	THORACOLUMBAR FRACTURE	3796	2.08
42.11	38.58	SPLEEN RUPTURE	160	2.85
41.83	37.86	SPLEEN LACERATION	1902	2.59
41.80	35.77	LACERATION OF GENITALS	256	2.73
41.21	38.18	SHOULDER DISLOCATED	593	2.00
40.43	37.56	FEMUR FRACTURE	3651	3.00
40.43	34.89	NECK SUPERFICIAL INJURY	47	2.00
40.00	37.60	KIDNEY RUPTURE	15	4.80
40.00	32.90	LIVER RUPTURE	30	4.57
39.92	37.05	HIP JOINT SEPARATION	491	2.93
39.82	45.54	THORACOLUMBAR CORD LACERATION	150	4.90
39.82	43.04	ABDOMEN ARTERY LACERATION	260	3.60
39.55	42.35	TORSO ARTERY TRANSECTION/RUPTURE	177	5.47
39.27	41.22	NECK FRACTURE	2773	2.31
39.19	39.41	LEG AMPUTATION	74	3.22
38.78	43.84	DIAPHRAGM CONTUSION	49	2.00
38.52	38.24	HIP DISLOCATION	350	2.99
38.10	38.87	EYE AVULSION/DETACHMENT	63	2.17
37.84	40.09	INTESTINAL CONTUSION	788	2.02
37.89	38.82	ARM AMPUTATION	95	2.45
37.70	41.50	LUNG LACERATION	1818	3.25
37.88	38.13	INTESTINAL LACERATION	854	2.98
37.65	40.61	NECK RUPTURED VERTEBRA	85	2.19
37.50	40.38	THORACOLUMBAR CORD CONTUSION	96	3.95
36.90	35.52	HEAD ARTERY LACERATION	84	3.99
36.77	38.02	LUNG CONTUSION	3834	2.52
36.70	44.86	CHEST LACERATED ARTERY	1818	4.52
36.69	38.39	BRAIN UNK INJURY	8058	3.65
36.36	34.50	ARM ARTERY LACERATION	22	2.86
36.19	34.15	KIDNEY LACERATION	525	2.50
36.03	44.24	HEART RUPTURE	85	5.76
35.79	35.65	FACIAL FRACTURE	4133	2.19
35.47	36.76	BRAIN CONTUSION	2120	3.41
35.40	38.56	DIAPHRAGM RUPTURE	163	3.25
34.83	41.11	DIAPHRAGM LACERATION	338	3.00
34.79	47.26	HEART LACERATION	1051	3.37
34.55	40.47	BRAIN STEM TRANSECTION	55	6.00
33.46	34.67	BRAIN LACERATION	923	4.88
31.95	34.27	SKULL FRACTURE	3593	2.90
31.79	33.94	HEAD CRUSHED	151	6.00
31.37	39.24	CHEST CRUSHED	51	5.94
31.25	45.09	CERVICAL CORD CONTUSION	320	4.08
30.43	34.67	EAR EXTERNAL INJURY	46	2.00
28.33	37.27	ARM JOINT LACERATION	60	2.02
25.81	34.48	LEG ARTERY LACERATION	31	2.84
25.43	32.81	BURNS	173	5.03
18.75	38.69	INTESTINAL RUPTURE	16	4.00

Rupture and also contusion of the genitals are high on the list, possibly indicating an anatomical vulnerability of females. Several relatively infrequent neck injuries are high on the list, but others are dispersed throughout the table – with neck fracture, the most frequent severe injury, near the middle. Pelvis fractures and injuries to abdominal organs such as the liver, spleen, and kidneys are high-to-middle on the list (consistent with Table 6-3). Chest fractures are fairly high on the list, but other thoracic injuries are widely dispersed. Most of the head injuries, except concussions, are relatively less common in females. Burns, a phenomenon characteristic of "men's crashes," are next-to-last on the list.

Both benchmark lines are fairly near the top of the chart. Females constitute 50.15 percent of all crash-involved occupants in CDS and 52.14 percent of the occupants with MAIS $\geq$ 2. But most of the injuries on the chart have less than 50 percent female victims, in some cases, substantially less. The high vulnerability of females to leg and arm injuries counterweighs their relatively lower risk of many other types of injuries. It even squanders females' advantage of engaging in less high-risk driving and having less severe crashes than males.

Table 7-4 ranks injuries to back-seat occupants by the percentage of female victims. Femur, arm, and lower-leg fractures are fairly high on the list, but leg and arm injuries do not dominate the top of the list as in Tables 7-3. Not an issue in the back seat: short-statured female drivers having to sit close to the steering wheel and the lower instrument panel in order to reach the pedals, possibly increasing the risk of leg and arm injuries. Fractures to the neck, chest (i.e., ribs and sternum), shoulder, and pelvis have high percentages of female victims. Injuries to thoracic and abdominal organs are widely dispersed on the list. As in the front seat, many of the head injuries, except concussions, are relatively less prevalent among females.

7.4   Injuries ranked by victims' average age – unrestrained versus belted

Table 7-5 ranks the AIS $\geq$ 2 injuries of <u>unrestrained</u> drivers and RF passengers by the average age of the victim, from oldest to youngest. Table 7-6 presents the corresponding rankings for <u>belted</u> drivers and RF passengers. Tables 7-5 and 7-6, like Table 7-1 are limited to injuries that occurred at least 20 times if the average AIS < 3.5 or at least 10 times if the average AIS $\geq$ 3.5. The great majority of injuries have fairly similar rankings for unrestrained and belted front-seat occupants (similar to the findings of the MCOD analysis in Tables 6-5 and 6-6). Chest fractures, heart lacerations, and "chest lacerated artery" are in both cases the highest on the list among injuries that occur frequently (i.e., N > 400 for unrestrained; N > 300 for belted). There are, however, two groups of injuries that occupy different places on the unrestrained and belted charts:

- Several abdominal or lower-torso injuries – diaphragm laceration, spleen rupture, diaphragm rupture, thoracolumbar fracture, and kidney contusion – have older-than-average belted victims but relatively young unrestrained victims. Perhaps some of these indicate belt forces on the abdomen exceeding the tolerance limits of older occupants, or that belts are not as protective for older occupants as for young adults.

TABLE 7-4: AIS 2-6 INJURIES OF REAR-OUTBOARD PASSENGERS IN CARS
RANKED BY PERCENTAGE OF VICTIMS THAT ARE FEMALES
(1988-2010 CDS, OCCUPANTS AGE 18-96)

PERCENT FEMALE	AVG AGE	INJURY	N	AVG AIS
66.67	46.17	TORSO ARTERY TRANSECTION/RUPTURE	6	5.67
65.79	37.39	FACIAL LACERATION	38	2.05
59.46	33.70	KNEE OR ANKLE SPRAIN	37	2.00
56.32	40.10	NECK FRACTURE	190	2.34
55.56	36.61	HEART CONTUSION	18	3.11
55.56	34.50	DIAPHRAGM LACERATION	18	3.00
55.42	37.11	FEMUR FRACTURE	249	3.00
54.02	44.35	CHEST FRACTURES	435	2.96
53.85	28.46	THORACOLUMBAR CORD CONTUSION	13	3.92
53.57	28.79	LIVER CONTUSION	28	2.18
53.51	36.63	SHOULDER FRACTURE	185	2.00
52.38	35.81	SPLEEN CONTUSION	21	2.05
52.23	35.02	ARM FRACTURE	358	2.39
52.00	38.60	LOWER LEG FRACTURE	250	2.15
51.61	33.13	weighted average of crash-involved backseat outboard passengers, CDS 1988-2010		
50.60	36.44	PELVIS FRACTURE	334	2.27
50.00	45.70	ABDOMEN ARTERY LACERATION	20	3.35
50.00	35.38	HEAD SUPERFICIAL INJURY	42	2.02
47.37	30.32	KIDNEY LACERATION	38	2.61
46.43	34.96	CERVICAL CORD CONTUSION	28	3.86
46.34	37.37	INTESTINAL CONTUSION	41	2.00
46.07	30.91	CONCUSSION	471	2.31
45.70	31.76	THORACOLUMBAR FRACTURE	291	2.12
45.45	37.64	THORACOLUMBAR CORD LACERATION	11	5.00
45.32	32.40	LIVER LACERATION	139	2.70
45.24	34.17	NECK DISLOCATED VERTEBRA	42	2.21
45.10	37.12	INTESTINAL LACERATION	51	2.49
43.59	32.77	SPLEEN LACERATION	117	2.61
42.86	41.08	CHEST LACERATED ARTERY	84	4.68
42.86	31.57	ARM DISLOCATED JOINT	14	2.07
42.50	37.85	HIP DISLOCATION	40	2.30
42.16	35.28	weighted average of backseat outboard passengers with MAIS ≥ 2, CDS 1988-2010		
42.11	35.84	ARM SUPERFICIAL INJURY	19	2.11
41.57	30.87	FACIAL FRACTURE	255	2.22
40.00	38.50	TORSO SUPERFICIAL INJURY	10	2.10
40.00	38.20	PELVIS CRUSHED	5	4.20
40.00	33.38	LUNG CONTUSION	210	3.38
39.67	33.39	BRAIN UNK INJURY	547	3.60
39.13	36.04	LUNG LACERATION	92	3.26
38.30	31.91	SHOULDER DISLOCATED	47	2.00
38.19	31.04	BRAIN CONTUSION	144	3.42
36.36	30.85	LEG SUPERFICIAL INJURY	33	2.21
34.38	38.09	HIP JOINT SEPARATION	32	3.00
34.15	28.88	KIDNEY CONTUSION	41	2.00
33.33	38.89	HEART LACERATION	57	4.05
31.56	29.26	SKULL FRACTURE	244	2.93
30.65	28.35	BRAIN LACERATION	62	4.82
30.00	30.00	DIAPHRAGM RUPTURE	10	3.80
28.57	26.21	BURNS	14	5.21
22.22	31.83	LEG DISLOCATED JOINT	18	2.33
22.22	29.11	HEAD CRUSHED	9	6.00
21.43	31.64	LACERATION OF GENITALS	14	2.57
12.50	28.50	CERVICAL CORD LACERATION	8	5.63

157

TABLE 7-5: AIS 2-6 INJURIES OF UNRESTRAINED CAR DRIVERS AND RF PASSENGERS
RANKED BY AVERAGE AGE OF THE VICTIM
(1988-2010 CDS, OCCUPANTS AGE 18-96)

AVG AGE	PERCENT FEMALE	INJURY	N	AVG AIS
45.30	10.00	INTESTINAL RUPTURE	10	4.00
45.69	37.25	THORACOLUMBAR DISLOCATION	51	2.20
45.07	31.21	HEART LACERATION	487	4.07
43.75	37.66	THORACOLUMBAR CORD LACERATION	77	4.39
43.75	36.48	CHEST FRACTURES	3410	2.98
42.94	31.78	CHEST LACERATED ARTERY	595	4.57
42.05	33.33	THORACOLUMBAR CORD CONTUSION	45	3.90
41.93	24.64	CERVICAL CORD CONTUSION	133	4.09
41.13	33.04	HEART CONTUSION	224	3.28
40.68	25.64	ABDOMEN ARTERY LACERATION	117	3.49
40.18	30.03	TORSO ARTERY TRANSECTION/RUPTURE	113	5.49
39.73	34.52	NECK DISLOCATED VERTEBRA	252	2.23
39.41	31.15	INTESTINAL CONTUSION	392	2.02
39.37	36.67	LEG AMPUTATION	30	3.23
38.85	34.81	LUNG LACERATION	585	3.27
38.24	27.58	CHEST CRUSHED	37	5.67
38.08	30.30	NECK RUPTURED VERTEBRA	33	2.18
38.89	46.66	ARM FRACTURE	2739	2.36
38.25	47.33	LOWER LEG FRACTURE	4126	2.19
38.33	41.94	CONTUSION OF GENITALS	31	2.00
38.21	37.50	NECK SUPERFICIAL INJURY	24	2.00
38.14	42.50	HEART RUPTURE	40	5.73
37.89	39.17	ARM SUPERFICIAL INJURY	217	2.05
37.69	43.61	weighted averages of unrestrained drivers and RF passengers with MAIS ≥ 2, CDS 1988-2010		
37.61	31.30	NECK FRACTURE	1227	2.32
37.36	38.27	CERVICAL CORD LACERATION	81	5.33
37.28	47.21	INTESTINAL LACERATION	284	2.70
37.18	37.00	ARM DISLOCATED JOINT	100	2.37
37.11	46.83	PELVIS FRACTURE	1890	2.32
37.07	38.22	SHOULDER FRACTURE	1059	2.00
36.70	25.00	BRAIN STEM TRANSECTION	20	6.00
36.51	37.07	FEMUR FRACTURE	1570	3.00
36.40	38.27	LIVER LACERATION	1100	2.88
36.26	54.22	HIP DISLOCATION	328	2.38
36.35	28.10	DIAPHRAGM LACERATION	184	3.00
36.35	37.27	SHOULDER DISLOCATED	220	2.00
36.12	30.84	BRAIN UNK INJURY	3489	3.65
36.11	48.04	LEG DISLOCATED JOINT	331	2.22
36.08	40.11	LIVER CONTUSION	177	2.12
35.89	33.83	SPLEEN CONTUSION	120	2.11
35.80	40.48	TORSO SUPERFICIAL INJURY	188	2.06
35.67	51.61	NECK ARTERY LACERATION	31	3.10
35.65	51.97	LUNG CONTUSION	1867	3.33
35.60	33.33	EYE AVULSION/DETACHMENT	42	2.24
35.43	34.65	HEAD SUPERFICIAL INJURY	254	2.04
35.04	37.70	FACIAL LACERATION	810	2.02
34.50	35.85	SPLEEN LACERATION	782	2.00
34.84	28.63	BRAIN CONTUSION	949	3.99
34.52	33.33	EAR EXTERNAL INJURY	33	2.00
34.41	48.39	LEG SUPERFICIAL INJURY	465	2.08
34.37	37.44	CONCUSSION	4124	2.01
34.24	32.18	FACIAL FRACTURE	2047	2.19
34.11	31.58	LIVER RUPTURE	13	4.38
33.99	34.50	THORACOLUMBAR FRACTURE	1527	2.18
33.92	28.52	BRAIN LACERATION	437	4.85
33.82	14.29	ARM JOINT LACERATION	28	2.00
33.51	58.42	DIAPHRAGM RUPTURE	59	5.60
33.46	42.54	weighted average of unrestrained crash-involved drivers and RF passengers, CDS 1988-2010		
33.01	25.85	KIDNEY LACERATION	305	2.57
32.85	27.27	ARM AMPUTATION	33	2.42
32.75	38.30	SPLEEN RUPTURE	71	2.82
32.72	45.28	PELVIS CRUSHED	53	4.84
32.67	43.85	LEG JOINT LACERATION	130	2.22
32.80	27.84	SKULL FRACTURE	1893	2.89
32.22	35.28	HIP JOINT SEPARATION	133	2.28
31.36	54.51	LACERATION OF GENITALS	100	2.50
31.65	38.13	KIDNEY CONTUSION	268	2.06
31.55	38.30	FACIAL DISLOCATION	47	2.02
31.05	30.77	BURNS	39	4.49
31.03	27.03	HEAD ARTERY LACERATION	37	4.05
30.64	34.18	HEAD CRUSHED	79	6.00
30.67	46.62	KNEE OR ANKLE SPRAIN	148	2.00
28.75	50.00	RUPTURE OF GENITALS	38	3.47

158

TABLE 7-6: AIS 2-6 INJURIES OF BELTED CAR DRIVERS AND RF PASSENGERS
RANKED BY AVERAGE AGE OF THE VICTIM
(1988-2010 CDS, OCCUPANTS AGE 18-96)

AVG AGE	PERCENT FEMALE	INJURY	N	AVG AIS
53.57	52.70	CHEST FRACTURES	3783	2.81
52.31	52.20	HEART CONTUSION	169	3.22
52.10	26.19	THORACOLUMBAR CORD LACERATION	42	5.00
51.88	40.24	HEART LACERATION	333	3.83
51.35	41.49	CERVICAL CORD CONTUSION	84	3.96
50.64	37.50	HEART RUPTURE	24	5.75
50.37	45.35	CHEST LACERATED ARTERY	419	4.50
47.43	40.00	TORSO ARTERY TRANSECTION/RUPTURE	35	5.34
47.90	56.41	ABDOMEN ARTERY LACERATION	74	3.43
46.93	41.46	DIAPHRAGM LACERATION	128	3.01
46.55	51.82	NECK FRACTURE	957	2.32
46.08	24.55	TORSO SUPERFICIAL INJURY	154	2.02
44.95	61.01	ARM FRACTURE	3452	2.39
46.32	82.93	SHOULDER FRACTURE	1380	2.00
44.86	50.51	LUNG LACERATION	432	3.20
44.14	47.68	SPLEEN RUPTURE	79	2.66
44.12	40.00	DIAPHRAGM RUPTURE	65	3.82
43.92	50.85	THORACOLUMBAR FRACTURE	1564	2.03
43.88	53.28	PELVIS FRACTURE	2572	2.30
43.53	49.84	INTESTINAL CONTUSION	301	2.02
43.58	61.93	LOWER LEG FRACTURE	4710	2.17
43.32	50.00	LIVER CONTUSION	164	2.09
43.27	58.57	CONTUSION OF GENITALS	56	2.00
43.20	57.68	NECK DISLOCATED VERTEBRA	190	2.18
43.01	46.03	ARM SUPERFICIAL INJURY	189	2.10
42.98	46.43	THORACOLUMBAR CORD CONTUSION	28	3.96
42.88	59.31	weighted average of belted drivers and RF passengers with MAIS ≥ 2, CDS 1988-2010		
42.83	65.13	LEG DISLOCATED JOINT	390	2.12
42.71	44.87	LUNG CONTUSION	1395	3.30
42.35	48.00	KIDNEY CONTUSION	225	2.02
42.40	49.42	HIP JOINT SEPARATION	172	3.00
42.31	44.44	CERVICAL CORD LACERATION	81	5.67
42.28	44.49	SHOULDER DISLOCATED	236	2.00
42.11	47.37	PELVIS CRUSHED	19	4.26
42.07	39.29	BRAIN STEM TRANSECTION	28	6.00
42.04	48.22	INTESTINAL LACERATION	338	2.84
41.86	60.00	ARM DISLOCATED JOINT	125	2.26
41.85	45.99	BRAIN UNK INJURY	2905	3.68
41.83	56.52	LEG AMPUTATION	23	3.17
41.82	51.35	SPLEEN CONTUSION	148	2.09
41.37	53.53	FACIAL LACERATION	299	2.02
41.12	53.93	LACERATION OF GENITALS	89	2.71
41.03	47.50	ARM AMPUTATION	40	2.43
40.77	51.87	LIVER LACERATION	750	2.51
40.73	48.61	SPLEEN LACERATION	763	2.67
40.64	46.15	NECK RUPTURED VERTEBRA	39	2.13
40.28	53.80	HEAD SUPERFICIAL INJURY	171	2.05
40.08	45.87	BRAIN CONTUSION	762	3.45
39.60	47.73	FEMUR FRACTURE	1257	3.00
39.56	59.26	ARM CRUSHED	27	2.48
39.51	33.33	HEAD CRUSHED	39	6.00
38.56	52.00	THORACOLUMBAR DISLOCATION	25	2.24
38.82	50.00	FACIAL DISLOCATION	28	2.00
38.67	48.28	HEAD ARTERY LACERATION	29	3.97
38.37	51.49	CONCUSSION	3294	2.23
38.36	42.05	FACIAL FRACTURE	1258	2.18
38.00	73.86	RUPTURE OF GENITALS	36	3.21
37.88	51.14	HIP DISLOCATION	176	2.38
37.84	59.95	LEG SUPERFICIAL INJURY	412	2.07
37.80	52.68	weighted average of belted crash-involved drivers and RF passengers, CDS 1988-2010		
37.67	33.59	SKULL FRACTURE	1123	2.91
36.67	37.92	BRAIN LACERATION	269	4.88
36.58	59.50	KNEE OR ANKLE SPRAIN	363	2.01
36.49	45.34	KIDNEY LACERATION	167	2.56
33.82	47.06	LEG JOINT LACERATION	119	2.14
32.78	22.50	BURNS	40	4.33

- Some relatively rare injuries in the unbelted table are entirely absent in the belted table – chest crushed, eye avulsion/detachment, ear external injury – presumably because belts are quite effective in preventing these injuries for occupants of all ages.

The similarity of the relative rankings for unrestrained and belted occupants implies that the rankings are largely intrinsic to the aging process and only secondarily affected by safety technologies. Older people are highly susceptible to multiple rib fractures and associated thoracic injuries, regardless of whether they are unrestrained or belted.

Even though the relative rankings are quite similar, the absolute ages of the belted casualties are approximately 5 to 10 years older than the unrestrained. For example, the average age of chest fractures is 43.75 for unrestrained and 53.07 for belted occupants. Two factors contribute to the difference, as may be discerned from the benchmark entries on the tables. Older people are historically more likely to buckle up, as evidenced by a 33.46 average age for unrestrained (and not necessarily injured) drivers and RF passengers on CDS, versus 37.80 for belted. In addition, because belts are a bit more effective for younger occupants, the average unrestrained occupant with MAIS $\geq$ 2 is 37.69 years old, but the average belted MAIS $\geq$ 2, 42.88.

As in Section 6.4, it might be worthwhile to look only at belted front-seat occupants in relatively late-model vehicles equipped with frontal air bags and compare the injury distribution to Table 7-6 (belted occupants of vehicles of all model years). Table 7-6a limits Table 7-6 to MY 1999 and later vehicles, all equipped with dual frontal air bags. It was necessary to go back as far as MY 1999 to garner adequate numbers of injury cases. Table 7-6a includes injuries that occurred at least 10 times if the average AIS < 3.5 or at least 5 times if the average AIS $\geq$ 3.5, but it still generates a shorter list of injuries than Table 7-6 (which required 10 cases and 5 cases, respectively).

The most visible difference between Tables 7-6 and 7-6a is the concentration of neck injuries near the top of Table 7-6a; specifically "cervical cord laceration" and "neck ruptured vertebra" have moved to considerably higher positions. This may reflect the relative success of frontal air bags, pretensioners, and load limiters in mitigating torso and head injuries of older occupants; it may also reflect a higher proportion of female drivers (vulnerable to neck injury) in late-model passenger cars. Conversely, some torso injuries such as heart rupture and thoracolumbar-cord laceration have moved well down the list, while others such as heart contusion and spleen rupture were omitted from Table 7-6a because they occurred so infrequently in the late-model cars.

Tables 7-7 and 7-8 rank the injuries by average age for unrestrained and belted back-seat occupants. Tables 7-7 and 7-8 (and, for that matter, Tables 7-5, 7-6, and 7-9 to 7-12) are limited to people at seats equipped with a 3-point belt. They exclude seats equipped with just a lap belt. Without that exclusion, CDS data stretching back to CY 1988 would have included a fairly large share of lap-belted back-seat occupants and even some lap-belted drivers and RF passengers. Because of limited data on back-seat occupants, Table 7-8 only provides a short list of injuries even though it includes injuries that occurred as few as 10 times if the average AIS < 3.5 or as few as 5 times if the average AIS $\geq$ 3.5.

TABLE 7-6a: MY 1999-2011 CARS WITH DUAL AIR BAGS
AIS 2-6 INJURIES OF BELTED CAR DRIVERS AND RF PASSENGERS RANKED BY AVERAGE AGE OF THE VICTIM
(1998-2010 CDS, OCCUPANTS AGE 18-96)

AVG AGE	PERCENT FEMALE	INJURY	N	AVG AIS
54.76	36.00	CERVICAL CORD CONTUSION	25	4.12
52.83	51.72	CHEST FRACTURES	1015	2.64
52.72	34.18	HEART LACERATION	79	3.85
51.09	50.00	CERVICAL CORD LACERATION	22	5.73
49.38	38.46	CONTUSION OF GENITALS	13	2.00
49.14	32.46	CHEST LACERATED ARTERY	114	4.41
48.15	55.87	NECK FRACTURE	349	2.36
45.82	51.49	SHOULDER FRACTURE	369	2.00
45.41	51.26	THORACOLUMBAR FRACTURE	873	2.08
45.10	60.00	ABDOMEN ARTERY LACERATION	20	3.45
45.00	64.44	LEG DISLOCATED JOINT	135	2.00
44.87	34.78	DIAPHRAGM LACERATION	23	3.00
44.10	60.35	ARM FRACTURE	966	2.39
44.09	54.55	NECK RUPTURED VERTEBRA	11	2.00
44.08	49.02	INTESTINAL CONTUSION	102	2.00
43.73	**62.98**	**weighted average of belted drivers and RF passengers on CDS with MAIS 2+, MY 1999-2011**		
43.18	57.89	ARM SUPERFICIAL INJURY	38	2.16
43.14	69.05	NECK DISLOCATED VERTEBRA	42	2.07
43.10	52.50	LIVER CONTUSION	40	2.10
43.08	39.76	SHOULDER DISLOCATED	83	2.00
42.92	66.67	RUPTURE OF GENITALS	12	3.08
42.60	34.29	DIAPHRAGM RUPTURE	35	4.00
42.59	37.27	LUNG LACERATION	110	3.33
42.58	59.06	PELVIS FRACTURE	762	2.34
42.57	50.00	KIDNEY CONTUSION	42	2.05
42.55	63.64	PELVIS CRUSHED	11	4.27
42.47	63.76	LOWER LEG FRACTURE	1363	2.19
42.46	46.15	ARM AMPUTATION	13	2.69
42.33	11.11	THORACOLUMBAR CORD CONTUSION	9	3.89
42.20	37.14	SPLEEN CONTUSION	35	2.09
41.58	46.14	BRAIN UNK INJURY	1010	3.64
41.55	47.62	INTESTINAL LACERATION	105	2.79
41.48	59.02	HIP JOINT SEPARATION	61	3.00
41.20	40.00	HEART RUPTURE	5	6.00
41.00	49.78	SPLEEN LACERATION	231	2.48
40.76	35.29	THORACOLUMBAR CORD LACERATION	17	5.00
40.71	50.00	HEAD ARTERY LACERATION	14	3.71
40.48	51.72	LACERATION OF GENITALS	29	2.62
40.38	42.73	LUNG CONTUSION	433	3.37
39.75	33.33	HEAD CRUSHED	12	6.00
39.67	50.22	LIVER LACERATION	225	2.56
39.65	34.78	FACIAL LACERATION	23	2.00
39.44	48.76	FEMUR FRACTURE	363	3.00
39.33	41.67	FACIAL DISLOCATION	12	2.00
39.12	34.12	BRAIN LACERATION	85	4.96
39.05	47.06	BRAIN CONTUSION	255	3.47
38.40	40.00	BRAIN STEM TRANSECTION	15	6.00
38.27	**55.04**	**weighted average of belted crash-involved drivers and RF passengers on CDS, MY 1999-2011**		
38.26	55.56	ARM DISLOCATED JOINT	27	2.00
37.68	49.59	CONCUSSION	740	2.16
37.22	42.64	FACIAL FRACTURE	265	2.21
36.86	64.93	KNEE OR ANKLE SPRAIN	134	2.00
36.60	46.75	KIDNEY LACERATION	77	2.52
35.73	38.64	SKULL FRACTURE	352	2.96
35.51	66.67	LEG SUPERFICIAL INJURY	87	2.06
34.52	48.48	LEG JOINT LACERATION	33	2.00
34.15	23.08	BURNS	13	4.62
32.69	51.43	HIP DISLOCATION	35	2.00
31.96	48.15	HEAD SUPERFICIAL INJURY	26	2.04

TABLE 7-7: AIS 2-6 INJURIES OF UNRESTRAINED REAR-OUTBOARD PASSENGERS IN CARS
RANKED BY AVERAGE AGE OF THE VICTIM
(1988-2010 CDS, OCCUPANTS AGE 18-96)

AVG AGE	PERCENT FEMALE	INJURY	N	AVG AIS
43.08	83.33	ABDOMEN ARTERY LACERATION	12	3.25
42.04	42.31	HEART LACERATION	26	4.12
40.83	75.00	INTESTINAL LACERATION	12	2.17
40.13	46.33	CHEST FRACTURES	177	3.08
38.60	44.44	LUNG LACERATION	45	3.18
38.32	48.25	LOWER LEG FRACTURE	114	2.15
38.23	50.78	FEMUR FRACTURE	128	3.00
37.45	43.18	CHEST LACERATED ARTERY	44	4.84
37.21	52.63	CERVICAL CORD CONTUSION	19	4.11
37.17	**46.85**	**wtd avg of unrestrained backseat outboard psgrs with MAIS ≥ 2, CDS 1988-2010**		
36.29	50.00	HIP JOINT SEPARATION	14	3.00
35.69	52.94	NECK FRACTURE	102	2.38
35.38	31.25	HIP DISLOCATION	16	2.25
34.51	54.35	SHOULDER FRACTURE	92	2.00
34.40	73.33	KNEE OR ANKLE SPRAIN	15	2.00
34.20	40.00	INTESTINAL CONTUSION	10	2.00
34.14	64.29	HEAD SUPERFICIAL INJURY	14	2.07
33.77	37.18	BRAIN UNK INJURY	277	3.60
33.71	50.29	ARM FRACTURE	173	2.36
33.50	47.24	PELVIS FRACTURE	163	2.28
33.37	42.86	SPLEEN LACERATION	63	2.52
32.22	27.78	KIDNEY CONTUSION	18	2.00
32.18	45.45	NECK DISLOCATED VERTEBRA	22	2.14
32.17	33.33	HEAD CRUSHED	6	6.00
31.92	32.39	BRAIN CONTUSION	71	3.45
31.80	50.00	SPLEEN CONTUSION	10	2.00
31.73	**49.71**	**wtd avg of crash-involved unrestrained backseat outboard psgrs, CDS 1988-2010**		
30.80	20.00	CERVICAL CORD LACERATION	5	5.80
30.78	38.38	LUNG CONTUSION	99	3.42
30.77	50.62	THORACOLUMBAR FRACTURE	162	2.09
30.48	42.25	LIVER LACERATION	71	2.68
30.17	45.00	CONCUSSION	180	2.37
29.36	27.27	SHOULDER DISLOCATED	22	2.00
28.88	45.83	KIDNEY LACERATION	24	2.46
28.70	41.40	FACIAL FRACTURE	199	2.30
28.42	27.27	BRAIN LACERATION	33	4.97
28.13	25.00	SKULL FRACTURE	128	2.90
27.93	20.00	LEG SUPERFICIAL INJURY	15	2.13
23.29	57.14	THORACOLUMBAR CORD CONTUSION	7	3.71
23.08	38.46	LIVER CONTUSION	13	2.15

162

TABLE 7-8: AIS 2-6 INJURIES OF BELTED REAR-OUTBOARD PASSENGERS IN CARS
RANKED BY AVERAGE AGE OF THE VICTIM
(1988-2010 CDS, OCCUPANTS AGE 18-96)

AVG AGE	PERCENT FEMALE	INJURY	N	AVG AIS
69.63	50.00	CHEST LACERATED ARTERY	8	4.75
60.63	83.33	NECK FRACTURE	24	2.29
54.35	67.37	CHEST FRACTURES	95	2.68
53.67	44.44	HEART LACERATION	9	3.89
50.46	46.15	INTESTINAL CONTUSION	13	2.00
45.93	64.29	INTESTINAL LACERATION	14	2.86
44.24	54.05	PELVIS FRACTURE	37	2.38
44.11	66.67	SHOULDER FRACTURE	27	2.00
40.85	48.15	LUNG CONTUSION	27	3.48
40.68	49.65	wtd avg of belted backseat outboard passengers with MAIS ≥ 2, CDS 1988-2010		
39.73	54.55	SPLEEN LACERATION	11	2.55
39.22	59.26	LOWER LEG FRACTURE	27	2.11
35.97	56.37	wtd avg of crash-involved belted backseat outboard passengers, CDS 1988-2010		
35.64	50.00	FEMUR FRACTURE	22	3.00
35.61	36.84	ARM FRACTURE	38	2.45
33.80	52.00	FACIAL FRACTURE	25	2.28
33.74	51.46	BRAIN UNK INJURY	103	3.79
33.61	39.39	SKULL FRACTURE	33	2.94
32.87	60.00	LIVER LACERATION	15	2.00
32.74	40.00	THORACOLUMBAR FRACTURE	50	2.14
31.32	53.23	CONCUSSION	62	2.23
31.00	33.33	BRAIN LACERATION	6	4.67
27.45	45.00	BRAIN CONTUSION	20	3.60

Most of the injuries listed in Table 7-8 (belted) have similar relative rankings as in Table 7-7 (unrestrained), with torso injuries more prevalent among older occupants, head injuries less. Neck fractures, however, are closer to the top of the belted list, possibly indicating that older occupants are especially vulnerable to strain on the neck when the head moves but the torso is restrained. Lower-leg and femur fractures are closer to the top of the unrestrained list, which may indicate that belts in the back seat protect even older occupants from consequences of leg contacts with the vehicle interior; similarly, lung lacerations are moderately frequent and moderately severe for unbelted occupants but are not frequent enough to make the belted list, indicating mitigation by belts.

7.5   Injuries ranked by percent female victims – unrestrained versus belted

Table 7-9 lists exactly the same injuries as Table 7-5 (unrestrained car drivers and RF passengers), but ranked by percent female rather than average age. Table 7-10 presents the corresponding rankings for belted front-seat occupants. The great majority of the injuries have similar ranks in the unrestrained and belted tables. There are a few possible exceptions:

- Neck fractures have a moderately high percentage of female victims in the belted table, but rather low for the unrestrained. Neck-dislocated-vertebra is quite high in the belted table, but in the middle of the Table 7-9. These findings parallel the results on fractures of the first and second cervical vertebrae in MCOD (Tables 6-9 and 6-10) as well as findings in Table 4-1d and 4-4d that belted females are highly susceptible to neck injury, especially in frontal crashes in vehicles without air bags.
- Abdominal artery laceration and intestinal laceration are located higher on the belted than the unrestrained table, perhaps indicating that belt forces on the abdomen may be above some females' tolerance level.
- Conversely, crushing of the pelvis, head, and chest are at lower levels or not even present on the belted list, indicating that unbelted females are especially vulnerable to these severe blunt-impact injuries, but belts offer protection.

Cord lacerations at two levels – cervical and thoracolumbar – may be relatively more prevalent for unrestrained than belted females. These may also be injuries involving severe blunt impact.

Table 7-10a is limited to belted front-seat occupants of MY 1999 and later cars, all equipped with dual frontal air bags. Most of the injuries occupy similar relative positions on Table 7-10 (all model years, including no vehicles without air bags) and Table 7-10a. As in the comparison of Tables 7-6 with 7-6a, some neck injuries, specifically dislocated vertebrae, ruptured vertebrae and cord lacerations rank higher on Table 7-10a than on Table 7-10, perhaps indicating the relative success of air bags and other recent technologies in mitigating females' head and torso injuries. Conversely, "head superficial injury" and facial lacerations became less characteristic of females. Trends are unclear for injuries to the lower torso: "pelvis crushed" and "hip joint separation" moved to higher positions in Table 7-10a, but "contusion of genitals" and spleen contusion are lower on the list than they were in Table 7-10.

PERCENT FEMALE	AVG AGE	INJURY	N	AVG AIS
51.61	35.87	NECK ARTERY LACERATION	31	3.10
50.00	29.76	RUPTURE OF GENITALS	36	3.47
48.39	34.41	LEG SUPERFICIAL INJURY	465	2.08
48.28	32.72	PELVIS CRUSHED	29	4.24
48.04	36.11	LEG DISLOCATED JOINT	331	2.22
47.33	38.25	LOWER LEG FRACTURE	4126	2.19
46.66	38.89	ARM FRACTURE	2739	2.36
46.63	37.11	PELVIS FRACTURE	1920	2.32
46.62	30.67	KNEE OR ANKLE SPRAIN	148	2.00
43.85	32.67	LEG JOINT LACERATION	130	2.22
43.81	37.69	weighted average of unrestrained drivers and RF passengers with MAIS ≥ 2, CDS 1988-2010		
42.54	33.48	weighted average of unrestrained crash-involved drivers and RF passengers, CDS 1988-2010		
41.94	38.23	CONTUSION OF GENITALS	31	2.00
40.49	35.08	TORSO SUPERFICIAL INJURY	128	2.06
40.11	36.06	LIVER CONTUSION	177	2.12
39.55	37.59	SPLEEN CONTUSION	129	2.11
39.27	35.40	LIVER LACERATION	1100	2.58
39.17	37.89	ARM SUPERFICIAL INJURY	217	2.05
39.09	40.16	TORSO ARTERY TRANSECTION/RUPTURE	110	5.40
38.30	31.55	FACIAL DISLOCATION	47	2.02
38.27	37.36	CERVICAL CORD LACERATION	81	5.83
38.22	37.07	SHOULDER FRACTURE	1038	2.00
37.70	35.04	FACIAL LACERATION	810	2.02
37.66	43.78	THORACOLUMBAR CORD LACERATION	77	4.99
37.50	38.21	NECK SUPERFICIAL INJURY	24	2.00
37.44	34.37	CONCUSSION	4124	2.34
37.27	38.35	SHOULDER DISLOCATED	220	2.00
37.25	45.88	THORACOLUMBAR DISLOCATION	51	2.20
37.07	36.51	FEMUR FRACTURE	1570	3.00
37.00	37.18	ARM DISLOCATED JOINT	100	2.37
36.67	39.37	LEG AMPUTATION	30	3.23
36.40	43.75	CHEST FRACTURES	3410	2.98
36.13	31.65	KIDNEY CONTUSION	206	2.05
35.80	34.96	SPLEEN LACERATION	782	2.69
35.23	32.22	HIP JOINT SEPARATION	109	2.56
34.91	31.98	LACERATION OF GENITALS	106	2.69
34.81	33.35	LUNG LACERATION	580	3.27
34.65	35.43	HEAD SUPERFICIAL INJURY	254	2.04
34.58	39.03	THORACOLUMBAR FRACTURE	1527	2.10
34.52	39.79	NECK DISLOCATED VERTEBRA	252	2.23
34.32	36.33	HIP DISLOCATION	383	2.33
34.18	30.84	HEAD CRUSHED	79	6.00
33.80	32.75	SPLEEN RUPTURE	71	2.82
33.53	42.06	THORACOLUMBAR CORD CONTUSION	49	3.36
33.33	35.60	EYE AVULSION/DETACHMENT	42	2.24
33.33	34.52	EAR EXTERNAL INJURY	39	2.00
33.04	41.16	HEART CONTUSION	224	3.29
32.56	38.14	HEART RUPTURE	43	5.74
32.19	34.24	FACIAL FRACTURE	2047	2.19
31.97	35.65	LUNG CONTUSION	1887	3.33
31.70	42.24	CHEST LACERATED ARTERY	595	4.57
31.58	34.11	LIVER RUPTURE	19	4.68
31.30	37.51	NECK FRACTURE	1227	2.32
31.21	45.07	HEART LACERATION	487	4.07
31.13	38.41	INTESTINAL CONTUSION	302	2.02
30.84	36.12	BRAIN UNK INJURY	3488	3.65
30.77	31.05	BURNS	39	4.49
30.36	39.06	NECK RUPTURED VERTEBRA	33	2.18
28.52	33.32	BRAIN LACERATION	437	4.85
28.19	36.36	DIAPHRAGM LACERATION	134	3.00
28.63	34.64	BRAIN CONTUSION	943	3.38
27.64	32.80	SKULL FRACTURE	1693	2.69
27.27	32.85	ARM AMPUTATION	33	2.42
27.21	37.28	INTESTINAL LACERATION	294	2.70
27.03	39.24	CHEST CRUSHED	37	5.82
27.03	31.03	HEAD ARTERY LACERATION	37	4.05
26.42	32.59	DIAPHRAGM RUPTURE	53	3.60
25.65	33.01	KIDNEY LACERATION	235	2.53
25.64	40.89	ABDOMEN ARTERY LACERATION	117	3.43
25.00	36.70	BRAIN STEM TRANSECTION	20	6.00
24.64	41.96	CERVICAL CORD CONTUSION	138	4.09
14.29	33.82	ARM JOINT LACERATION	28	2.00
10.00	45.80	INTESTINAL RUPTURE	10	4.00

TABLE 7-10: AIS 2-6 INJURIES OF BELTED CAR DRIVERS AND RF PASSENGERS
RANKED BY PERCENTAGE OF VICTIMS THAT ARE FEMALES
(1988-2010 CDS, OCCUPANTS AGE 18-96)

PERCENT FEMALE	AVG AGE	INJURY	N	AVG AIS
73.68	38.00	RUPTURE OF GENITALS	38	3.21
65.13	42.83	LEG DISLOCATED JOINT	390	2.12
61.93	43.58	LOWER LEG FRACTURE	4710	2.17
61.01	44.95	ARM FRACTURE	3452	2.39
60.93	43.58	PELVIS FRACTURE	2572	2.30
60.00	41.86	ARM DISLOCATED JOINT	125	2.26
59.95	37.84	LEG SUPERFICIAL INJURY	412	2.07
59.50	36.58	KNEE OR ANKLE SPRAIN	363	2.01
59.31	42.88	weighted average of belted drivers and RF passengers with MAIS ≥ 2, CDS 1988-2010		
59.26	39.56	ARM CRUSHED	27	2.48
57.89	43.20	NECK DISLOCATED VERTEBRA	190	2.18
56.52	41.83	LEG AMPUTATION	23	3.17
56.41	47.30	ABDOMEN ARTERY LACERATION	74	3.48
54.55	48.06	TORSO SUPERFICIAL INJURY	154	2.02
53.38	41.12	LACERATION OF GENITALS	68	2.71
53.80	40.28	HEAD SUPERFICIAL INJURY	171	2.05
53.57	43.37	CONTUSION OF GENITALS	55	2.00
53.53	41.37	FACIAL LACERATION	289	2.02
52.83	44.92	SHOULDER FRACTURE	1380	2.00
52.70	53.07	CHEST FRACTURES	3763	2.81
52.66	37.80	weighted average of belted crash-involved drivers and RF passengers, CDS 1988-2010		
52.20	32.31	HEART CONTUSION	150	3.22
52.00	53.38	THORACOLUMBAR DISLOCATION	25	2.24
51.87	46.77	LIVER LACERATION	750	2.61
51.82	46.95	NECK FRACTURE	957	2.32
51.48	38.37	CONCUSSION	3294	2.23
51.35	41.52	SPLEEN CONTUSION	148	2.09
51.14	37.88	HIP DISLOCATION	176	2.28
50.33	43.82	THORACOLUMBAR FRACTURE	1524	2.09
50.60	43.32	LIVER CONTUSION	164	2.03
50.00	39.32	FACIAL DISLOCATION	28	2.00
49.42	42.48	HIP JOINT SEPARATION	172	3.00
48.82	42.04	INTESTINAL LACERATION	336	2.84
48.61	40.73	SPLEEN LACERATION	769	2.07
48.28	38.97	HEAD ARTERY LACERATION	29	3.97
48.09	42.55	KIDNEY CONTUSION	230	2.02
47.95	46.14	SPLEEN RUPTURE	79	2.56
47.73	39.60	FEMUR FRACTURE	1257	3.00
47.50	41.03	ARM AMPUTATION	40	2.43
47.57	42.11	PELVIS CRUSHED	15	4.26
47.06	33.82	LEG JOINT LACERATION	119	2.14
46.54	43.58	INTESTINAL CONTUSION	301	2.02
46.43	42.96	THORACOLUMBAR CORD CONTUSION	28	3.86
46.15	40.64	NECK RUPTURED VERTEBRA	39	2.13
46.03	43.01	ARM SUPERFICIAL INJURY	189	2.10
45.99	41.85	BRAIN UNK INJURY	2905	3.86
45.89	38.46	KIDNEY LACERATION	197	2.50
45.87	40.08	BRAIN CONTUSION	782	3.45
45.95	50.27	CHEST LACERATED ARTERY	419	4.50
44.07	42.79	LUNG CONTUSION	1905	3.30
44.12	44.11	SHOULDER DISLOCATED		2.00
44.44	42.31	CERVICAL CORD LACERATION	81	5.67
42.05	35.38	FACIAL FRACTURE	1258	2.18
41.49	51.36	CERVICAL CORD CONTUSION	94	3.96
41.46	46.66	DIAPHRAGM LACERATION	129	3.01
40.98	51.68	HEART LACERATION	332	3.93
40.51	44.68	LUNG LACERATION	432	3.50
40.00	47.40	TORSO ARTERY TRANSECTION/RUPTURE	35	5.34
40.00	46.12	DIAPHRAGM RUPTURE	85	3.82
39.53	37.87	SKULL FRACTURE	1126	2.91
39.29	42.07	BRAIN STEM TRANSECTION	28	6.00
38.10	52.10	THORACOLUMBAR CORD LACERATION	42	5.00
37.92	36.87	BRAIN LACERATION	263	4.98
37.50	60.54	HEART RUPTURE	24	5.75
33.33	38.51	HEAD CRUSHED	38	6.00
22.50	32.78	BURNS	40	4.33

PERCENT FEMALE	AVG AGE	INJURY	N	AVG AIS
69.05	43.14	NECK DISLOCATED VERTEBRA	42	2.07
66.67	42.92	RUPTURE OF GENITALS	12	3.08
66.67	35.51	LEG SUPERFICIAL INJURY	87	2.06
64.93	36.86	KNEE OR ANKLE SPRAIN	134	2.00
64.44	45.00	LEG DISLOCATED JOINT	135	2.00
63.76	42.47	LOWER LEG FRACTURE	1363	2.19
63.64	42.55	PELVIS CRUSHED	11	4.27
62.98	43.73	weighted average of belted drivers and RF passengers on CDS with MAIS 2+, MY 1999-2011		
60.35	44.10	ARM FRACTURE	966	2.39
60.00	45.10	ABDOMEN ARTERY LACERATION	20	3.45
59.06	42.58	PELVIS FRACTURE	762	2.34
59.02	41.48	HIP JOINT SEPARATION	61	3.00
57.89	43.18	ARM SUPERFICIAL INJURY	38	2.16
55.87	46.15	NECK FRACTURE	349	2.36
55.56	38.26	ARM DISLOCATED JOINT	27	2.00
55.04	38.27	weighted average of belted crash-involved drivers and RF passengers on CDS, MY 1999-2011		
54.55	44.09	NECK RUPTURED VERTEBRA	11	2.00
52.50	43.10	LIVER CONTUSION	40	2.10
51.72	52.83	CHEST FRACTURES	1015	2.84
51.72	40.48	LACERATION OF GENITALS	29	2.62
51.49	45.82	SHOULDER FRACTURE	369	2.00
51.43	32.69	HIP DISLOCATION	35	2.00
51.26	45.41	THORACOLUMBAR FRACTURE	673	2.08
50.22	39.67	LIVER LACERATION	225	2.56
50.00	51.09	CERVICAL CORD LACERATION	22	5.73
50.00	42.57	KIDNEY CONTUSION	42	2.05
50.00	40.71	HEAD ARTERY LACERATION	14	3.71
49.78	41.00	SPLEEN LACERATION	231	2.48
49.59	37.68	CONCUSSION	740	2.16
49.02	44.08	INTESTINAL CONTUSION	102	2.00
48.76	39.44	FEMUR FRACTURE	363	3.00
48.48	34.52	LEG JOINT LACERATION	33	2.00
47.62	41.55	INTESTINAL LACERATION	105	2.79
47.06	39.05	BRAIN CONTUSION	255	3.47
46.75	36.60	KIDNEY LACERATION	77	2.52
46.15	42.46	ARM AMPUTATION	13	2.69
46.15	31.96	HEAD SUPERFICIAL INJURY	26	2.04
46.14	41.56	BRAIN UNK INJURY	1010	3.64
42.73	40.38	LUNG CONTUSION	433	3.37
42.64	37.22	FACIAL FRACTURE	265	2.21
41.67	39.33	FACIAL DISLOCATION	12	2.00
40.00	41.20	HEART RUPTURE	5	6.00
40.00	38.40	BRAIN STEM TRANSECTION	15	6.00
39.76	43.08	SHOULDER DISLOCATED	83	2.00
38.64	35.73	SKULL FRACTURE	352	2.96
38.46	49.38	CONTUSION OF GENITALS	13	2.00
37.27	42.58	LUNG LACERATION	110	3.33
37.14	42.20	SPLEEN CONTUSION	35	2.09
36.00	54.76	CERVICAL CORD CONTUSION	25	4.12
35.29	40.76	THORACOLUMBAR CORD LACERATION	17	5.00
34.78	44.57	DIAPHRAGM LACERATION	23	3.00
34.78	39.65	FACIAL LACERATION	23	2.00
34.29	42.60	DIAPHRAGM RUPTURE	35	4.00
34.18	52.72	HEART LACERATION	79	3.85
34.12	39.12	BRAIN LACERATION	85	4.96
33.33	39.75	HEAD CRUSHED	12	6.00
32.46	49.14	CHEST LACERATED ARTERY	114	4.41
23.08	34.15	BURNS	13	4.62
11.11	42.33	THORACOLUMBAR CORD CONTUSION	9	3.89

Tables 7-11 and 7-12 rank the injuries for unrestrained and belted <u>back-seat</u> occupants by the percentage of victims that are females. Most of the injuries frequent enough to make the list in Table 7-12 occupy similar relative positions in the unrestrained rankings, but there are some discrepancies:

- Neck fractures are no. 1 on the belted list, by a wide margin. A remarkable 83.33 percent of the belted victims are females (20 of 24). Neck fractures are also fairly high on the unbelted list, but here only 52.94 percent of the victims are women. Table 4-4d noted that belted females were especially vulnerable to neck injury in the front seat of vehicles without frontal air bags. In the back seat, of course, there are no frontal air bags.
- Chest fractures are no. 2 on the belted list but just in the middle of the unbelted list, possibly indicating that belts in the back seat apply loads that may exceed tolerance levels of some females.
- Two types of abdominal injuries – liver and spleen lacerations – are also relatively more prevalent for belted females, possibly indicating excessive loads for the lap portion of the belt. This parallels the MCOD findings in Tables 6-11 and 6-12.
- Conversely, thoracolumbar fractures and arm fractures are two fairly frequent injuries that occupy higher positions in the unbelted table. Perhaps belts are especially effective for females, whose shorter stature would tend to limit contact with interior surfaces.

TABLE 7-11: AIS 2-6 INJURIES OF UNRESTRAINED REAR-OUTBOARD PASSENGERS IN CARS
RANKED BY PERCENTAGE OF VICTIMS THAT ARE FEMALES
(1988-2010 CDS, OCCUPANTS AGE 18-96)

PERCENT FEMALE	AVG AGE	INJURY	N	AVG AIS
83.33	43.08	ABDOMEN ARTERY LACERATION	12	3.25
75.00	40.83	INTESTINAL LACERATION	12	2.17
73.33	34.40	KNEE OR ANKLE SPRAIN	15	2.00
64.29	34.14	HEAD SUPERFICIAL INJURY	14	2.07
57.14	23.29	THORACOLUMBAR CORD CONTUSION	7	3.71
54.35	34.51	SHOULDER FRACTURE	92	2.00
52.94	35.69	NECK FRACTURE	102	2.38
52.63	37.21	CERVICAL CORD CONTUSION	19	4.11
50.78	38.23	FEMUR FRACTURE	128	3.00
50.62	30.77	THORACOLUMBAR FRACTURE	162	2.09
50.29	33.71	ARM FRACTURE	173	2.36
50.00	36.29	HIP JOINT SEPARATION	14	3.00
50.00	31.80	SPLEEN CONTUSION	10	2.00
49.71	31.73	wtd avg of crash-involved unrestrained backseat outboard psgrs, CDS 1988-2010		
48.25	38.32	LOWER LEG FRACTURE	114	2.15
47.24	33.50	PELVIS FRACTURE	163	2.26
46.85	37.17	wtd avg of unrestrained backseat outboard psgrs with MAIS ≥ 2, CDS 1988-2010		
46.33	40.13	CHEST FRACTURES	177	3.08
45.83	26.88	KIDNEY LACERATION	24	2.46
45.45	32.18	NECK DISLOCATED VERTEBRA	22	2.14
45.00	30.17	CONCUSSION	180	2.37
44.44	38.60	LUNG LACERATION	45	3.18
43.18	37.45	CHEST LACERATED ARTERY	44	4.64
42.86	33.37	SPLEEN LACERATION	63	2.52
42.31	42.04	HEART LACERATION	26	4.12
42.25	30.49	LIVER LACERATION	71	2.86
41.46	28.76	FACIAL FRACTURE	123	2.20
40.00	34.20	INTESTINAL CONTUSION	10	2.00
38.46	23.08	LIVER CONTUSION	13	2.15
38.38	30.78	LUNG CONTUSION	99	3.42
37.18	33.77	BRAIN UNK INJURY	277	3.60
33.33	32.17	HEAD CRUSHED	6	6.00
32.39	31.92	BRAIN CONTUSION	71	3.45
31.25	35.38	HIP DISLOCATION	16	2.25
27.78	32.22	KIDNEY CONTUSION	18	2.00
27.27	29.36	SHOULDER DISLOCATED	22	2.00
27.27	28.42	BRAIN LACERATION	33	4.97
25.00	28.13	SKULL FRACTURE	128	2.90
20.00	30.80	CERVICAL CORD LACERATION	5	5.80
20.00	27.93	LEG SUPERFICIAL INJURY	15	2.13

169

TABLE 7-12: AIS 2-6 INJURIES OF BELTED REAR-OUTBOARD PASSENGERS IN CARS
RANKED BY PERCENTAGE OF VICTIMS THAT ARE FEMALES
(1988-2010 CDS, OCCUPANTS AGE 18-96)

PERCENT FEMALE	AVG AGE	INJURY	N	AVG AIS
83.33	60.63	NECK FRACTURE	24	2.29
67.37	54.35	CHEST FRACTURES	95	2.68
66.67	44.11	SHOULDER FRACTURE	27	2.00
64.29	45.93	INTESTINAL LACERATION	14	2.86
60.00	32.87	LIVER LACERATION	15	2.00
59.26	39.22	LOWER LEG FRACTURE	27	2.11
56.37	35.97	wtd avg of crash-involved belted backseat outboard passengers, CDS 1988-2010		
54.55	39.73	SPLEEN LACERATION	11	2.55
54.05	44.24	PELVIS FRACTURE	37	2.38
53.23	31.32	CONCUSSION	62	2.23
52.00	33.80	FACIAL FRACTURE	25	2.28
51.46	33.74	BRAIN UNK INJURY	103	3.79
50.00	69.63	CHEST LACERATED ARTERY	8	4.75
50.00	35.64	FEMUR FRACTURE	22	3.00
49.65	40.68	wtd avg of belted backseat outboard passengers with MAIS ≥ 2, CDS 1988-2010		
48.15	40.85	LUNG CONTUSION	27	3.48
46.15	50.46	INTESTINAL CONTUSION	13	2.00
45.00	27.45	BRAIN CONTUSION	20	3.60
44.44	53.67	HEART LACERATION	9	3.89
40.00	32.74	THORACOLUMBAR FRACTURE	50	2.14
39.39	33.61	SKULL FRACTURE	33	2.94
36.84	35.61	ARM FRACTURE	38	2.45
33.33	31.00	BRAIN LACERATION	6	4.67

CHAPTER 8

VEHICLE COMPONENTS PREVALENT AS SOURCES OF INJURY TO OLDER OCCUPANTS AND WOMEN – ANALYSES OF 1988-2010 CDS DATA

8.0   Summary

NASS-CDS tries to identify the injury source (vehicle component contacted) for each individual fatal or nonfatal injury in the database. For any specific injury source, a high average age of the victims with AIS $\geq$ 2 injuries attributed to that source may indicate that older occupants are vulnerable to injury from that source; high percentages of female victims indicate that females are vulnerable. Injury sources with high average ages of victims include: the air bag in frontal impacts, the belt system in frontal and far-side impacts, and the side interior surface in nearside impacts. The findings do not imply that air bags or belts are harmful for older people; instead, they suggest that when air bags or belts save a young person's life, he or she may walk away from the crash with little or no injury, but when they save an older occupant's life there may still be injuries ranging from moderate to severe. One source of leg injuries has a high proportion of female victims: the floor and its components, including the toe pan and pedals. Arm injuries involving contact with the steering assembly, the air bag, or the A-pillar have high percentages of female victims. So do torso injuries from the armrest or side hardware in nearside impacts.

8.1 Ranking the injury sources by victims' average age or percent females

The NASS-CDS file, as discussed in Section 5.1, describes the individual AIS $\geq$ 2 injuries of each occupant of a vehicle involved in a towaway crash: nonfatal as well as fatal crashes. Whenever possible, CDS identifies the vehicle component contacted that is the source of the injury – e.g., the steering wheel, instrument panel, or armrest. For any specific injury source it is possible to compute the average victim age for the various AIS $\geq$ 2 injuries [or AIS $\geq$ 3, or AIS $\geq$ 4] attributed to that source and the percent that are female. The injury sources can be ranked by average age from oldest to youngest or by percent female victims from highest to lowest. A high average age possibly indicates that the component poses some type of challenge for older people. One such challenge could be force levels exceeding an older person's tolerance. Another is that the component contacts older occupants differently than young people, because of a difference in posture. A high percentage female victims means women are especially vulnerable because of lower tolerance and/or because their different height, weight, or anatomy affects how the component contacts them. The rankings can be treated as "control charts," flagging the components where the average age of the victim or the percent female is significantly higher than for the "typical" injury.

Data on injury sources is limited. The analysis is exploratory. A flag is a starting point for further investigation, not definitive evidence that a component poses a challenge to older occupants or women. There is much opportunity for Type 1 error – failing to identify challenging components due to limited numbers of cases – and Type 2 error – false alarms. As in Chapters 5 and 7, the analyses are based on unweighted CDS data to avoid the misleading results if a few especially old or young occupants with high sampling weights are injured by a particular source.

One difference from Chapter 7 is the need to be more specific. Chapter 7 is primarily an inquiry into the intrinsic vulnerability of older occupants or women to certain types of injury. For that purpose, it was acceptable to combine frontals with side impacts or drivers with passengers because the origin of the physical insult was not so important, just the outcome. This chapter investigates the performance of the vehicle. Details matter. For example, the role of the steering assembly as a source of injury is relevant only for drivers, not passengers; the role of frontal air bags is relevant only for vehicles equipped with air bags and involved in frontal crashes. In other words, separate analyses must be conducted by type of crash and type of occupant protection, further subdividing the limited data.

Separate analyses are performed for each **cell** created by four variables: seat position (driver, RF passenger, or backseat-outboard passenger), type of impact (frontal, nearside, far-side, rollover, or rear/other), type of occupant protection, and body region (head, neck, torso, arm, or leg; or burns regardless of body region). "Rollovers" include not only single-vehicle, first-event rollovers but any vehicle for which a rollover occurred and was judged the most harmful event (as evidenced by GAD1 = T). The types of occupant protection considered in frontal impacts are: 3-point belted + air bag available; 3-point belted, no air bag; unbelted, but air bag available; no belt and no air bag. The types of occupant protection considered in all other impacts are: 3-point belted; unrestrained (there were not enough cases to consider side air bags). The body region of the injury is identified by the value of BODYREG: head injuries include H and F (head and face); N = neck injury; torso includes B, C, M, P, and S (back, chest, abdomen, pelvis, and shoulder); arms include A, E, R, W, and X (upper arm, elbow, forearm, wrist/hand, entire arm); and legs include K, L, Q, T, and Y (knee, lower leg, ankle/foot, thigh, entire leg). All burns (LESION = B) are listed separately, regardless of the body region(s) where the burn occurred.

One set of analyses is performed for AIS 2 to 6 injuries, another for AIS 3 to 6, and another for AIS 4 to 6. The data combines passenger cars and LTVs (with MY $\geq$ 1960) to maximize the number of cases and also includes all adults, age 18 to 96.

Grouping of injury sources: Since its inception in 1988, CDS has used a variable with the same name, INJSOU, to identify the injury source. The numerical values, however, changed slightly in 1993 and substantially in 1995. Various codes have been added, deleted, or modified between 1995 and 2010. INJSOU is defined at the injury level, with possibly different codes for each of an occupant's various injuries.

The numerous individual codes have been grouped into 18 categories of related components, in order to provide large enough numbers of injuries for meaningful statistical analyses. The 18 categories are:

Air bag	A-pillar	Roof & its components
Belt system	B-pillar	Floor & its components
Steering assembly	Side hardware or armrest	Seat & its components
Instrument panel	Side interior surface	Exterior
Other frontal component	Other side component	Fire

Other occs, self, cargo   Non-contact   Other

Appendix D lists which numeric codes are included in each category in CY 1998-1992, 1993-1994, and 1995-2010. The numeric codes are almost all unambiguous and allow definitions of the categories that are consistent from year to year.

Analysis method: For any given cell – e.g., drivers, frontal impacts, belted with air bags, head injuries – the tables that follow lists each of the 18 categories that occurs at least the threshold number of times in the 1988-2010 CDS data. For the AIS ≥ 2 analyses, the threshold is at least 15 occurrences; for the AIS ≥ 3 analyses, 10 occurrences; and for the AIS ≥ 4 analyses, five occurrences. For example, here are the results for two cells for drivers at the AIS ≥ 2 level:

IMPACT TYPE	OCCUPANT PROTECTION	BODY REGION	AVERAGE AGE	PERCENT FEMALE	N	INJURY SOURCE
FRONTAL	BELTED, AIR BAG	HEAD	46.63	47.37	19	BELT SYSTEM
			45.65	46.15	26	INSTRUMENT PANEL
			43.37	51.30	4255	average injured occupant
			43.29	35.29	34	EXTERIOR
			43.00	38.89	18	SEAT & ITS COMPONENTS
			42.62	43.08	130	AIR BAG
			40.68	45.61	57	B-PILLAR
			40.52	54.02	174	NON-CONTACT
			40.25	35.85	53	OTHER FRONTAL COMPONENT
			39.71	42.68	82	A-PILLAR
			39.43	37.65	170	ROOF & ITS COMPONENTS
			39.40	26.67	15	OTHER OCCS, SELF, CARGO
			39.27	45.06	324	STEERING ASSEMBLY
			38.64	16.00	25	OTHER
			37.86	54.55	22	OTHER SIDE COMPONENT
FRONTAL	BELTED, AIR BAG	TORSO	62.37	60.00	60	AIR BAG
			51.71	45.83	48	NON-CONTACT
			48.63	48.02	683	BELT SYSTEM
			47.09	48.18	110	SIDE HARDWARE OR ARMREST
			45.69	42.55	463	STEERING ASSEMBLY
			44.80	42.86	133	SIDE INTERIOR SURFACE
			44.23	40.91	22	B-PILLAR
			43.37	51.30	4255	average injured occupant
			42.51	46.15	234	INSTRUMENT PANEL
			41.96	41.38	116	SEAT & ITS COMPONENTS
			40.42	35.82	67	FLOOR & ITS COMPONENTS

The first cell comprises AIS ≥ 2 head injuries to belted drivers in frontal impacts of vehicles with air bags. Fourteen of the 18 injury-source categories contribute 15 or more injuries in 1988-2010 CDS (all except the side interior surface, side hardware, floor, and fire). The number of injuries ranges from 324 involving the steering assembly down to the bare-minimum 15 due to contacts with other occupants. The average age of the victims ranges from 46.63 for the belt system down to 37.86 for other side components. The benchmark or comparison value is 43.37, the average age of the 4,255 occupants with MAIS ≥ 2 in a frontal impact, belted, with air bags. It is printed in red. Two of the 14 injury sources have average age higher than 43.37, 12 have lower.

The second cell tabulates the torso injuries of the same occupants. Here, only 10 injury-source categories are represented, with seven having higher average age than the benchmark and three, lower. Note that the benchmark age is the same in both cells, because it is defined at the occupant level (occupants with MAIS ≥ 2), not the injury level, and is the same for all body regions. However, the benchmark would change for other seat locations, impact types, restraint-use conditions, or injury-severity levels.

Significance testing: The key to the analysis is a statistical comparison of each injury-source category within the cell to the benchmark value. For example, is the 46.63 average age of the 19 head injuries from the belt system significantly higher than the 43.37 benchmark? For that comparison, the standard deviation of the average age is computed for the 4,255 benchmark cases and it is treated as the population standard deviation (because the benchmark sample of 4,255 far exceeds any of the sample sizes for the individual categories). The standard deviation is 18.07. A Z-score is computed for the difference of the belted cases (N=19) and the benchmark cases (N=4,255), assuming the population standard deviation computed from the benchmark cases:

$$Z = (46.63 - 43.37)/\sqrt{18.07^2 *((1/19)+(1/4255))} = 0.78$$

These Z-scores are normally distributed. The customary threshold of statistical significance (two-sided $\alpha = .05$) is crossed when Z is 1.96 or more – i.e., when the average age for a particular injury source is 1.96 or more standard deviations above the benchmark value. However, in these analyses, a somewhat more stringent level of 3 standard deviations (two-sided $\alpha = .002$) may be more meaningful throughout this chapter because of the multiple applications of the statistical tests (which may create false positives) as well as the additional uncertainty due to cluster sampling in CDS and the use of unweighted case counts. Furthermore, a solitary test result of > 3 standard deviations deserves only limited credence. Most important are the injury sources that repeatedly test out at > 3 standard deviations at different AIS levels, types of restraint use, seat positions, and/or impact types. In the preceding example, because Z = 0.78, which is less than 1.96, the 46.63 based on 19 cases is not significantly higher than 43.37. But in the next cell, the average age 62.37 for the 60 cases of torso injury involving contact with the air bag is significantly higher than the 43.37 benchmark (Z = 8.09).

The results of the statistical testing are added to the preceding listings of two cells. The cells are tabulated like control charts, with injury sources whose average age is three or more standard deviations above the benchmark printed bold and marked with two stars. These injury sources would be the initial candidates for further investigation to obtain a better understanding of why older occupants appear to be vulnerable. Injury sources whose average age is 1.96 to 3 standard deviations above the benchmark are marked with one star, as possible follow-up candidates for further investigation. No statistical tests are needed if the average age is below the benchmark, and these injury sources are not marked with stars.

IMPACT TYPE	OCCUPANT PROTECTION	BODY REGION	AVERAGE AGE	PERCENT FEMALE	N	INJURY SOURCE
FRONTAL	BELTED, AIR BAG	HEAD	46.63	47.37	19	BELT SYSTEM
			45.65	46.15	26	INSTRUMENT PANEL
			43.37	51.30	4255	average injured occupant
			43.29	35.29	34	EXTERIOR
			43.00	38.89	18	SEAT & ITS COMPONENTS
			42.62	43.08	130	AIR BAG
			40.68	45.61	57	B-PILLAR
			40.52	54.02	174	NON-CONTACT
			40.25	35.85	53	OTHER FRONTAL COMPONENT
			39.71	42.68	82	A-PILLAR
			39.43	37.65	170	ROOF & ITS COMPONENTS
			39.40	26.67	15	OTHER OCCS, SELF, CARGO
			39.27	45.06	324	STEERING ASSEMBLY
			38.64	16.00	25	OTHER
			37.86	54.55	22	OTHER SIDE COMPONENT
FRONTAL	BELTED, AIR BAG	TORSO	** 62.37	60.00	60	AIR BAG
			** 51.71	45.83	48	NON-CONTACT
			** 48.63	48.02	683	BELT SYSTEM
			* 47.09	48.18	110	SIDE HARDWARE OR ARMREST
			* 45.69	42.55	463	STEERING ASSEMBLY
			44.80	42.86	133	SIDE INTERIOR SURFACE
			44.23	40.91	22	B-PILLAR
			43.37	51.30	4255	average injured occupant
			42.51	46.15	234	INSTRUMENT PANEL
			41.96	41.38	116	SEAT & ITS COMPONENTS
			40.42	35.82	67	FLOOR & ITS COMPONENTS

**3+ standard deviations higher than average injured occupant (p < .002).
*1.96-3 standard deviations higher than average injured occupant (p < .05).

None of the 14 sources of head injury have average victims' age significantly higher than the benchmark. Only two of the sources have higher age than the benchmark, but it is just slightly higher. By contrast, the first three sources of torso injury have average age 3+ standard deviations higher than the benchmark and the next two, 1.96 to 3 standard deviations higher. Of course, the results are consistent with the literature and preceding chapters' findings that older occupants are especially vulnerable to torso injury, not so much to head injury. But what is unique about this analysis is that it identifies the specific vehicle components prevalent as injury sources to older occupants – e.g., the air bag, non-contact injury, and the belt system.

The identical method produces control charts for the percent of victims that are females. Here, for example is the cell of torso injuries to belted occupants in frontal impacts of vehicles with air bags – the same data as the second cell in the preceding example, but now ranked by percent of victims that are females:

IMPACT TYPE	OCCUPANT PROTECTION	BODY REGION	PERCENT FEMALE	AVERAGE AGE	N	INJURY SOURCE
FRONTAL	BELTED, AIR BAG	TORSO	60.00	62.37	60	AIR BAG
			51.30	43.37	4255	average injured occupant
			48.18	47.09	110	SIDE HARDWARE OR ARMREST
			48.02	48.63	683	BELT SYSTEM
			46.15	42.51	234	INSTRUMENT PANEL
			45.83	51.71	48	NON-CONTACT
			42.86	44.80	133	SIDE INTERIOR SURFACE
			42.55	45.69	463	STEERING ASSEMBLY
			41.38	41.96	116	SEAT & ITS COMPONENTS
			40.91	44.23	22	B-PILLAR
			35.82	40.42	67	FLOOR & ITS COMPONENTS

The benchmark percentage of females is 51.30. Only one of the 10 injury sources has more than 51.30 percent female victims, but none of them is significantly higher than 51.30. The analysis of this cell does not identify any sources of torso injury where belted females in frontal crashes of vehicles with air bags show exceptional vulnerability.

8.2   Injury sources with high average victims' ages

Table 8-1, which covers the next six pages, ranks the sources of AIS $\geq$ 2 injuries to drivers of cars and LTVs by the average age of the victim, from oldest to youngest – with separate rankings by type of impact, type of occupant protection, and body region. In each cell, the benchmark age, printed in red, is the average age of drivers with MAIS $\geq$ 2 for that type of impact and occupant protection in 1988-2010 CDS (i.e., the average at the occupant level, not the injury level). Table 8-1 displays the average age of the victims (left column), the percent female (second column), the number of times the injury source is reported in CDS (next-to-last column), and the name of the injury source. Table 8-1 is limited to injury sources that occurred at least 15 times.

It is evident in Table 8-1 that for the overwhelming majority of injury sources the average age of the victim is not significantly older than the benchmark age – the age of the "typical" injured driver. To begin with, approximately half of the injury sources have an average victim age younger than the benchmark – e.g., many sources of head injury and contacts with objects exterior to the vehicle. Of the remainder, many have average age close to the benchmark, while others do not occur often enough for differences to be significant.

It is difficult to work through the many pages of mostly non-significant results in search of the injury sources that consistently result in injuries with older-than-average victims. Therefore, the remaining tables of this chapter are limited to the injury sources where the average age of the victim [or the percent female victims] is at least 1.96 standard deviations higher than the benchmark. However, the complete tables including all the injury sources may be found in Appendix E.

TABLE 8-1: AVERAGE AGE OF DRIVERS AGE 18-96 WITH AIS 2+ INJURIES, CDS 1988-2010

IMPACT TYPE	OCCUPANT PROTECTION	BODY REGION		AVERAGE AGE	PERCENT FEMALE	N	INJURY SOURCE
FRONTAL	BELTED, AIR BAG	HEAD		46.63	47.37	19	BELT SYSTEM
				45.65	46.15	26	INSTRUMENT PANEL
				43.37	51.30	4255	average injured occupant
				43.29	35.29	34	EXTERIOR
				43.00	38.89	18	SEAT & ITS COMPONENTS
				42.62	43.08	130	AIR BAG
				40.68	45.61	57	B-PILLAR
				40.52	54.02	174	NON-CONTACT
				40.25	35.85	53	OTHER FRONTAL COMPONENT
				39.71	42.68	82	A-PILLAR
				39.43	37.65	170	ROOF & ITS COMPONENTS
				39.40	26.67	15	OTHER OCCS, SELF, CARGO
				39.27	45.06	324	STEERING ASSEMBLY
				38.64	16.00	25	OTHER
				37.86	54.55	22	OTHER SIDE COMPONENT
FRONTAL	BELTED, AIR BAG	NECK	**	56.00	66.67	21	BELT SYSTEM
			*	55.11	47.37	19	SEAT & ITS COMPONENTS
			**	52.00	64.58	48	NON-CONTACT
			*	51.53	52.63	19	A-PILLAR
			*	49.26	46.81	47	STEERING ASSEMBLY
				45.82	37.25	51	ROOF & ITS COMPONENTS
				43.37	51.30	4255	average injured occupant
FRONTAL	BELTED, AIR BAG	TORSO	**	62.37	60.00	80	AIR BAG
			**	51.71	45.83	48	NON-CONTACT
			**	48.63	48.02	683	BELT SYSTEM
			*	47.09	48.18	110	SIDE HARDWARE OR ARMREST
			*	45.69	42.55	463	STEERING ASSEMBLY
				44.80	42.86	133	SIDE INTERIOR SURFACE
				44.23	40.91	22	B-PILLAR
				43.37	51.30	4255	average injured occupant
				42.51	46.15	234	INSTRUMENT PANEL
				41.96	41.38	116	SEAT & ITS COMPONENTS
				40.42	35.82	67	FLOOR & ITS COMPONENTS
FRONTAL	BELTED, AIR BAG	ARM		46.64	50.00	98	OTHER FRONTAL COMPONENT
			*	46.08	62.40	258	STEERING ASSEMBLY
				45.59	37.93	29	EXTERIOR
				44.83	80.00	150	AIR BAG
				44.02	52.53	158	SIDE INTERIOR SURFACE
				43.85	51.65	333	INSTRUMENT PANEL
				43.37	51.30	4255	average injured occupant
				43.33	69.88	83	A-PILLAR
				42.42	47.37	19	SIDE HARDWARE OR ARMREST
				41.98	59.26	54	ROOF & ITS COMPONENTS
FRONTAL	BELTED, AIR BAG	LEG		43.69	63.36	1209	FLOOR & ITS COMPONENTS
				43.43	47.62	21	SIDE HARDWARE OR ARMREST
				43.37	51.30	4255	average injured occupant
				42.46	49.55	1112	INSTRUMENT PANEL
				42.28	27.78	54	SIDE INTERIOR SURFACE
				35.26	33.87	62	STEERING ASSEMBLY
FRONTAL	BELTED, NO AIR BAG	HEAD		42.17	43.18	3175	average injured occupant
				41.75	31.58	95	ROOF & ITS COMPONENTS
				39.98	35.29	102	A-PILLAR
				39.76	52.94	17	B-PILLAR
				39.61	41.67	36	OTHER SIDE COMPONENT
				39.23	46.67	30	EXTERIOR
				38.94	43.32	247	OTHER FRONTAL COMPONENT
				38.91	43.76	713	STEERING ASSEMBLY
				38.15	40.74	27	NON-CONTACT
				36.33	35.00	60	INSTRUMENT PANEL
FRONTAL	BELTED, NO AIR BAG	NECK	*	52.55	45.00	20	NON-CONTACT
			*	48.13	46.03	63	STEERING ASSEMBLY
				42.17	43.18	3175	average injured occupant
FRONTAL	BELTED, NO AIR BAG	TORSO		47.54	53.66	41	NON-CONTACT
			**	47.24	41.06	899	STEERING ASSEMBLY
				47.02	30.23	43	SEAT & ITS COMPONENTS
			**	45.98	45.12	492	BELT SYSTEM
				45.66	26.32	76	SIDE INTERIOR SURFACE
				42.38	38.68	106	INSTRUMENT PANEL
				42.17	43.18	3175	average injured occupant
				41.09	33.33	33	SIDE HARDWARE OR ARMREST
				37.10	42.86	21	FLOOR & ITS COMPONENTS

**3+ standard deviations higher than average injured occupant (p < .002).
*1.96-3 standard deviations higher than average injured occupant (p < .05).

177

IMPACT TYPE	OCCUPANT PROTECTION	BODY REGION	AVERAGE AGE	PERCENT FEMALE	N	INJURY SOURCE
FRONTAL	BELTED, NO AIR BAG	ARM	48.13	31.25	16	OTHER FRONTAL COMPONENT
			46.87	42.22	45	SIDE INTERIOR SURFACE
			43.08	53.45	232	STEERING ASSEMBLY
			42.17	43.18	3175	average injured occupant
			41.84	42.86	259	INSTRUMENT PANEL
FRONTAL	BELTED, NO AIR BAG	LEG	42.17	43.18	3175	average injured occupant
			41.34	54.21	629	FLOOR & ITS COMPONENTS
			40.16	43.77	658	INSTRUMENT PANEL
			38.28	38.89	36	SIDE INTERIOR SURFACE
			38.18	28.07	57	STEERING ASSEMBLY
FRONTAL	NO BELT, AIR BAG	HEAD	39.85	35.90	39	NON-CONTACT
			38.78	25.24	103	A-PILLAR
			37.55	58.33	60	AIR BAG
			37.24	32.63	2400	average injured occupant
			36.95	19.93	301	OTHER FRONTAL COMPONENT
			35.36	22.87	188	ROOF & ITS COMPONENTS
			35.32	32.57	175	STEERING ASSEMBLY
			33.81	20.93	43	INSTRUMENT PANEL
			32.52	28.00	25	OTHER
			31.77	20.87	115	EXTERIOR
FRONTAL	NO BELT, AIR BAG	NECK	* 43.00	39.13	46	STEERING ASSEMBLY
			42.19	26.19	42	OTHER FRONTAL COMPONENT
			39.78	22.22	18	A-PILLAR
			39.35	17.65	17	INSTRUMENT PANEL
			38.52	16.67	42	ROOF & ITS COMPONENTS
			37.24	32.63	2400	average injured occupant
			33.69	25.00	36	EXTERIOR
FRONTAL	NO BELT, AIR BAG	TORSO	* 44.91	42.86	35	AIR BAG
			40.83	33.33	36	FLOOR & ITS COMPONENTS
			** 40.22	31.76	677	STEERING ASSEMBLY
			38.66	27.55	98	SIDE INTERIOR SURFACE
			38.06	16.67	36	SEAT & ITS COMPONENTS
			37.24	32.63	2400	average injured occupant
			37.13	27.06	255	INSTRUMENT PANEL
			35.21	35.71	42	SIDE HARDWARE OR ARMREST
			33.20	25.00	84	EXTERIOR
			30.96	26.09	23	ROOF & ITS COMPONENTS
FRONTAL	NO BELT, AIR BAG	ARM	* 46.84	68.42	19	AIR BAG
			40.19	31.91	47	SIDE INTERIOR SURFACE
			* 40.15	28.64	206	INSTRUMENT PANEL
			38.05	45.45	22	A-PILLAR
			37.24	32.63	2400	average injured occupant
			36.50	41.18	34	OTHER FRONTAL COMPONENT
			35.87	42.86	91	STEERING ASSEMBLY
			33.83	33.33	18	ROOF & ITS COMPONENTS
			31.17	33.33	30	EXTERIOR
FRONTAL	NO BELT, AIR BAG	LEG	37.58	48.57	523	FLOOR & ITS COMPONENTS
			37.32	32.22	568	INSTRUMENT PANEL
			37.24	32.63	2400	average injured occupant
			36.64	26.17	107	STEERING ASSEMBLY
			34.58	23.08	26	SIDE INTERIOR SURFACE
			32.08	25.00	24	EXTERIOR
FRONTAL	NO BELT, NO AIR BAG	HEAD	40.47	26.67	15	SIDE INTERIOR SURFACE
			37.93	34.68	173	INSTRUMENT PANEL
			37.61	30.55	5081	average injured occupant
			37.36	18.85	191	ROOF & ITS COMPONENTS
			36.91	35.27	601	STEERING ASSEMBLY
			36.65	23.53	17	NON-CONTACT
			36.58	38.00	50	OTHER SIDE COMPONENT
			35.81	28.68	1440	OTHER FRONTAL COMPONENT
			35.25	18.75	160	EXTERIOR
			34.55	28.64	206	A-PILLAR
			31.16	15.79	19	OTHER
FRONTAL	NO BELT, NO AIR BAG	NECK	* 44.29	14.29	42	ROOF & ITS COMPONENTS
			** 44.10	32.17	115	OTHER FRONTAL COMPONENT
			39.99	32.53	83	STEERING ASSEMBLY
			38.60	40.00	20	INSTRUMENT PANEL
			37.61	30.55	5081	average injured occupant
			37.50	16.67	36	EXTERIOR
			35.96	20.83	24	A-PILLAR

**3+ standard deviations higher than average injured occupant (p < .002).
*1.96-3 standard deviations higher than average injured occupant (p < .05).

178

IMPACT TYPE	OCCUPANT PROTECTION	BODY REGION	AVERAGE AGE	PERCENT FEMALE	N	INJURY SOURCE
FRONTAL	NO BELT, NO AIR BAG	TORSO	** 50.56	38.89	18	NON-CONTACT
			** 42.39	28.57	119	SIDE INTERIOR SURFACE
			** 41.88	28.08	1631	STEERING ASSEMBLY
			41.50	34.21	38	FLOOR & ITS COMPONENTS
			** 40.53	22.28	359	INSTRUMENT PANEL
			39.52	33.33	27	OTHER FRONTAL COMPONENT
			38.79	31.58	19	SIDE HARDWARE OR ARMREST
			37.61	30.55	5081	average injured occupant
			35.95	15.00	20	ROOF & ITS COMPONENTS
			35.04	32.00	50	SEAT & ITS COMPONENTS
			33.26	25.58	86	EXTERIOR
FRONTAL	NO BELT, NO AIR BAG	ARM	** 40.91	33.69	374	INSTRUMENT PANEL
			39.50	35.00	20	A-PILLAR
			39.15	41.26	286	STEERING ASSEMBLY
			38.14	31.25	64	SIDE INTERIOR SURFACE
			37.65	38.78	49	OTHER FRONTAL COMPONENT
			37.61	30.55	5081	average injured occupant
			34.40	17.14	35	EXTERIOR
FRONTAL	NO BELT, NO AIR BAG	LEG	39.16	10.53	38	SIDE INTERIOR SURFACE
			38.15	43.90	779	FLOOR & ITS COMPONENTS
			37.83	33.69	1021	INSTRUMENT PANEL
			37.61	30.55	5081	average injured occupant
			36.90	36.99	146	STEERING ASSEMBLY
			33.68	44.00	25	EXTERIOR
NEARSIDE IMPACT	BELTED	HEAD	43.94	48.72	2974	average injured occupant
			43.77	53.19	47	STEERING ASSEMBLY
			43.75	52.46	183	OTHER SIDE COMPONENT
			43.59	64.06	64	SIDE INTERIOR SURFACE
			42.88	50.00	140	A-PILLAR
			42.83	47.08	308	B-PILLAR
			42.34	49.32	148	EXTERIOR
			40.82	48.96	96	NON-CONTACT
			39.57	61.90	21	AIR BAG
			39.06	38.89	18	OTHER OCCS, SELF, CARGO
			37.48	34.39	221	ROOF & ITS COMPONENTS
			34.87	60.00	30	BELT SYSTEM
			34.54	53.85	26	OTHER FRONTAL COMPONENT
			33.92	31.82	66	OTHER
NEARSIDE IMPACT	BELTED	NECK	* 50.05	53.13	64	B-PILLAR
			47.44	58.97	39	NON-CONTACT
			44.45	36.36	33	EXTERIOR
			43.94	48.72	2974	average injured occupant
			41.16	52.00	25	SIDE INTERIOR SURFACE
			39.97	34.38	32	ROOF & ITS COMPONENTS
NEARSIDE IMPACT	BELTED	TORSO	* 47.70	41.04	212	B-PILLAR
			* 47.40	45.10	204	BELT SYSTEM
			** 46.49	46.53	1255	SIDE INTERIOR SURFACE
			* 46.10	52.77	741	SIDE HARDWARE OR ARMREST
			44.63	59.26	27	INSTRUMENT PANEL
			44.29	47.20	125	STEERING ASSEMBLY
			43.94	48.72	2974	average injured occupant
			41.39	50.00	18	OTHER
			40.88	49.33	75	SEAT & ITS COMPONENTS
			40.33	49.45	91	FLOOR & ITS COMPONENTS
			37.82	58.82	17	OTHER OCCS, SELF, CARGO
NEARSIDE IMPACT	BELTED	ARM	48.04	63.24	68	STEERING ASSEMBLY
			44.39	55.10	49	INSTRUMENT PANEL
			43.94	48.72	2974	average injured occupant
			42.02	46.82	173	SIDE INTERIOR SURFACE
			41.00	56.25	16	B-PILLAR
			38.93	32.14	28	SIDE HARDWARE OR ARMREST
NEARSIDE IMPACT	BELTED	LEG	46.91	59.05	105	FLOOR & ITS COMPONENTS
			43.94	48.72	2974	average injured occupant
			43.10	50.75	134	INSTRUMENT PANEL
			40.04	43.56	163	SIDE INTERIOR SURFACE
			38.81	51.19	84	SIDE HARDWARE OR ARMREST
			32.93	33.33	15	STEERING ASSEMBLY

**3+ standard deviations higher than average injured occupant (p < .002).
*1.96-3 standard deviations higher than average injured occupant (p < .05).

TABLE 8-1 (CONTINUED): AVERAGE AGE OF DRIVERS AGE 18-96 WITH AIS 2+ INJURIES, CDS 1988-2010

IMPACT TYPE	OCCUPANT PROTECTION	BODY REGION	AVERAGE AGE	PERCENT FEMALE	N	INJURY SOURCE
NEARSIDE IMPACT	NO BELT	HEAD	44.17	56.67	30	SIDE INTERIOR SURFACE
			39.41	37.62	101	OTHER SIDE COMPONENT
			38.63	32.43	111	B-PILLAR
			37.94	46.88	32	NON-CONTACT
			37.83	33.71	1771	average injured occupant
			37.16	32.58	132	A-PILLAR
			36.10	34.33	67	OTHER FRONTAL COMPONENT
			34.50	34.78	207	EXTERIOR
			33.32	26.09	161	ROOF & ITS COMPONENTS
			30.56	17.74	62	OTHER
			26.29	39.29	28	STEERING ASSEMBLY
NEARSIDE IMPACT	NO BELT	NECK	45.80	40.00	15	SIDE INTERIOR SURFACE
			41.21	31.58	19	B-PILLAR
			39.35	26.09	46	EXTERIOR
			37.83	33.71	1771	average injured occupant
			35.55	18.42	38	ROOF & ITS COMPONENTS
NEARSIDE IMPACT	NO BELT	TORSO	40.50	10.00	20	INSTRUMENT PANEL
			* 40.47	43.46	306	SIDE HARDWARE OR ARMREST
			** 40.42	37.61	694	SIDE INTERIOR SURFACE
			40.35	41.18	17	A-PILLAR
			39.54	33.33	81	B-PILLAR
			38.03	45.45	33	SEAT & ITS COMPONENTS
			37.99	35.00	160	STEERING ASSEMBLY
			37.83	33.71	1771	average injured occupant
			37.63	37.50	48	FLOOR & ITS COMPONENTS
			35.85	26.88	93	EXTERIOR
			31.79	21.05	19	OTHER OCCS, SELF, CARGO
NEARSIDE IMPACT	NO BELT	ARM	40.71	48.57	35	STEERING ASSEMBLY
			39.61	26.09	23	INSTRUMENT PANEL
			37.97	36.13	119	SIDE INTERIOR SURFACE
			37.83	33.71	1771	average injured occupant
			32.47	32.56	43	EXTERIOR
NEARSIDE IMPACT	NO BELT	LEG	41.34	33.78	74	FLOOR & ITS COMPONENTS
			37.83	33.71	1771	average injured occupant
			36.89	25.30	83	INSTRUMENT PANEL
			36.79	22.64	53	SIDE HARDWARE OR ARMREST
			34.78	33.58	134	SIDE INTERIOR SURFACE
			34.55	22.58	31	EXTERIOR
FAR-SIDE IMPACT	BELTED	HEAD	47.72	38.89	18	EXTERIOR
			45.47	42.86	49	OTHER FRONTAL COMPONENT
			43.12	48.48	33	INSTRUMENT PANEL
			43.05	47.40	1310	average injured occupant
			42.65	52.38	63	STEERING ASSEMBLY
			40.62	45.28	53	B-PILLAR
			39.14	28.57	21	SEAT & ITS COMPONENTS
			38.33	36.11	36	OTHER OCCS, SELF, CARGO
			37.96	41.67	24	OTHER SIDE COMPONENT
			37.09	49.58	119	SIDE INTERIOR SURFACE
			36.26	41.86	43	NON-CONTACT
			36.16	28.89	90	ROOF & ITS COMPONENTS
			34.53	64.71	17	AIR BAG
FAR-SIDE IMPACT	BELTED	NECK	43.05	47.40	1310	average injured occupant
			42.32	60.00	25	NON-CONTACT
			37.92	56.00	25	SIDE INTERIOR SURFACE
			37.07	29.63	27	ROOF & ITS COMPONENTS
FAR-SIDE IMPACT	BELTED	TORSO	** 53.00	51.79	56	STEERING ASSEMBLY
			* 52.58	50.00	24	INSTRUMENT PANEL
			** 51.16	55.17	203	BELT SYSTEM
			* 49.87	43.59	39	OTHER
			47.50	44.44	18	NON-CONTACT
			44.53	52.71	129	FLOOR & ITS COMPONENTS
			43.05	47.40	1310	average injured occupant
			40.70	49.60	125	SIDE INTERIOR SURFACE
			39.82	39.00	100	SEAT & ITS COMPONENTS
			39.39	39.47	38	SIDE HARDWARE OR ARMREST
			38.78	26.83	41	OTHER OCCS, SELF, CARGO
			35.80	30.00	20	B-PILLAR
FAR-SIDE IMPACT	BELTED	ARM	* 51.08	55.26	38	INSTRUMENT PANEL
			* 50.94	72.00	50	STEERING ASSEMBLY
			43.05	47.40	1310	average injured occupant
			36.66	48.78	41	SIDE INTERIOR SURFACEE

**3+ standard deviations higher than average injured occupant (p < .002).
*1.96-3 standard deviations higher than average injured occupant (p < .05)

IMPACT TYPE	OCCUPANT PROTECTION	BODY REGION	AVERAGE AGE	PERCENT FEMALE	N	INJURY SOURCE
FAR-SIDE IMPACT	BELTED	LEG	** 51.51	64.20	81	FLOOR & ITS COMPONENTS
			47.25	58.02	81	INSTRUMENT PANEL
			43.05	47.40	1310	average injured occupant
FAR-SIDE IMPACT	NO BELT	HEAD	44.17	33.33	18	SIDE HARDWARE OR ARMREST
			42.94	32.08	53	INSTRUMENT PANEL
			42.37	29.59	98	A-PILLAR
			41.42	38.46	52	OTHER SIDE COMPONENT
			39.65	35.53	197	OTHER FRONTAL COMPONENT
			38.99	36.24	149	SIDE INTERIOR SURFACE
			38.60	31.21	1535	average injured occupant
			35.78	32.65	49	B-PILLAR
			35.54	24.14	145	ROOF & ITS COMPONENTS
			34.83	22.22	144	EXTERIOR
			34.07	30.00	30	STEERING ASSEMBLY
			33.74	16.13	31	OTHER
FAR-SIDE IMPACT	NO BELT	NECK	** 51.57	23.81	21	A-PILLAR
			* 46.84	36.36	44	SIDE INTERIOR SURFACE
			41.33	30.95	42	OTHER FRONTAL COMPONENT
			38.60	31.21	1535	average injured occupant
			37.24	22.22	45	EXTERIOR
			33.74	18.42	38	ROOF & ITS COMPONENTS
FAR-SIDE IMPACT	NO BELT	TORSO	44.60	13.33	15	B-PILLAR
			* 44.43	26.09	23	OTHER
			42.27	28.07	114	INSTRUMENT PANEL
			40.77	32.75	284	SIDE INTERIOR SURFACE
			40.38	30.59	85	STEERING ASSEMBLY
			39.46	25.42	59	SEAT & ITS COMPONENTS
			38.96	37.04	27	OTHER OCCS, SELF, CARGO
			38.60	31.21	1535	average injured occupant
			36.39	25.42	59	SIDE HARDWARE OR ARMREST
			36.31	29.85	67	FLOOR & ITS COMPONENTS
			34.81	29.59	98	EXTERIOR
			33.00	10.00	20	ROOF & ITS COMPONENTS
FAR-SIDE IMPACT	NO BELT	ARM	* 45.94	38.30	47	INSTRUMENT PANEL
			41.40	36.54	52	SIDE INTERIOR SURFACE
			38.67	51.85	27	STEERING ASSEMBLY
			38.60	31.21	1535	average injured occupant
			38.31	25.71	35	EXTERIOR
FAR-SIDE IMPACT	NO BELT	LEG	** 47.09	50.00	56	FLOOR & ITS COMPONENTS
			43.13	46.27	67	INSTRUMENT PANEL
			38.60	31.21	1535	average injured occupant
			34.64	27.27	22	EXTERIOR
			34.52	21.74	23	SIDE INTERIOR SURFACE
ROLLOVER FIRST/WORST EVENT	BELTED	HEAD	38.70	48.15	27	B-PILLAR
			38.00	37.75	1224	average injured occupant
			37.62	36.00	400	ROOF & ITS COMPONENTS
			36.77	50.00	22	STEERING ASSEMBLY
			35.35	38.24	34	EXTERIOR
			34.62	46.15	26	OTHER FRONTAL COMPONENT
			29.50	50.00	18	NON-CONTACT
ROLLOVER FIRST/WORST EVENT	BELTED	NECK	* 42.01	27.59	145	ROOF & ITS COMPONENTS
			38.00	37.75	1224	average injured occupant
ROLLOVER FIRST/WORST EVENT	BELTED	TORSO	* 43.59	26.09	46	B-PILLAR
			* 42.83	38.46	65	SEAT & ITS COMPONENTS
			42.70	45.65	46	SIDE HARDWARE OR ARMREST
			* 42.15	38.94	113	BELT SYSTEM
			41.37	26.32	19	FLOOR & ITS COMPONENTS
			39.29	30.88	68	ROOF & ITS COMPONENTS
			39.17	41.94	124	SIDE INTERIOR SURFACE
			38.00	37.75	1224	average injured occupant
			36.07	26.32	76	STEERING ASSEMBLY
ROLLOVER FIRST/WORST EVENT	BELTED	ARM	45.21	42.11	19	INSTRUMENT PANEL
			41.21	50.68	73	ROOF & ITS COMPONENTS
			39.86	43.94	66	EXTERIOR
			39.08	47.50	40	SIDE INTERIOR SURFACE
			38.00	37.75	1224	average injured occupant
			28.25	45.00	20	STEERING ASSEMBLY

3+ standard deviations higher than average injured occupant (p < .002).
*1.96-3 standard deviations higher than average injured occupant (p < .05).

181

IMPACT TYPE	OCCUPANT PROTECTION	BODY REGION	AVERAGE AGE	PERCENT FEMALE	N	INJURY SOURCE
ROLLOVER FIRST/WORST EVENT	BELTED	LEG	38.00	37.75	1224	average injured occupant
			35.83	52.50	40	FLOOR & ITS COMPONENTS
			35.41	39.22	51	INSTRUMENT PANEL
			34.24	47.06	17	STEERING ASSEMBLY
ROLLOVER FIRST/WORST EVENT	NO BELT	HEAD	32.82	29.41	34	A-PILLAR
			32.53	24.02	1815	average injured occupant
			32.04	17.65	306	ROOF & ITS COMPONENTS
			31.88	29.41	17	OTHER SIDE COMPONENT
			31.66	26.70	442	EXTERIOR
			30.86	26.25	80	OTHER FRONTAL COMPONENT
			27.00	30.00	20	STEERING ASSEMBLY
ROLLOVER FIRST/WORST EVENT	NO BELT	NECK	* 35.43	18.18	99	ROOF & ITS COMPONENTS
			33.06	35.38	130	EXTERIOR
			32.53	24.02	1815	average injured occupant
ROLLOVER FIRST/WORST EVENT	NO BELT	TORSO	* 39.19	9.38	32	SEAT & ITS COMPONENTS
			* 38.61	27.78	18	FLOOR & ITS COMPONENTS
			* 35.47	17.56	131	STEERING ASSEMBLY
			34.25	20.83	24	INSTRUMENT PANEL
			33.86	17.65	119	ROOF & ITS COMPONENTS
			33.11	40.74	27	SIDE HARDWARE OR ARMREST
			32.53	24.02	1815	average injured occupant
			32.27	26.56	128	SIDE INTERIOR SURFACE
			31.46	29.22	503	EXTERIOR
			29.19	6.25	16	OTHER SIDE COMPONENT
ROLLOVER FIRST/WORST EVENT	NO BELT	ARM	* 38.53	42.11	19	INSTRUMENT PANEL
			32.80	29.92	127	EXTERIOR
			32.62	20.59	34	ROOF & ITS COMPONENTS
			32.53	24.02	1815	average injured occupant
			31.74	36.84	19	SIDE INTERIOR SURFACE
ROLLOVER FIRST/WORST EVENT	NO BELT	LEG	33.52	26.09	46	INSTRUMENT PANEL
			33.37	31.71	41	FLOOR & ITS COMPONENTS
			32.77	36.89	122	EXTERIOR
			32.53	24.02	1815	average injured occupant
			26.80	6.67	15	STEERING ASSEMBLY
REAR & OTHER	BELTED	HEAD	46.06	31.25	16	OTHER
			44.81	47.22	36	B-PILLAR
			43.41	49.37	79	SEAT & ITS COMPONENTS
			42.43	44.63	475	average injured occupant
			41.56	38.89	36	STEERING ASSEMBLY
			40.39	27.27	33	ROOF & ITS COMPONENTS
			39.15	40.00	20	NON-CONTACT
REAR & OTHER	BELTED	NECK	47.57	34.29	35	SEAT & ITS COMPONENTS
			42.43	44.63	475	average injured occupant
REAR & OTHER	BELTED	TORSO	* 49.03	48.39	31	BELT SYSTEM
			45.27	50.53	95	SEAT & ITS COMPONENTS
			44.41	37.93	29	STEERING ASSEMBLY
			43.30	30.00	20	B-PILLAR
			42.43	44.63	475	average injured occupant
			38.37	40.07	15	SIDE INTERIOR CONTROL
REAR & OTHER	BELTED	LEG	43.10	55.00	20	FLOOR & ITS COMPONENTS
			42.43	44.63	475	average injured occupant
			40.83	63.33	30	INSTRUMENT PANEL
REAR & OTHER	NO BELT	HEAD	40.94	18.75	16	B-PILLAR
			38.89	42.86	35	SEAT & ITS COMPONENTS
			38.27	13.64	22	EXTERIOR
			36.55	28.51	242	average injured occupant
			30.28	13.79	29	ROOF & ITS COMPONENTS
REAR & OTHER	NO BELT	NECK	40.80	20.00	20	SEAT & ITS COMPONENTS
			36.55	28.51	242	average injured occupant
REAR & OTHER	NO BELT	TORSO	40.12	21.43	42	SEAT & ITS COMPONENTS
			39.57	21.74	23	EXTERIOR
			36.55	28.51	242	average injured occupant
			36.54	14.29	28	STEERING ASSEMBL

**3+ standard deviations higher than average injured occupant (p < .002).
*1.96-3 standard deviations higher than average injured occupant (p < .05).

A key finding of Tables 4-1a and 5-2a as well as the literature is that head-injury risk increases relatively slowly with age, torso-injury risk more rapidly. Another way to say this is that head-injury risk increases slower than the risk of the "average" injury; torso-injury risk increases faster. Table 8-1 reflects these trends in that most of the individual sources of head injury have average age lower than the benchmark (the "average" or "typical" injury) whereas most sources of torso injury have higher average age than the benchmark; likewise for neck injuries.

Two factors cause the "benchmark" ages to vary throughout Table 8-1. One is exposure. Older drivers, for example, are relatively less likely to be involved in rollover crashes and more likely to buckle up. Thus, the benchmark ages are lower in rollovers than in frontal or side impacts and higher for belted than unbelted drivers. The second factor is that belts, although beneficial at all ages, are somewhat more effective for younger drivers. That further skews the injury distribution of belted drivers toward older occupants.

The significance of the difference between the average age for a particular injury source and the benchmark depends on the magnitude of the observed difference and the number of CDS cases with that injury source. Because of smaller numbers of cases, a higher age may be non-significant while a lower age is significant. For example, the analysis of frontal impacts, belted, without air bags, resulting in torso injuries (the last cell on the first page of Table 8-1) includes eight injury sources that occurred at least 15 times each. The first of these sources is non-contact injury; the second source is the steering assembly. The average age for the 41 non-contact injuries, 47.54 is not significantly higher than the benchmark of 42.17, whereas the 47.24 for steering assembly, based on 699 cases, is statistically significant. It is important to consider the absolute magnitude of the difference from the benchmark as well as its statistical significance. For example, if an injury source averages 3 years older than the benchmark, it does not mean too much, even if the difference is significant due to very large numbers of cases, but if the source averages 20 years older than the benchmark, based on a reasonably large sample and a significant result, it means quite a bit.

The average age of the victims for an injury source reflects two factors: the intrinsic vulnerability of the occupant and the specific role of the injury source. Because older people are inherently vulnerable to torso injury, the average ages for most of the sources of torso injury will inevitably be higher than for most of the head-injury sources, as discussed above. Nevertheless, the fact that some of the torso-injury sources have substantially higher average age than others shows that the vehicle component also matters, that some components are especially likely to injure older people and/or not injure young people.

Table 8-2 lists all the drivers' injury sources with average victims' ages significantly (> 1.96 standard deviations) higher than the benchmark. The first section of Table 8-2, analyzing injuries with AIS 2 to 6, recapitulates Table 8-1, but all injury sources with non-significant results are omitted; furthermore, if none of the results in a cell are significant, the cell is omitted. The last two sections of Table 8-2 present the corresponding significant results for AIS 3-to-6 and AIS 4-to-6 injuries.

Table 8-2: Drivers' Injury Sources With High Average Victims' Ages
(Cars and LTVs, CDS 1988-2010)

IMPACT TYPE	OCCUPANT PROTECTION	BODY REGION	AVERAGE AGE	PERCENT FEMALE	N	INJURY SOURCE

AIS ≥ 2

IMPACT TYPE	OCCUPANT PROTECTION	BODY REGION	AVERAGE AGE	PERCENT FEMALE	N	INJURY SOURCE
FRONTAL	BELTED, AIR BAG	NECK	** 56.00	66.67	21	BELT SYSTEM
			* 55.11	47.37	19	SEAT & ITS COMPONENTS
			** 52.00	64.58	48	NON-CONTACT
			* 51.53	52.63	19	A-PILLAR
			* 49.26	46.81	47	STEERING ASSEMBLY
			43.37	51.30	4255	average injured occupant
FRONTAL	BELTED, AIR BAG	TORSO	** 62.37	60.00	60	AIR BAG
			** 51.71	45.83	48	NON-CONTACT
			** 48.63	48.02	683	BELT SYSTEM
			* 47.09	48.18	110	SIDE HARDWARE OR ARMREST
			* 45.69	42.55	463	STEERING ASSEMBLY
			43.37	51.30	4255	average injured occupant
FRONTAL	BELTED, AIR BAG	ARM	* 46.08	62.40	258	STEERING ASSEMBLY
			43.37	51.30	4255	average injured occupant
FRONTAL	BELTED, NO AIR BAG	NECK	* 52.55	45.00	20	NON-CONTACT
			* 48.13	46.03	63	STEERING ASSEMBLY
			42.17	43.18	3175	average injured occupant
FRONTAL	BELTED, NO AIR BAG	TORSO	** 47.24	41.06	699	STEERING ASSEMBLY
			** 45.96	45.12	492	BELT SYSTEM
			42.17	43.18	3175	average injured occupant
FRONTAL	NO BELT, AIR BAG	NECK	* 43.00	39.13	46	STEERING ASSEMBLY
			37.24	32.63	2400	average injured occupant
FRONTAL	NO BELT, AIR BAG	TORSO	* 44.91	42.86	35	AIR BAG
			** 40.22	31.76	677	STEERING ASSEMBLY
			37.24	32.63	2400	average injured occupant
FRONTAL	NO BELT, AIR BAG	ARM	* 46.84	68.42	19	AIR BAG
			* 40.15	28.64	206	INSTRUMENT PANEL
			37.24	32.63	2400	average injured occupant
FRONTAL	NO BELT, NO AIR BAG	NECK	* 44.29	14.29	42	ROOF & ITS COMPONENTS
			** 44.10	32.17	115	OTHER FRONTAL COMPONENT
			37.61	30.55	5081	average injured occupant
FRONTAL	NO BELT, NO AIR BAG	TORSO	** 50.56	38.89	18	NON-CONTACT
			** 42.39	28.57	119	SIDE INTERIOR SURFACE
			** 41.88	28.06	1631	STEERING ASSEMBLY
			** 40.53	22.28	359	INSTRUMENT PANEL
			37.61	30.55	5081	average injured occupant
FRONTAL	NO BELT, NO AIR BAG	ARM	** 40.91	33.69	374	INSTRUMENT PANEL
			37.61	30.55	5081	average injured occupant
NEARSIDE IMPACT	BELTED	NECK	* 50.05	53.13	64	B-PILLAR
			43.94	48.72	2974	average injured occupant
NEARSIDE IMPACT	BELTED	TORSO	* 47.70	41.04	212	B-PILLAR
			* 47.40	45.10	204	BELT SYSTEM
			** 46.46	46.53	1255	SIDE INTERIOR SURFACE
			* 46.10	52.77	741	SIDE HARDWARE OR ARMREST
			43.94	48.72	2974	average injured occupant
NEARSIDE IMPACT	NO BELT	TORSO	* 40.47	43.46	306	SIDE HARDWARE OR ARMREST
			** 40.42	37.61	694	SIDE INTERIOR SURFACE
			37.83	33.71	1771	average injured occupant

**3+ standard deviations higher than average injured occupant (p < .002).
*1.96-3 standard deviations higher than average injured occupant (p < .05).

Table 8-2 (Continued): Drivers' Injury Sources With High Average Victims' Ages (Cars and LTVs, CDS 1988-2010)

IMPACT TYPE	OCCUPANT PROTECTION	BODY REGION	AVERAGE AGE	PERCENT FEMALE	N	INJURY SOURCE

AIS ≥ 2 (Continued)

IMPACT TYPE	OCCUPANT PROTECTION	BODY REGION	AVERAGE AGE	PERCENT FEMALE	N	INJURY SOURCE
FAR-SIDE IMPACT	BELTED	TORSO	** 53.00	51.79	56	STEERING ASSEMBLY
			* 52.58	50.00	24	INSTRUMENT PANEL
			** 51.16	55.17	203	BELT SYSTEM
			* 49.87	43.59	39	OTHER
			43.05	47.40	1310	average injured occupant
FAR-SIDE IMPACT	BELTED	ARM	* 51.08	55.26	38	INSTRUMENT PANEL
			* 50.94	72.00	50	STEERING ASSEMBLY
			43.05	47.40	1310	average injured occupant
FAR-SIDE IMPACT	BELTED	LEG	** 51.51	64.20	81	FLOOR & ITS COMPONENTS
			43.05	47.40	1310	average injured occupant
FAR-SIDE IMPACT	NO BELT	NECK	** 51.57	23.81	21	A-PILLAR
			* 46.84	36.36	44	SIDE INTERIOR SURFACE
			38.60	31.21	1535	average injured occupant
FAR-SIDE IMPACT	NO BELT	TORSO	* 44.43	26.09	23	OTHER
			38.60	31.21	1535	average injured occupant
FAR-SIDE IMPACT	NO BELT	ARM	* 45.94	38.30	47	INSTRUMENT PANEL
			38.60	31.21	1535	average injured occupant
FAR-SIDE IMPACT	NO BELT	LEG	** 47.09	50.00	58	FLOOR & ITS COMPONENTS
			38.60	31.21	1535	average injured occupant
ROLLOVER FIRST/WORST EVENT	BELTED	NECK	* 42.01	27.59	145	ROOF & ITS COMPONENTS
			38.00	37.75	1224	average injured occupant
ROLLOVER FIRST/WORST EVENT	BELTED	TORSO	* 43.59	26.09	46	B-PILLAR
			* 42.83	38.46	65	SEAT & ITS COMPONENTS
			* 42.15	38.94	113	BELT SYSTEM
			38.00	37.75	1224	average injured occupant
ROLLOVER FIRST/WORST EVENT	NO BELT	NECK	* 35.43	18.18	99	ROOF & ITS COMPONENTS
			32.53	24.02	1815	average injured occupant
ROLLOVER FIRST/WORST EVENT	NO BELT	TORSO	* 39.19	9.38	32	SEAT & ITS COMPONENTS
			* 38.61	27.78	18	FLOOR & ITS COMPONENTS
			* 35.47	17.56	131	STEERING ASSEMBLY
			32.53	24.02	1815	average injured occupant
ROLLOVER FIRST/WORST EVENT	NO BELT	ARM	* 38.53	42.11	19	INSTRUMENT PANEL
			32.53	24.02	1815	average injured occupant
REAR & OTHER	BELTED	TORSO	* 49.03	48.39	31	BELT SYSTEM
			42.43	44.63	475	average injured occupant

AIS ≥ 3

IMPACT TYPE	OCCUPANT PROTECTION	BODY REGION	AVERAGE AGE	PERCENT FEMALE	N	INJURY SOURCE
FRONTAL	BELTED, AIR BAG	NECK	* 53.33	76.19	21	NON-CONTACT
			44.45	48.73	2048	average injured occupant
FRONTAL	BELTED, AIR BAG	TORSO	** 64.45	69.70	33	AIR BAG
			* 47.17	50.20	245	BELT SYSTEM
			44.45	48.73	2048	average injured occupant
FRONTAL	BELTED, AIR BAG	ARM	* 49.25	86.84	76	AIR BAG
			44.45	48.73	2048	average injured occupant

**3+ standard deviations higher than average injured occupant (p < .002).
*1.96-3 standard deviations higher than average injured occupant (p < .05).

185

Table 8-2 (Continued): Drivers' Injury Sources With High Average Victims' Ages (Cars and LTVs, CDS 1988-2010)

AIS ≥ 3 (Continued)

IMPACT TYPE	OCCUPANT PROTECTION	BODY REGION	AVERAGE AGE	PERCENT FEMALE	N	INJURY SOURCE
FRONTAL	BELTED, NO AIR BAG	TORSO	* 48.14	43.75	144	BELT SYSTEM
			** 47.44	38.07	415	STEERING ASSEMBLY
			43.81	41.97	1413	average injured occupant
FRONTAL	NO BELT, AIR BAG	HEAD	* 49.55	63.64	11	AIR BAG
			38.15	29.71	1518	average injured occupant
FRONTAL	NO BELT, AIR BAG	TORSO	* 47.11	52.63	19	AIR BAG
			* 40.58	32.06	496	STEERING ASSEMBLY
			38.15	29.71	1518	average injured occupant
FRONTAL	NO BELT, NO AIR BAG	NECK	* 46.52	37.50	48	OTHER FRONTAL COMPONENT
			39.66	28.26	2788	average injured occupant
FRONTAL	NO BELT, NO AIR BAG	TORSO	* 48.10	35.00	20	FLOOR & ITS COMPONENTS
			** 42.83	25.29	1103	STEERING ASSEMBLY
			39.66	28.26	2788	average injured occupant
NEARSIDE IMPACT	BELTED	HEAD	* 56.00	50.00	14	STEERING ASSEMBLY
			44.77	46.34	1830	average injured occupant
NEARSIDE IMPACT	BELTED	NECK	* 53.04	50.00	26	B-PILLAR
			44.77	46.34	1830	average injured occupant
NEARSIDE IMPACT	BELTED	TORSO	* 46.46	45.94	849	SIDE INTERIOR SURFACE
			44.77	46.34	1830	average injured occupant
NEARSIDE IMPACT	NO BELT	TORSO	* 40.79	36.05	516	SIDE INTERIOR SURFACE
			38.39	31.95	1280	average injured occupant
FAR-SIDE IMPACT	BELTED	TORSO	** 55.05	44.87	78	BELT SYSTEM
			44.05	42.75	648	average injured occupant
FAR-SIDE IMPACT	BELTED	ARM	* 54.06	83.33	18	STEERING ASSEMBLY
			44.05	42.75	648	average injured occupant
FAR-SIDE IMPACT	NO BELT	HEAD	* 51.14	22.73	22	INSTRUMENT PANEL
			39.74	29.25	988	average injured occupant
FAR-SIDE IMPACT	NO BELT	TORSO	* 44.74	27.03	74	INSTRUMENT PANEL
			39.74	29.25	988	average injured occupant
FAR-SIDE IMPACT	NO BELT	LEG	* 52.33	41.67	12	STEERING ASSEMBLY
			39.74	29.25	988	average injured occupant
ROLLOVER FIRST/WORST EVENT	BELTED	ARM	* 51.00	50.00	12	INSTRUMENT PANEL
			39.13	38.32	669	average injured occupant
ROLLOVER FIRST/WORST EVENT	NO BELT	TORSO	* 41.11	0.00	18	SEAT & ITS COMPONENTS
			33.20	23.67	1293	average injured occupant
REAR & OTHER	BELTED	HEAD	* 56.00	54.55	11	B-PILLAR
			44.24	45.74	188	average injured occupant
REAR & OTHER	BELTED	NECK	* 54.62	30.77	13	SEAT & ITS COMPONENTS
			44.24	45.74	188	average injured occupant

**3+ standard deviations higher than average injured occupant (p < .002).
*1.96-3 standard deviations higher than average injured occupant (p < .05).

Table 8-2 (Continued): Drivers' Injury Sources With High Average Victims' Ages
(Cars and LTVs, CDS 1988-2010)

IMPACT TYPE	OCCUPANT PROTECTION	BODY REGION	AVERAGE AGE	PERCENT FEMALE	N	INJURY SOURCE
			AIS ≥ 4			
FRONTAL	BELTED, AIR BAG	TORSO	** 71.00	68.75	18	AIR BAG
			47.79	40.68	649	average injured occupant
FRONTAL	NO BELT, NO AIR BAG	TORSO	* 43.78	25.04	603	STEERING ASSEMBLY
			41.46	25.02	1431	average injured occupant
NEARSIDE IMPACT	NO BELT	TORSO	* 42.02	34.53	278	SIDE INTERIOR SURFACE
			39.00	30.52	842	average injured occupant
FAR-SIDE IMPACT	BELTED	TORSO	* 69.40	80.00	5	STEERING ASSEMBLY
			42.52	37.01	354	average injured occupant
FAR-SIDE IMPACT	NO BELT	NECK	* 59.00	40.00	5	OTHER FRONTAL COMPONENT
			39.34	25.69	654	average injured occupant
FAR-SIDE IMPACT	NO BELT	TORSO	* 45.97	25.00	32	INSTRUMENT PANEL
			39.34	25.69	654	average injured occupant

**3+ standard deviations higher than average injured occupant (p < .002).
*1.96-3 standard deviations higher than average injured occupant (p < .05).

Perhaps the injury source that stands out the most is the air bag in torso injuries. It is 3+ standard deviations older than the benchmark, for belted drivers, at the AIS $\geq$ 2, AIS $\geq$ 3, and AIS $\geq$ 4 levels and 1.96-to-3 higher for unbelted drivers at the AIS $\geq$ 2 and AIS $\geq$ 3 levels. Moreover, the average ages are substantially higher than the benchmark in absolute terms: 62.37 versus 43.37 at belted AIS $\geq$ 2, 64.45 versus 44.45 at AIS $\geq$ 3, and 71.00 versus 47.79 at AIS $\geq$ 4.

These findings do not imply that air bags are harmful for older drivers. On the contrary, they are quite effective in saving lives, as will be shown in Section 9.7. Instead, they suggest that when air bags save a young driver's life, that driver may walk away from the crash with little or no injury, but when they save an older driver's life there may still be injuries ranging from moderate to severe. Of course, the first priority for air bags is to continue saving lives of drivers of all ages. However, a distant second priority could be further analysis to understand and possibly mitigate these residual injuries to older drivers.

The belt system is cited as an injury source with victims' ages 3+ standard deviations higher than the benchmark for torso injury in frontal and far-side impacts as well as neck injury in frontals, plus additional cites at the lower level of significance. The literature states that the effect of aging is more severe in belt loading than in blunt impact force.[59] Again, these findings do not suggest that belts are ineffective, let alone harmful for older drivers – just that they are somewhat less effective than for young drivers and, even when they save lives, they may leave residual AIS 2+ injuries.

Non-contact injuries of the neck or torso show up three times in bold type on the first page of Table 8-2 (frontals, AIS $\geq$ 2 injuries). The finding does not reflect on any specific injury source, but rather that older occupants are intrinsically more vulnerable to this type of injury. However, it is conceivable that improvements to restraint systems could mitigate the injuries.

The steering assembly shows up many times as a source of neck, arm, or especially torso injury, a fair number of them in bold print. However, the absolute difference from the benchmark in victims' age is usually five years or less: the large numbers of injuries makes the differences statistically significant. Nevertheless, the results suggest that contact with the steering assembly is more likely to result in forces exceeding tolerance limits when the occupants are older. Similarly, the side interior surface for torso injuries in nearside impacts, the floor components for leg injuries in far-side impacts, and the instrument panel for various types of injuries have a significant preponderance of older victims.

Table 8-3 presents the corresponding findings for right-front passengers: injury sources with average victims' ages significantly (> 1.96 standard deviations) higher than the benchmark. Because there are only about ⅓ as many RF passengers as drivers, fewer differences are statistically significant and all of Table 8-3 fits on one page.

---

[59] Zhou, Rouhana, & Melvin (1996).

Table 8-3: Right-Front Passengers' Injury Sources With High Average Victims' Ages (Cars and LTVs, CDS 1988-2010)

IMPACT TYPE	OCCUPANT PROTECTION	BODY REGION	AVERAGE AGE	PERCENT FEMALE	N	INJURY SOURCE

AIS ≥ 2

IMPACT TYPE	OCCUPANT PROTECTION	BODY REGION	AVERAGE AGE	PERCENT FEMALE	N	INJURY SOURCE
FRONTAL	BELTED, AIR BAG	TORSO	* 55.75	70.00	20	AIR BAG
			** 51.75	78.37	237	BELT SYSTEM
			45.76	69.37	790	average injured occupant
FRONTAL	BELTED, AIR BAG	ARM	* 57.13	87.50	16	OTHER FRONTAL COMPONENT
			45.76	69.37	790	average injured occupant
FRONTAL	BELTED, NO AIR BAG	TORSO	** 51.76	73.49	298	BELT SYSTEM
			45.93	70.46	765	average injured occupant
FRONTAL	BELTED, NO AIR BAG	ARM	* 51.46	73.95	119	INSTRUMENT PANEL
			45.93	70.46	765	average injured occupant
FRONTAL	NO BELT, NO AIR BAG	NECK	** 48.78	62.96	27	INSTRUMENT PANEL
			** 48.78	45.00	40	OTHER FRONTAL COMPONENT
			36.53	51.34	1498	average injured occupant
FRONTAL	NO BELT, NO AIR BAG	TORSO	** 40.99	52.40	479	INSTRUMENT PANEL
			36.53	51.34	1498	average injured occupant
FRONTAL	NO BELT, NO AIR BAG	ARM	* 39.83	61.42	254	INSTRUMENT PANEL
			36.53	51.34	1498	average injured occupant
NEARSIDE IMPACT	BELTED	TORSO	* 61.56	62.50	16	OTHER
			* 53.73	79.10	67	BELT SYSTEM
			47.45	68.73	339	average injured occupant
NEARSIDE IMPACT	NO BELT	TORSO	* 41.15	41.03	39	INSTRUMENT PANEL
			34.49	44.48	353	average injured occupant
FAR-SIDE IMPACT	NO BELT	TORSO	* 37.61	50.00	234	SIDE INTERIOR SURFACE
			34.46	48.42	632	average injured occupant

AIS ≥ 3

IMPACT TYPE	OCCUPANT PROTECTION	BODY REGION	AVERAGE AGE	PERCENT FEMALE	N	INJURY SOURCE
FRONTAL	BELTED, NO AIR BAG	TORSO	* 55.52	70.92	141	BELT SYSTEM
			49.86	68.00	375	average injured occupant
FRONTAL	NO BELT, NO AIR BAG	TORSO	** 42.74	52.07	290	INSTRUMENT PANEL
			38.82	50.65	768	average injured occupant
NEARSIDE IMPACT	BELTED	TORSO	* 60.81	77.78	27	BELT SYSTEM
			49.34	67.48	163	average injured occupant
FAR-SIDE IMPACT	NO BELT	TORSO	* 39.25	51.18	170	SIDE INTERIOR SURFACE
			35.43	48.52	439	average injured occupant

AIS ≥ 4

IMPACT TYPE	OCCUPANT PROTECTION	BODY REGION	AVERAGE AGE	PERCENT FEMALE	N	INJURY SOURCE
FRONTAL	BELTED, AIR BAG	HEAD	* 69.20	100.00	5	AIR BAG
			47.91	61.74	115	average injured occupant

**3+ standard deviations higher than average injured occupant (p < .002).
*1.96-3 standard deviations higher than average injured occupant (p < .05).

189

As in Table 8-2, the victims of torso injury from the air bag or from the belt system are significantly older than the benchmark (but only the belt system reaches significance of 3+ standard deviations). The principal difference from Table 8-2, of course, is that the steering assembly is not an important injury source for RF passengers. In its place, the instrument panel repeatedly shows up as a source of torso, neck, or arm injuries with older-than-average victims.

Table 8-4 analyzes <u>backseat-outboard passengers</u>. There is even less data than in Table 8-3; few differences reach statistical significance. The "seat and its components" – i.e., the back of the front seat – shows up twice for unbelted passengers in frontal impacts as a source of leg injuries with older-than-average victims.

Table 8-4: Backseat Outboard Passengers' Injury Sources With High Average Victims' Ages
(Cars and LTVs, CDS 1988-2010)

IMPACT TYPE	OCCUPANT PROTECTION	BODY REGION	AVERAGE AGE	PERCENT FEMALE	N	INJURY SOURCE
			$AIS \geq 2$			
FRONTAL	BELTED	TORSO	* 50.11	75.38	65	BELT SYSTEM
			42.79	66.17	133	average injured occupant
FRONTAL	NO BELT	LEG	** 43.71	68.57	70	SEAT & ITS COMPONENTS
			35.51	48.32	358	average injured occupant
FAR-SIDE IMPACT	BELTED	TORSO	* 51.27	65.38	26	SIDE INTERIOR SURFACE
			40.03	56.76	74	average injured occupant
			$AIS \geq 3$			
FRONTAL	NO BELT	LEG	* 45.53	67.44	43	SEAT & ITS COMPONENTS
			36.10	47.76	201	average injured occupant
			$AIS \geq 4$			
NONE						

**3+ standard deviations higher than average injured occupant (p < .002).
*1.96-3 standard deviations higher than average injured occupant (p < .05).

190

8.3   Injury sources with high percentages of female victims

Table 8-5 lists all the drivers' injury sources for which the percentage of victims who are females is significantly (> 1.96 standard deviations) higher than the benchmark percentage of injured occupants who are female. Table 8-5 is distilled from Tables E-10, E-11, and E-12 in Appendix E, which list all injury sources, including those that do not have significantly higher-than-average percentages of female victims. The three sections of Table 8-5 analyze AIS ≥ 2, AIS ≥ 3 and AIS ≥ 4 injuries, respectively.

Because gender is a dichotomous variable whereas age is a continuous variable without extreme values, it takes a somewhat larger N of cases to show a significant difference in the percent of female victims than to show it for average age. That is one reason Table 8-5 is only three pages long, versus four pages for Table 8-2.

The salient feature of Table 8-5 is the strong presence of arm and leg injuries relative to torso, head, and neck injuries. The results are consistent with Table 5-2d showing high risk of arm and leg injuries for female occupants. One injury source dominates Table 8-5, consistently showing a high proportion of female victims at both the AIS ≥ 2 and the AIS ≥ 3 levels: the floor and its components, including the toe pan and pedals. Arm injuries have a more eclectic list of sources with high percentages of female victims at the AIS ≥ 2 and AIS ≥ 3 levels: the steering assembly, the air bag, and the A-pillar. All of these components can apply enough force to break (AIS 2) or shatter (AIS 3) a woman's arm bones more easily than a man's. The percentage of female victims is often far above the benchmark, and in several cases above 80 percent, peaking at 86.84 percent of the AIS ≥ 3 arm injuries attributed to air bags (66 of the 76 observed injuries).

Neck injuries are not conspicuous on the lists, considering that Tables 4-1d and 4-4d showed a high vulnerability of females to neck injury in fatal crashes and Table 5-2d also showed a fairly strong effect in the mostly nonfatal crashes of CDS. But one source of neck injury is notable: at the AIS > 3 level with belts and air bags, 76.19 percent of the victims of non-contact neck injury are females, as compared to the benchmark percentage of 48.73.

The side hardware or armrest appears five times as a cause of torso injury, four of them belted or unbelted nearside impacts. The armrest is perhaps positioned at a level where it presents more of a threat to females, whose stature and posture may be different from males – or, perhaps, females do not have men's typically larger arms and shoulders to keep the armrest away from their bodies. These CDS statistics, however, are based almost entirely on vehicles without side air bags; results might be different for newer vehicles.

Females are a high percentage of the victims of head injury from contacting the side interior surface in nearside impacts: four significant results in Table 8-5, one in bold print. Here, too, the smaller stature of females may be an issue. If they had been taller, their heads might have hit just the side window, which generally shatters easily and presents little risk of serious injury.

Table 8-5: Drivers' Injury Sources With High Percentages of Female Victims
(Cars and LTVs, CDS 1988-2010)

IMPACT TYPE	OCCUPANT PROTECTION	BODY REGION	PERCENT FEMALE	AVERAGE AGE	N	INJURY SOURCE

AIS ≥ 2

IMPACT TYPE	OCCUPANT PROTECTION	BODY REGION	PERCENT FEMALE	AVERAGE AGE	N	INJURY SOURCE
FRONTAL	BELTED, AIR BAG	ARM	** 60.00	44.83	150	AIR BAG
			** 69.88	43.33	83	A-PILLAR
			** 62.40	46.08	258	STEERING ASSEMBLY
			51.30	43.37	4255	average injured occupant
FRONTAL	BELTED, AIR BAG	LEG	** 63.36	43.69	1209	FLOOR & ITS COMPONENTS
			51.30	43.37	4255	average injured occupant
FRONTAL	BELTED, NO AIR BAG	ARM	** 53.45	43.08	232	STEERING ASSEMBLY
			43.18	42.17	3175	average injured occupant
FRONTAL	BELTED, NO AIR BAG	LEG	** 54.21	41.34	629	FLOOR & ITS COMPONENTS
			43.18	42.17	3175	average injured occupant
FRONTAL	NO BELT, AIR BAG	HEAD	** 58.33	37.55	60	AIR BAG
			32.63	37.24	2400	average injured occupant
FRONTAL	NO BELT, AIR BAG	ARM	** 68.42	46.84	19	AIR BAG
			* 42.86	35.87	91	STEERING ASSEMBLY
			32.63	37.24	2400	average injured occupant
FRONTAL	NO BELT, AIR BAG	LEG	** 48.57	37.58	523	FLOOR & ITS COMPONENTS
			32.63	37.24	2400	average injured occupant
FRONTAL	NO BELT, NO AIR BAG	HEAD	* 35.27	36.91	601	STEERING ASSEMBLY
			30.55	37.61	5081	average injured occupant
FRONTAL	NO BELT, NO AIR BAG	ARM	** 41.26	39.15	286	STEERING ASSEMBLY
			30.55	37.61	5081	average injured occupant
FRONTAL	NO BELT, NO AIR BAG	LEG	** 43.90	38.15	779	FLOOR & ITS COMPONENTS
			* 33.69	37.83	1021	INSTRUMENT PANEL
			30.55	37.61	5081	average injured occupant
NEARSIDE IMPACT	BELTED	HEAD	* 64.06	43.59	64	SIDE INTERIOR SURFACE
			48.72	43.94	2974	average injured occupant
NEARSIDE IMPACT	BELTED	TORSO	* 52.77	46.10	741	SIDE HARDWARE OR ARMREST
			48.72	43.94	2974	average injured occupant
NEARSIDE IMPACT	BELTED	ARM	* 63.24	48.04	68	STEERING ASSEMBLY
			48.72	43.94	2974	average injured occupant
NEARSIDE IMPACT	BELTED	LEG	* 59.05	46.91	105	FLOOR & ITS COMPONENTS
			48.72	43.94	2974	average injured occupant
NEARSIDE IMPACT	NO BELT	HEAD	* 56.67	44.17	30	SIDE INTERIOR SURFACE
			33.71	37.83	1771	average injured occupant
NEARSIDE IMPACT	NO BELT	TORSO	** 43.46	40.47	306	SIDE HARDWARE OR ARMREST
			33.71	37.83	1771	average injured occupant
FAR-SIDE IMPACT	BELTED	TORSO	* 55.17	51.16	203	BELT SYSTEM
			47.40	43.05	1310	average injured occupant
FAR-SIDE IMPACT	BELTED	ARM	** 72.00	50.94	50	STEERING ASSEMBLY
			47.40	43.05	1310	average injured occupant
FAR-SIDE IMPACT	BELTED	LEG	* 64.20	51.51	81	FLOOR & ITS COMPONENTS
			47.40	43.05	1310	average injured occupant

**3+ standard deviations higher than average injured occupant (p < .002).
*1.96-3 standard deviations higher than average injured occupant (p < .05).

Table 8-5 (Continued): Drivers' Injury Sources With High Percentages of Female Victims (Cars and LTVs, CDS 1988-2010)

IMPACT TYPE	OCCUPANT PROTECTION	BODY REGION	PERCENT FEMALE	AVERAGE AGE	N	INJURY SOURCE

AIS ≥ 2 (Continued)

IMPACT TYPE	OCCUPANT PROTECTION	BODY REGION	PERCENT FEMALE	AVERAGE AGE	N	INJURY SOURCE
FAR-SIDE IMPACT	NO BELT	ARM	* 51.85	38.67	27	STEERING ASSEMBLY
			31.21	38.60	1535	average injured occupant
FAR-SIDE IMPACT	NO BELT	LEG	* 50.00	47.09	56	FLOOR & ITS COMPONENTS
			* 46.27	43.13	67	INSTRUMENT PANEL
			31.21	38.60	1535	average injured occupant
ROLLOVER FIRST/WORST EVENT	BELTED	ARM	* 50.68	41.21	73	ROOF & ITS COMPONENTS
			37.75	38.00	1224	average injured occupant
ROLLOVER FIRST/WORST EVENT	NO BELT	NECK	* 35.38	33.06	130	EXTERIOR
			24.02	32.53	1815	average injured occupant
ROLLOVER FIRST/WORST EVENT	NO BELT	TORSO	* 40.74	33.11	27	SIDE HARDWARE OR ARMREST
			* 29.22	31.46	503	EXTERIOR
			24.02	32.53	1815	average injured occupant
ROLLOVER FIRST/WORST EVENT	NO BELT	LEG	** 36.89	32.77	122	EXTERIOR
			24.02	32.53	1815	average injured occupant
REAR & OTHER	BELTED	LEG	* 63.33	40.83	30	INSTRUMENT PANEL
			44.63	42.43	475	average injured occupant

AIS ≥ 3

IMPACT TYPE	OCCUPANT PROTECTION	BODY REGION	PERCENT FEMALE	AVERAGE AGE	N	INJURY SOURCE
FRONTAL	BELTED, AIR BAG	NECK	* 76.19	53.33	21	NON-CONTACT
			48.73	44.45	2048	average injured occupant
FRONTAL	BELTED, AIR BAG	TORSO	* 69.70	64.45	33	AIR BAG
			48.73	44.45	2048	average injured occupant
FRONTAL	BELTED, AIR BAG	ARM	** 86.84	49.25	76	AIR BAG
			** 71.60	41.00	80	STEERING ASSEMBLY
			* 67.57	43.68	37	A-PILLAR
			48.73	44.45	2048	average injured occupant
FRONTAL	BELTED, AIR BAG	LEG	** 64.45	45.37	256	FLOOR & ITS COMPONENTS
			48.73	44.45	2048	average injured occupant
FRONTAL	BELTED, NO AIR BAG	ARM	* 54.22	46.57	83	STEERING ASSEMBLY
			41.97	43.81	1413	average injured occupant
FRONTAL	BELTED, NO AIR BAG	LEG	** 59.09	42.50	132	FLOOR & ITS COMPONENTS
			41.97	43.81	1413	average injured occupant
FRONTAL	NO BELT, AIR BAG	HEAD	* 63.64	49.55	11	AIR BAG
			29.71	38.15	1518	average injured occupant
FRONTAL	NO BELT, AIR BAG	TORSO	* 52.63	47.11	19	AIR BAG
			29.71	38.15	1518	average injured occupant
FRONTAL	NO BELT, AIR BAG	ARM	* 63.64	41.09	11	OTHER FRONTAL COMPONENT
			* 55.56	42.06	18	SIDE INTERIOR SURFACE
			29.71	38.15	1518	average injured occupant
FRONTAL	NO BELT, AIR BAG	LEG	** 47.86	40.55	117	FLOOR & ITS COMPONENTS
			29.71	38.15	1518	average injured occupant

**3+ standard deviations higher than average injured occupant (p < .002).
*1.96-3 standard deviations higher than average injured occupant (p < .05).

IMPACT TYPE	OCCUPANT PROTECTION	BODY REGION	PERCENT FEMALE	AVERAGE AGE	N	INJURY SOURCE

AIS $\geq$ 3 (Continued)

IMPACT TYPE	OCCUPANT PROTECTION	BODY REGION	PERCENT FEMALE	AVERAGE AGE	N	INJURY SOURCE
FRONTAL	NO BELT, NO AIR BAG	HEAD	* 42.19	39.72	64	INSTRUMENT PANEL
			28.26	39.66	2788	average injured occupant
FRONTAL	NO BELT, NO AIR BAG	ARM	** 42.27	41.05	97	STEERING ASSEMBLY
			* 37.40	40.82	131	INSTRUMENT PANEL
			28.26	39.66	2788	average injured occupant
FRONTAL	NO BELT, NO AIR BAG	LEG	** 47.19	37.88	178	FLOOR & ITS COMPONENTS
			28.26	39.66	2788	average injured occupant
NEARSIDE IMPACT	BELTED	HEAD	* 63.64	43.61	33	SIDE INTERIOR SURFACE
			46.34	44.77	1830	average injured occupant
NEARSIDE IMPACT	BELTED	TORSO	* 52.66	45.84	395	SIDE HARDWARE OR ARMREST
			46.34	44.77	1830	average injured occupant
NEARSIDE IMPACT	NO BELT	HEAD	** 75.00	44.31	16	SIDE INTERIOR SURFACE
			31.95	38.39	1280	average injured occupant
NEARSIDE IMPACT	NO BELT	TORSO	* 39.88	40.53	173	SIDE HARDWARE OR ARMREST
			31.95	38.39	1280	average injured occupant
NEARSIDE IMPACT	NO BELT	ARM	* 58.82	43.35	17	STEERING ASSEMBLY
			31.95	38.39	1280	average injured occupant
FAR-SIDE IMPACT	BELTED	ARM	** 83.33	54.06	18	STEERING ASSEMBLY
			* 75.00	38.75	16	SIDE INTERIOR SURFACE
			42.75	44.05	648	average injured occupant
FAR-SIDE IMPACT	BELTED	LEG	* 66.67	53.33	18	FLOOR & ITS COMPONENTS
			42.75	44.05	648	average injured occupant
ROLLOVER FIRST/WORST EVENT	NO BELT	LEG	* 36.00	33.07	75	EXTERIOR
			23.67	33.20	1293	average injured occupant

AIS $\geq$ 4

IMPACT TYPE	OCCUPANT PROTECTION	BODY REGION	PERCENT FEMALE	AVERAGE AGE	N	INJURY SOURCE
FRONTAL	BELTED, AIR BAG	HEAD	* 62.96	36.81	27	A-PILLAR
			* 57.63	43.41	59	STEERING ASSEMBLY
			40.68	47.79	649	average injured occupant
FRONTAL	BELTED, AIR BAG	TORSO	* 68.75	71.00	16	AIR BAG
			40.68	47.79	649	average injured occupant
FRONTAL	NO BELT, AIR BAG	NECK	* 80.00	43.20	5	STEERING ASSEMBLY
			25.20	38.94	742	average injured occupant
FRONTAL	NO BELT, NO AIR BAG	HEAD	* 43.24	35.05	37	INSTRUMENT PANEL
			25.02	41.46	1431	average injured occupant
FRONTAL	NO BELT, NO AIR BAG	TORSO	* 40.63	33.69	32	EXTERIOR
			25.02	41.46	1431	average injured occupant
NEARSIDE IMPACT	BELTED	HEAD	* 70.00	43.05	20	SIDE INTERIOR SURFACE
			* 61.11	43.44	36	A-PILLAR
			43.34	45.92	1006	average injured occupant
NEARSIDE IMPACT	NO BELT	HEAD	* 66.67	51.33	9	SIDE INTERIOR SURFACE
			30.52	39.00	842	average injured occupant
FAR-SIDE IMPACT	BELTED	TORSO	* 80.00	69.40	5	STEERING ASSEMBLY
			37.01	42.52	354	average injured occupant

**3+ standard deviations higher than average injured occupant (p < .002).
*1.96-3 standard deviations higher than average injured occupant (p < .05).

194

Table 8-6 lists <u>RF passengers</u>' injury sources with high percentages of female victims. It is a much shorter list than for drivers (Table 8-5). Not a single injury source has a percentage of females 3+ standard deviations above the benchmark. This is, of course, partly because of the far smaller number of RF passengers in CDS and because the dichotomous nature of the gender variable requires more data to show significant differences. But there also appear to be some "real" differences between RF passengers and drivers.

Table 8-6: Right-Front Passengers' Injury Sources With High Percentages of Female Victims (Cars and LTVs, CDS 1988-2010)

IMPACT TYPE	OCCUPANT PROTECTION	BODY REGION	PERCENT FEMALE	AVERAGE AGE	N	INJURY SOURCE
			AIS ≥ 2			
FRONTAL	BELTED, AIR BAG	TORSO	* 76.37	51.75	237	BELT SYSTEM
			69.37	45.76	790	average injured occupant
FRONTAL	BELTED, AIR BAG	LEG	* 80.30	45.83	132	FLOOR & ITS COMPONENTS
			69.37	45.76	790	average injured occupant
FRONTAL	NO BELT, AIR BAG	LEG	* 65.93	35.24	91	FLOOR & ITS COMPONENTS
			52.07	36.19	484	average injured occupant
FRONTAL	NO BELT, NO AIR BAG	ARM	* 61.42	39.83	254	INSTRUMENT PANEL
			51.34	36.53	1498	average injured occupant
FRONTAL	NO BELT, NO AIR BAG	LEG	* 59.52	36.87	168	FLOOR & ITS COMPONENTS
			51.34	36.53	1498	average injured occupant
NEARSIDE IMPACT	NO BELT	ARM	* 68.18	41.50	22	INSTRUMENT PANEL
			44.48	34.49	353	average injured occupant
			AIS ≥ 3			
FRONTAL	NO BELT, AIR BAG	ARM	* 76.67	43.00	30	INSTRUMENT PANEL
			55.87	38.10	281	average injured occupant
FRONTAL	NO BELT, NO AIR BAG	ARM	* 62.26	40.39	106	INSTRUMENT PANEL
			50.65	38.82	768	average injured occupant
NEARSIDE IMPACT	NO BELT	ARM	* 83.33	45.33	12	INSTRUMENT PANEL
			43.06	35.45	209	average injured occupant t
			AIS ≥ 4			

NONE

3+ standard deviations higher than average injured occupant (p < .002).
*1.96-3 standard deviations higher than average injured occupant (p < .05).

195

The component most cited in Table 8-6 is the instrument panel as a source of arm injuries: five of the nine entries in Table 8-6, including all three at the AIS ≥ 3 level. But all of them pertain to <u>unbelted</u> passengers. Leg injuries from the floor and its components are cited three times, but two of these are also unbelted passengers. Belted female RF passengers do not have significant excess vulnerability to arm and leg injuries. That is a principal difference between the driver's and RF passenger's seats: a driver of short stature must sit close to the steering wheel and lower instrument panel in order to reach the pedals, but the RF passenger can move the seat back and, if she buckles up, keep her arms and legs out of harm's way.

Table 8-7 presents corresponding results for <u>backseat-outboard passengers</u>. The "seat and its components" – i.e., the back of the front seat – shows up twice for unbelted passengers in frontal impacts as a source of leg injuries. It also showed up twice in Table 8-4, indicating a shared vulnerability of females and older occupants to these leg injuries.

Table 8-7: Backseat Outboard Passengers' Injury Sources With High Percentages of Female Victims (Cars and LTVs, CDS 1988-2010)

IMPACT TYPE	OCCUPANT PROTECTION	BODY REGION	PERCENT FEMALE	AVERAGE AGE	N	INJURY SOURCE
			AIS ≥ 2			
FRONTAL	NO BELT	LEG	** 68.57	43.71	70	SEAT & ITS COMPONENTS
			48.32	35.51	358	average injured occupant
			AIS ≥ 3			
FRONTAL	NO BELT	LEG	* 67.44	45.53	43	SEAT & ITS COMPONENTS
			47.76	36.10	201	average injured occupant
FAR-SIDE IMPACT	NO BELT	TORSO	* 72.73	36.18	11	EXTERIOR
			40.00	34.40	125	average injured occupant
			AIS ≥ 4			
NONE						

**3+ standard deviations higher than average injured occupant (p < .002).
*1.96-3 standard deviations higher than average injured occupant (p < .05).

CHAPTER 9

CRASHWORTHINESS TECHNOLOGIES: FATALITY REDUCTION FOR OLDER OCCUPANTS AND WOMEN – ANALYSES OF 1975-2010 FARS DATA

9.0   Summary

NHTSA's analyses of the fatality-reducing effectiveness of seat belts and other crashworthiness technologies can be performed separately for occupants of different age groups or for males and females. All of these technologies in current-model vehicles have at least some benefit for adults of all age groups and of either gender; none of them are harmful for a particular age group or gender. Nevertheless, the analyses show some significant variations in effectiveness by age and gender. Seat belts, with the possible exception of the latest types equipped with pretensioners and load-limiters, have been somewhat less effective for older occupants than for young adults. Seat belts appear to be more effective for female drivers than for male drivers; for passengers, the reverse appears to be true, except possibly when the belts have pretensioners and load limiters. Non-inflatable non-belt technologies, such as energy-absorbing steering assemblies or side door beams tend to be less effective for older people than young adults; frontal air bags are about equally effective across all age groups; side air bags with head protection may be especially effective for older occupants. All of these non-belt technologies – with and without air bags – are helping females just as much and quite possibly more than they protect males.

9.1   Crashworthiness technologies that save lives

Chapter 3 estimated the overall increase in fatality risk attributable to aging one year and the risk for a female relative to a male of the same age. Then it investigated how the effects have changed over time. The increases for aging and females may have both intensified slightly from vehicles of the 1960s up to about 1990; since then, the added risk for females has substantially diminished, probably to less than half, while the increase for aging may also have decreased, but by a much smaller amount. A possible explanation of those results is that the crashworthiness of the vehicle fleet at first improved somewhat more for young occupants and men than for older occupants and women (intensifying the relative difference in risk); later on, crashworthiness improved for female and possibly older occupants even more than it continued to improve for young, male occupants. The observed trends over time combine the net effects of everything that might have changed in vehicles:

- Introductions of major new safety technologies or equipment – whose fatality reduction has been demonstrated by statistical analysis of crash data – such as frontal air bags
- Other safety technology gradually introduced in small increments, usually not required by any specific FMVSS, such as incremental revisions in design to improve performance on crash ratings
- Revised materials or design to refresh the product or modernize production, with consequent effects on safety, such as the gradual shift to higher-strength steels or new types of plastic

- Evolving market trends with safety consequences, such as shifts from cars to SUVs or from two-door to four-door cars

This chapter focuses on the first of those four groups: major individual technologies introduced since the 1950s to protect occupants in crashes:

- Seat belts, including 3-point belts, lap belts, and 2-point automatic belts – for drivers, RF passengers, and back-seat passengers, in cars and LTVs;
- Frontal air bags, including barrier-certified and sled-certified, for drivers and RF passengers, in cars and LTVs (FARS does not yet have enough cases with CAC air bags – phased in during MY 2004-2007 – to analyze their effectiveness by occupants' gender or age);
- Side air bags and curtain air bags;
- Energy-absorbing steering columns introduced in cars in the 1960s;
- Side impact protection without air bags, such as side-door beams in the 1970s and side-impact upgrades to meet the dynamic FMVSS No. 214 test in the 1990s; and
- The head-impact upgrade of FMVSS No. 201 in 1999-2003.

Furthermore, the chapter focuses directly on how effective each of these technologies is for older and female occupants, unlike Chapter 3 which merely tracks the overall effects of aging and gender over time and implies that any change could reflect, among many other factors, the influence of new safety technologies that are more effective (or less effective) for older occupants and/or women than for young males. NHTSA has already published evaluations of the technologies listed above. Each evaluation defines a statistical procedure to estimate fatality reduction for that device: the overall fatality reduction, comprising occupants of all ages, male and female. Here, the investigation consists of repeating those statistical procedures, with the latest FARS data, but now estimating effectiveness separately for males and females and for various age groups of occupants. Statistical tests will show if effectiveness is significantly different for older and younger occupants or for males and females.

9.2   Analysis method for non-belt technologies

NHTSA's effectiveness analyses based on FARS data for frontal and side air bags and for static built-in protection such as energy-absorbing steering columns usually consist of a 2x2 table comprising four cells, which are four counts of occupant fatalities on FARS. The rows of the table are vehicles produced immediately after versus immediately before the technology was introduced. The left column is a type of crash or impact where the technology is likely to be effective and the right column is a control group of crashes or impacts where little or no effectiveness should be expected.

In 1981, for example, NHTSA evaluated energy-absorbing (EA) steering assemblies meeting FMVSS Nos. 203 and 204.[60] At that time FARS data was available from CY 1975 through approximately nine months of CY 1979 (slightly more in some States, less in others). The

---

[60] Kahane, C. J. (1981). *An Evaluation of Federal Motor Vehicle Safety Standards for Passenger Car Steering Assemblies.* (Report No. DOT HS 805 705, Table 5-32 on p. 202). Washington, DC: National Highway Traffic Safety Administration. Available at www-nrd.nhtsa.dot.gov/Pubs/805705.PDF

technology was introduced by some manufacturers in MY 1967 and by others in 1968. It is designed to cushion and limit the force, at least to some extent, when a driver contacts the steering assembly in a frontal impact. It would probably make little difference in a side or rear impact or a first-event rollover: those crash involvements are the control group. NHTSA's 1981 report compared the driver fatality counts from the existing FARS data of MY 1965-1966 cars (none of which have EA steering assemblies) to MY 1968-1969 cars (all of which are equipped with the technology) – in frontal impacts (IMPACT2 = 11, 12, or 1) to the control group of non-frontal impacts (IMPACT2 = 2 to 10). The report skipped MY 1967 vehicles because some were equipped with the new technology and some were not:

```
EFFECTIVENESS OF ENERGY-ABSORBING STEERING ASSEMBLIES IN REDUCING FRONTAL FATALITIES
1981 NHTSA Report (1975 to late-1979 FARS)
Table of FMVSS203 by FRONTAL (Driver Fatality Counts)


FMVSS203        FRONTAL

Frequency      │FRONTAL │SIDE OR │  Total
               │        │REAR    │
         ──────┼────────┼────────┼
MY 1968-1969   │  5518  │  3233  │   8751
         ──────┼────────┼────────┼
MY 1965-1966   │  3793  │  1977  │   5770
         ──────┼────────┼────────┼
Total             9311     5210     14521
```

The overall reduction of frontal-impact fatalities for drivers is:

$$1 - [(5518/3233) / (3793/1977)] = 11.0 \text{ percent}$$

It is a statistically significant reduction because chi-square for the table is 10.86 (which exceeds the 3.84 required for statistical significance at the two-sided .05 level).

Thirty years' hindsight suggests a few improvements in the original analysis:

- Limit the data to cars manufactured by AMC, Chrysler, Ford, or GM, where NHTSA is certain that the technology was introduced at the beginning of MY 1967 (AMC, Chrysler, and GM) or at the beginning of MY 1968 (Ford).
- Include MY 1967 cars in the analysis; do not skip that year.
- Compare cars of the first 2 MY with EA steering assemblies to the last 2 MY without them – i.e., 1967-1968 versus 1965-1966 AMC, Chrysler, and GM; 1968-1969 versus 1966-1967 Ford.
- Assign first-event rollovers to the non-frontal control group, even if IMPACT2 = 11, 12, or 1.

With these modifications, the effectiveness estimate based on FARS data from 1975 through September 1979 rises to 16 percent, perhaps because the analysis is better focused, or perhaps just by chance:

EFFECTIVENESS OF ENERGY-ABSORBING STEERING ASSEMBLIES IN REDUCING FRONTAL FATALITIES
FARS JANUARY 1975-SEPTEMBER 1979, FIRST 2 MY WITH EA COLUMNS VERSUS LAST 2 WITHOUT
Table of FMVSS203 by FRONTAL

FMVSS203   FRONTAL

Frequency	FRONTAL	NON-FRONTAL	Total
EA COLUMN	4194	3161	7355
NO EA COLUMN	3506	2224	5730
Total	7700	5385	13085

The statistically significant ($\chi^2 = 23.06$) overall reduction of frontal-impact fatalities for drivers is:

$$1 - [(4194/3161) / (3506/2224)] = 15.8 \text{ percent.}$$

In the meantime, many additional years of FARS data have become available to update the analyses. However, there is little point in using FARS data past 1985, because cars of the mid-1960s did not have the lifespan of today's cars and LTVs and were mostly retired by 1985. When the data are extended to CY 1985, effectiveness drops back to 12 percent, similar to the estimate in NHTSA's 1981 report, but continues to be statistically significant ($\chi^2 = 18.64$):

EFFECTIVENESS OF ENERGY-ABSORBING STEERING ASSEMBLIES IN REDUCING FRONTAL FATALITIES
FARS 1975-1985, FIRST 2 MY WITH EA COLUMNS VERSUS LAST 2 WITHOUT
Table of FMVSS203 by FRONTAL

FMVSS203   FRONTAL

Frequency	FRONTAL	NON-FRONTAL	Total
EA COLUMN	6004	4424	10428
NO EA COLUMN	4718	3047	7765
Total	10722	7471	18193

$$1 - [(6004/4424) / (4718/3047)] = 12.4 \text{ percent}$$

This last table, based on the near-maximum available 18,193 fatality cases, is the starting point for the analyses by gender and age. For the analysis by gender, separate tables are created for the male and female driver cases:

EFFECTIVENESS OF ENERGY-ABSORBING STEERING ASSEMBLIES IN REDUCING FRONTAL FATALITIES
FARS 1975-1985, FIRST 2 MY WITH EA COLUMNS VS. LAST 2 WITHOUT

MALES

Frequency	FRONTAL	NON-FRONTAL	Total
EA COLUMN	4951	3509	8460
NO EA COLUMN	3906	2451	6357
Total	8857	5960	14817

FEMALES

Frequency	FRONTAL	NON-FRONTAL	Total
EA COLUMN	1053	915	1968
NO EA COLUMN	812	596	1408
Total	1865	1511	3376

Fatality reduction is $1 - [(4951/3509) / (3906/2451)] = 11.5$ percent for the male drivers and it is $1 - [(1053/915) / (812/596)] = 15.5$ percent for the female drivers. (Note the high proportion of males in fatal crashes 30 to 40 years ago.) Evidently, effectiveness is similar – or at least not greatly different – for males and females. However, the SAS procedure CATMOD permits a statistical test of whether the observed effectiveness estimates for males and females are significantly different. CATMOD is a form of logistic regression that allows all the variables to be categorical rather than continuous and also accepts the data in the form of tables. In this case, FRONTAL is treated as the dependent variable (= 1 if the impact is frontal, = 2 if not). The two independent variables are FMVSS203 (= 1 if equipped with EA steering assembly, = 2 if not) and MALE (= 1 if male, = 2 if female). The 18,193 fatality cases in the preceding tables (14,817 males plus 3,376 females) are the 18,193 independent data points in the regression, each one receiving the values of FRONTAL, FMVSS203, and MALE according to which cell it inhabits. The logistic regression model for the dependent variable FRONTAL includes not only the two independent variables FMVSS203 and MALE but also their interaction term FMVSS203 * MALE. The regression coefficients and their statistics are:

Analysis of Maximum Likelihood Estimates

Parameter	Value	Estimate	Standard Error	Chi-Square	Pr > ChiSq
Intercept		0.3150	0.0195	260.14	<.0001
FMVSS203	EA COLUMN	-0.0726	0.0195	13.83	0.0002
MALE	MALE	0.0901	0.0195	21.30	<.0001
FMVSS203*MALE	EA COLUMN & MALE	0.0118	0.0195	0.36	0.5472

The positive intercept indicates that more fatalities are frontal (10,722) than non-frontal (7,471). The statistically significant negative (-0.0726) coefficient for FMVSS203 ($\chi^2 = 13.83$) says that EA steering assemblies reduce the odds that a fatality will be frontal – i.e., they save lives in frontal crashes relative to the non-frontal control group. The positive coefficient for MALE indicates that male drivers have a higher proportion of their fatalities in frontal crashes (8,857 of 14,817, which is 60%) than do female drivers (1,865 of 3,376, which is 55%). But the crucial statistics are the last line, FMVSS203*MALE. The positive coefficient 0.0118 indicates that the observed effectiveness of EA steering assemblies is slightly lower for males than for females (because more negative means more effective). However, with a chi-square of 0.36, this last effect is **not statistically significant**. We cannot conclude that the technology is more effective for females than males.

Actually, the regression coefficients themselves are not particularly important; it is their chi-square values that matter. It is sufficient to look at the simpler table of just the chi-squares and, as a matter of fact, even there, only the last row, the chi-square of FMVSS203*MALE really matters:

```
        Maximum Likelihood Analysis of Variance


Source              DF    Chi-Square    Pr > ChiSq
_____

Intercept            1       260.14       <.0001
FMVSS203             1        13.83       0.0002
MALE                 1        21.30       <.0001
FMVSS203*MALE        1         0.36       0.5472
```

The analysis of effectiveness by driver age is more complex because age is a continuous variable, not a dichotomy like gender. Furthermore, there is no particular reason to expect the fatality reduction to change linearly with age. It might be just as intuitively reasonable for effectiveness to remain fairly constant up to a certain age and then drop off rapidly. Thus, it appears unwarranted to use a mixed logistic regression model with age as a linear term and FMVSS203 as a categorical variable.

Instead the two-fold exploratory approach consists of, first, computing separate effectiveness estimates for five non-overlapping age groups of drivers: age 29 and younger, 30 to 54, 55 to 69, 70 to 79, and 80 to 96. The five estimates will provide a sense of whether or not there is any kind of age-related trend in the data. Next, three separate but partially overlapping CATMOD analyses will statistically test for a difference in effectiveness between a younger and an older group of drivers. In all three tests, the younger group consists of drivers age 50 or less. In the first test, the older group includes drivers 55+; in the second test, 70+; and in the third test, 80+. Note that the first older group includes all of the second and the second includes the third. These are not entirely independent statistical tests. If all three tests show directionally similar FMVSS203*age terms, at least two or even all three statistically significant, that would be strong evidence that effectiveness varies by age. On the other hand, if just one of the three test results is significant, it is difficult to judge whether that is a chance occurrence, a weak overall trend, or a strong effect at some ages but not at others.

Here are the effectiveness analyses of EA steering assemblies for the five age groups:

EFFECTIVENESS OF ENERGY-ABSORBING STEERING ASSEMBLIES IN REDUCING FRONTAL FATALITIES
FARS 1975-1985, FIRST 2 MY WITH EA COLUMNS VS. LAST 2 WITHOUT

AGEGP=UP TO 29: FATALITY REDUCTION 16%

Frequency	FRONTAL	NON-FRONTAL	Total
EA COLUMN	2957	2462	5419
NO EA COLUMN	2260	1574	3834
Total	5217	4036	9253

AGEGP=70-79: FATALITY REDUCTION 12%

Frequency	FRONTAL	NON-FRONTAL	Total
EA COLUMN	356	307	663
NO EA COLUMN	357	270	627
Total	713	577	1290

AGEGP=30-54: FATALITY REDUCTION 9%

Frequency	FRONTAL	NON-FRONTAL	Total
EA COLUMN	1749	1046	2795
NO EA COLUMN	1354	734	2088
Total	3103	1780	4883

AGEGP=80+: FATALITY REDUCTION -20%

Frequency	FRONTAL	NON-FRONTAL	Total
EA COLUMN	204	158	362
NO EA COLUMN	178	166	344
Total	382	324	706

AGEGP=55-69: FATALITY REDUCTION 13%

FMVSS203   FRONTAL

Frequency	FRONTAL	NON-FRONTAL	Total
EA COLUMN	738	451	1189
NO EA COLUMN	569	303	872
Total	1307	754	2061

The effectiveness estimates for ages 16 to 29, 30 to 54, 55 to 69, and 70 to 79 are all quite similar: 16, 9, 13, and 12 percent fatality reduction. The observed estimate for 80+, based on fewer cases than any of the others, is a 20 percent increase. It is difficult to say what kind of a trend, if any, is in the data. There is no particular reason to believe an EA steering assembly would increase fatality risk for any particular age group, because it is a static device that would not exert more force on the driver than the rigid steering assemblies in pre-standard vehicles. On the other hand, effectiveness might gradually decline with age, because the level of force needed to compress the EA devices is relatively high and perhaps better tolerated by a younger driver with a stronger ribcage. Thus, the observed negative effect might just be an indicator of close-to-zero effectiveness for the oldest drivers. The fairly constant results up to age 79 could indicate similar effectiveness, but they could also conceivably mask a gradual declining trend that started at an earlier age, because each of these estimates is based on relatively limited data and imprecise.

The first CATMOD analysis limits the data to drivers 49 and younger or 55 and older. The independent variable FIVEFIVE is defined: = 1 if 55 or older, = 2 if 49 or younger. The analysis is similar to the preceding CATMOD analysis by gender, except FIVEFIVE replaces MALE and the cases with drivers age 50 to 54 are excluded. Here are the two contingency tables and the chi-square statistics for the CATMOD analysis:

```
EA COLUMN EFFECTIVENESS FOR DRIVERS
AGE < 50 VS. AGE 55+
```

```
AGE < 50:  FATALITY REDUCTION 16%            AGE 55+:  FATALITY REDUCTION  5%
```

Frequency	FRONTAL	NON-FRON TAL	Total
EA COLUMN	4423	3369	7792
NO EA COLUMN	3410	2179	5589
Total	7833	5548	13381

Frequency	FRONTAL	NON-FRON TAL	Total
EA COLUMN	1298	916	2214
NO EA COLUMN	1104	739	1843
Total	2402	1655	4057

```
        Maximum Likelihood Analysis of Variance
```

Source	DF	Chi-Square	Pr > ChiSq
Intercept	1	400.45	<.0001
FMVSS203	1	9.67	0.0019
FIVEFIVE	1	0.17	0.6839
FMVSS203*FIVEFIVE	1	2.80	0.0946

The observed fatality reduction for EA steering assemblies is greater for drivers age < 50 (16%) than for drivers age 55+ (5%). The CATMOD analysis finds a chi-square of 2.80 for the FMVSS203*FIVEFIVE coefficient. When chi-square is between 2.71 and 3.84, it indicates statistical significance at the one-sided .05 level. That falls short of the usual criterion for statistical significance in NHTSA evaluations, the two-sided .05 level ($\chi^2 > 3.84$). It is a borderline significance that indicates the data are at least leaning in a certain direction. It is useful additional information to make note of results with one-sided significance; they flesh out the pattern based on the fully significant results – but it is important not to jump to any conclusion on the basis of a single one-sided result.

The second CATMOD analysis limits the data to drivers age 49 and younger or 70 and older. SEVENTY, replacing FIVEFIVE in the preceding regression, is defined: = 1 if 70 or older, = 2 if 49 or younger. Cases with drivers 50 to 69 are excluded.

```
EA COLUMN EFFECTIVENESS FOR DRIVERS
AGE < 50 VS. AGE 70+
```

```
AGE < 50:   FATALITY REDUCTION 16%              AGE 70+:   FATALITY REDUCTION   2%
```

Frequency	FRONTAL	NON-FRON TAL	Total		Frequency	FRONTAL	NON-FRON TAL	Total
EA COLUMN	4423	3369	7792		EA COLUMN	560	465	1025
NO EA COLUMN	3410	2179	5589		NO EA COLUMN	535	436	971
Total	7833	5548	13381		Total	1095	901	1996

```
            Maximum Likelihood Analysis of Variance
```

Source	DF	Chi-Square	Pr > ChiSq
Intercept	1	131.57	<.0001
FMVSS203	1	4.03	0.0447
SEVENTY	1	11.58	0.0007
FMVSS203*SEVENTY	1	2.63	0.1051

The observed fatality reduction for EA steering assemblies is again greater for drivers age < 50 (16%) than for drivers 70+ (2%). This time, however, the CATMOD chi-square of 2.63 for the FMVSS203*SEVENTY coefficient falls short of significance even at the one-sided .05 level (because the analysis is based on fewer cases). The last CATMOD analysis limits the data to drivers 49 and younger or 80 and older. EIGHTY replaces SEVENTY as an independent variable.

```
EA COLUMN EFFECTIVENESS FOR DRIVERS
AGE < 50 VS. AGE 80+
```

```
AGE < 50:   FATALITY REDUCTION 16%              AGE 80+:   FATALITY REDUCTION  -20%
```

Frequency	FRONTAL	NON-FRON TAL	Total		Frequency	FRONTAL	NON-FRON TAL	Total
EA COLUMN	4423	3369	7792		EA COLUMN	204	158	362
NO EA COLUMN	3410	2179	5589		NO EA COLUMN	178	166	344
Total	7833	5548	13381		Total	382	324	706

```
            Maximum Likelihood Analysis of Variance
```

Source	DF	Chi-Square	Pr > ChiSq
Intercept	1	45.25	<.0001
FMVSS203	1	0.00	0.9483
EIGHTY	1	6.45	0.0111
FMVSS203*EIGHTY	1	5.41	0.0200

The effect of EA steering assemblies appears to be substantially different for drivers age < 50 (16% fatality reduction) than for drivers 80+ (20% increase). In fact, the CATMOD chi-square of 5.41 for the FMVSS203*EIGHTY coefficient indicates the difference is statistically significant at the two-sided .05 level.

In summary, the three CATMOD analyses have directionally similar results, namely, that EA steering assemblies are less effective for the older than for the younger groups of drivers. One of these three is statistically significant (at the two-sided .05 level); one is borderline-significant (at the one-sided .05 level); and one is not significant. Together, they are moderately strong but not unequivocal evidence that the EA steering assemblies are less effective for older than for young occupants.

9.3   Analysis method for seat belts

NHTSA's effectiveness estimates for the various types of seat belts are based on <u>double-pair comparison</u> analyses of FARS data. Chapters 2 to 5 use double-pair comparison to estimate the effects of aging and gender or fatality risk, but the method was originally developed by Partyka and Evans to estimate fatality reduction by occupant protection requiring activation (namely, buckling up), such as child safety seats or seat belts.[61] Here is an example of double-pair comparison analysis to estimate the overall fatality reduction by seat belts for drivers, based on FARS cases of crash-involved vehicles occupied by a driver and an RF passenger, at least one of them a fatality. The presentation is extracted from NHTSA's 2000 evaluation of belts, substituting data used in this report. (The 2000 study also describes how to estimate belt effectiveness for the RF passenger from the same data – by exchanging the driver and passenger numbers and then repeating the computations).[62]

The starting point for the example is FARS data for CY 1975 to 1985. Records of passenger cars (BODY_TYP = 1-9) of MY 1974 to 1982 are extracted (1974 is the first model year that all cars were equipped with 3-point belt systems rather than separate lap belts and shoulder harnesses). The analysis is limited to:

- Cars with a driver and a RF passenger (and perhaps other passengers). When two or more people occupy the same seat, according to FARS, only the occupant with the lowest PER_NO is included.
- The driver, or the RF passenger, or both were fatally injured.
- The driver and the RF passenger both have known reported belt use: MAN_REST has to be 0 (unrestrained) or 1, 2, 3, 8 or 13 (belted, perhaps incorrectly). Volkswagen Rabbits with automatic 2-point belts are excluded from the analysis.

---

[61] Partyka, S. C. (1984). *Restraint Use and Fatality Risk for Infants and Toddlers*. Washington, DC: National Highway Traffic Safety Administration; Evans (1986a); Evans, L. (1986b). The Effectiveness of Safety Belts in Preventing Fatalities. *Accident Analysis and Prevention, 18*, pp. 229-241; Kahane, C. J. (1986). *An Evaluation of Child Passenger Safety: The Effectiveness and Benefits of Safety Seats*. (Report. DOT HS 806 890, Chapter 4). Washington, DC: National Highway Traffic Safety Administration Available at www-nrd.nhtsa.dot.gov/Pubs/806890.PDF ; Partyka, S. C. (1988). "Belt Effectiveness in Pickup Trucks and Passenger Cars by Crash Direction and Accident Year," *Papers on Adult Seat Belts – Effectiveness and Use*. (Report No. DOT HS 807 285). Washington, DC: National Highway Traffic Safety Administration.

[62] Kahane (2000), pp. 5-8.

- The driver and the RF passenger are both 13 to 96 years old. Cases are excluded if either occupant's gender is not reported.

There are 34,805 cars in CY 1975 to 1985 with a driver and a RF passenger, at least one fatal, both with known belt use and 13 to 96 years old. The vehicle cases tabulate as follows, based on each occupant's belt use and survival:

Vehicles	Driver Died RF Survived	Driver Survived RF Died	Both Died
Both unrestrained	12,861	13,272	6,051
Driver unrestrained, RF belted	290	154	67
Driver belted, RF unrestrained	203	516	97
Both belted	467	597	230

This can be re-tabulated as fatality counts rather than vehicle cases, by adding the "both died" column to each of the preceding columns. There are 41,250 fatalities (20,266 drivers and 20,984 RF passengers), classified as follows:

Fatalities	Driver Fatalities	RF Fatalities	Driver/RF Risk Ratio
Both unrestrained	18,912	19,323	0.979
Driver unrestrained, RF belted	357	221	1.615
Driver belted, RF unrestrained	300	613	0.489
Both belted	697	827	0.843

In CY 1975 to 1985, it is clear that (1) the overwhelming majority of people killed in crashes were unrestrained; (2) unrestrained drivers and RF passengers are at nearly equal risk in the same crash; and (3) whoever buckled up substantially reduced their risk.

The four rows of data allow a total of four double-pair comparisons, two for computing the effectiveness of belts for drivers, and two for RF passengers (not shown here). The first comparison for the driver is based on the first and third rows of data:

		Driver Fatalities	RF Fatalities	Driver/RF Risk Ratio
Driver unrestrained	RF unrestrained	18,912	19,323	0.979
Driver belted	RF unrestrained	300	613	0.489

In both pairs, the driver's fatality risk is compared to the same control group: the unrestrained RF passenger. The unrestrained driver has essentially the same fatality risk as the unrestrained RF passenger in the same crash, the belted driver about half. The fatality reduction for belts is

$$1 - (0.489/0.979) = 50.0 \text{ percent.}$$

The other comparison for the driver is based on the second and fourth rows of data:

		Driver Fatalities	RF Fatalities	Driver/RF Risk Ratio
Driver unrestrained	RF belted	357	221	1.615
Driver belted	RF belted	697	827	0.843

Here, the control group is the belted RF passenger. The unrestrained driver has higher fatality risk than the belted RF passenger in the same crash, the belted driver, lower. The fatality reduction is:

$$1 - (0.843/1.615) = 47.8 \text{ percent.}$$

It is important that the effectiveness estimates are quite similar with the two control groups: it suggests the estimates are robust and not affected by the choice of control group. The next task is to develop a weighting procedure that combines the two driver estimates into a single number, and likewise for the two RF passenger estimates. In the 1975-85 FARS data, the actual number of driver fatalities is:

$$\text{Actual driver fatalities} = 18,912 + 357 + 300 + 697 = 20,266$$

The first two numbers in that sum are unrestrained drivers, the last two, belted. However, if every driver had been unrestrained, that sum would have increased to:

$$\text{All-unrestrained driver fatalities} = 18,912 + 357 + (0.979 \times 613) + (1.615 \times 827) = 21,205$$

(Here, 613 was the number of unrestrained RF passenger fatalities that accompanied the 300 belted drivers and 0.979 is the risk ratio of unrestrained driver to unrestrained RF passenger fatalities; 827 is the number of belted RF passenger fatalities that accompanied the 697 belted drivers and 1.615 is the risk ratio of unrestrained drivers to belted RF passenger fatalities.) On the other hand, if every driver had buckled up, the sum would have dropped to:

$$\text{All-belted driver fatalities} = (0.489 \times 19,323) + (0.843 \times 221) + 300 + 697 = 10,632.$$

The overall effectiveness of 3-point belts for drivers in MY 1974-1982 cars is:

$$(21,205 - 10,632) / 21,205 = 49.82 \text{ percent,}$$

which is between the results of the two separate double-pair comparisons for drivers (50.0 and 47.8 percent).

This is the starting point for the analyses by gender and age. For the analysis by gender, separate double-pair comparison computations are performed for the subset of cases involving male drivers and the subset of female drivers. (The accompanying RF passenger may be of either gender in either of those subsets.) Results: belts reduce fatality risk by an estimated 48.1 percent for male drivers and by 54.6 percent for female drivers. The SAS procedure CATMOD permits a statistical test of whether the observed effectiveness estimates for males and females should be considered "approximately the same" or whether 3-point belts are significantly more effective for females than for males in MY 1974-1982 cars.

CATMOD, as discussed in the preceding section, is a form of logistic regression that allows all the variables to be categorical rather than continuous. Here, the 41,250 driver and RF passenger fatality cases from our example each have known values for the following four variables: DRIVER is treated as the dependent variable (= 1 if this specific fatality case is a driver, = 2 if it is an RF passenger). The three independent variables are BELT1 (= 1 if the driver of the <u>vehicle</u> in which the fatality occurred is belted, = 2 if unrestrained); BELT3 (= 1 if the RF passenger of the <u>vehicle</u> in which the fatality occurred is belted, = 2 if unrestrained); and MALE (= 1 if the driver of the <u>vehicle</u> in which the fatality occurred is male, = 2 if female).[63] The logistic regression model for the dependent variable DRIVER includes not only the three independent variables BELT1, BELT3, and MALE but also the three two-way interaction terms BELT1 * BELT3, BELT1 * MALE, and BELT3 * MALE. However, the three-way interaction term BELT1 * BELT3 * MALE is omitted. The regression coefficients and their statistics are:

Analysis of Maximum Likelihood Estimates

Parameter			Estimate	Standard Error	Chi-Square	Pr > ChiSq
Intercept			-0.0836	0.0335	6.22	0.0126
BELT1	BELTED		-0.3645	0.0347	110.09	<.0001
BELT3	BELTED		0.2959	0.0359	67.81	<.0001
MALE	MALE		-0.0331	0.0264	1.57	0.2097
BELT3*MALE	BELTED	MALE	-0.0672	0.0363	3.43	0.0641
BELT3*BELT1	BELTED	BELTED	0.00117	0.0311	0.00	0.9699
BELT1*MALE	BELTED	MALE	0.0509	0.0328	2.41	0.1204

As stated above, the dependent variable is DRIVER: given a fatally-injured occupant in these vehicles where drivers and RF passengers sit together, is this occupant a driver or an RF passenger? The results for the intercept and the three main terms (BELT1, BELT3, and MALE) are straightforward: the negative intercept indicates that, all else being equal, fatality risk is slightly lower for the driver than for the RF passenger. The statistically significant, strongly negative (- .3645) coefficient for BELT1 (χ^2 = 110.09) says that belt use by the driver reduces the odds that the fatality will be the driver – i.e., belts save drivers' lives. The significant <u>positive</u> (0.2959) coefficient for BELT3 (χ^2 = 67.81) says that belt use by the RF passenger increases the odds that, in this particular database, the fatality will be the driver – i.e., belts save passengers'

---

[63] In diagrams of passenger cars, the driver's seat is traditionally labeled no. 1, [the center-front seat, no. 2,] and the RF passenger seat no. 3 – thus, the variables for driver and RF passenger belt use are called BELT1 and BELT3, respectively.

lives; in this database where either the driver or the passenger (but only occasionally both) is a fatality, the survival of the passenger in this crash means that the driver must have been a fatality (the "zero-sum game" aspect of FARS discussed in Section 2.4 and elsewhere). The non-significant negative coefficient for MALE indicates that, all else being equal, male drivers are at slightly lower risk than females.

That leaves the three interaction terms. The crucial statistics are the last line, BELT1*MALE. The positive coefficient 0.0509 indicates that the observed effectiveness of belts is slightly lower for male drivers than for female drivers (because more negative means more fatality reduction). However, with a chi-square of 2.41, this last effect is **not statistically significant**. We cannot conclude that belts are more effective for female drivers than males in MY 1974-1982 cars. The other two interaction terms are difficult to interpret and they are also not particularly relevant except to the extent that they account for some of the variance in the data – and in this example, they are not statistically significant. For example, BELT3*MALE is the interaction between the passenger's belt use, the driver's gender, and the driver's fatality risk (the dependent variable). It says that when the driver is a male, the "zero-sum game" effect of passenger belt use increasing the driver's risk is not quite as strong as when the driver is a female; that could happen for various reasons involving how the variables interact, none of them obvious.

Actually, none of the regression coefficients themselves are particularly important; it is their chi-square values that matter. It is sufficient to look at the simpler table of just the chi-squares and, as a matter of fact, even there, only the chi-square of BELT1*MALE really matters:

```
          Maximum Likelihood Analysis of Variance


   Source          DF    Chi-Square     Pr > ChiSq
   ─────────────────────────────────────────────────
   Intercept        1         6.22         0.0126
   BELT3            1        67.81        <.0001
   BELT1            1       110.09        <.0001
   MALE             1         1.57         0.2097
   BELT3*MALE       1         3.43         0.0641
   BELT3*BELT1      1         0.00         0.9699
   BELT1*MALE       1         2.41         0.1204


   Likelihood Ratio  1         1.80         0.1802
```

This table, however, introduces one new term: a "Likelihood Ratio," which is found to be non-significant ($\chi^2 = 1.80$). It says that the three-way interaction term BELT1*BELT3*MALE, which is the only term omitted from the regression model, would not have been statistically significant if it had been included, and that it is therefore acceptable to omit it from the regression. That is good news, because in the absence of BELT1*BELT3*MALE, the two-way term BELT1*MALE directly measures the difference in belt effectiveness for males and females; its chi-square tells us if the difference is significant or not. But if the three-way term had been included, the two-way term BELT1*MALE would have lost its straightforward interpretation.

The analysis of effectiveness by driver age, as in the preceding section, consists of, first, obtaining point estimates for five non-overlapping age groups of drivers: 29 and younger, 30 to

210

54, 55 to 69, 70 to 79, and 80 to 96. (The accompanying RF passenger may be of any age 13 and older in any of those five subsets.) The five estimates will provide a sense of whether or not there is any kind of age-related trend in the data. Next, three CATMOD analyses will statistically test for a difference in effectiveness between a younger and an older group of drivers. In all three tests, the younger group consists of drivers 50 or younger. In the first test, the older group includes drivers 55+; in the second test, 70+; and in the third test, 80+. The five point estimates of effectiveness are:

Age Group	Fatality Reduction (%)
Up to 29	52
30-54	50
55-69	48
70-79	51
80 & older	- 16

The results happen to be similar to those for EA steering assemblies, with effectiveness staying almost constant up through age 79 and then dropping to negative for the oldest group (but where there is more uncertainty because there is less data).

The first CATMOD analysis limits the data to drivers 49 and younger or 55 and older (RF passengers may be any age, 13 and up). The independent variable FIVEFIVE, as in the preceding section, is defined: = 1 if 55 or older, = 2 if 49 or younger. The analysis is similar to the CATMOD analysis of belt effectiveness by gender, except FIVEFIVE replaces MALE and the cases with drivers 50 to 54 are excluded. Estimated effectiveness is 51 percent for the drivers age < 50 and 44 percent for the drivers 55+. Here are the chi-square statistics for the CATMOD analysis:

```
        Maximum Likelihood Analysis of Variance

Source              DF    Chi-Square    Pr > ChiSq

Intercept            1       19.52        <.0001
BELT3                1       43.71        <.0001
BELT1                1       76.73        <.0001
FIVEFIVE             1        7.40        0.0065
BELT3*FIVEFIVE       1        3.61        0.0574
BELT3*BELT1          1        0.29        0.5897
BELT1*FIVEFIVE       1        2.03        0.1542

Likelihood Ratio     1        0.79        0.3748
```

Because chi-square for the BELT1*FIVEFIVE interaction term is 2.03, the observed difference in effectiveness (51% for the younger drivers and 44% for the drivers age 55+) is not statistically significant. In the second CATMOD analysis, estimated effectiveness is 51 percent for the drivers age < 50 and 38 percent for the drivers 70+. Here are the chi-square statistics:

```
         Maximum Likelihood Analysis of Variance


Source              DF    Chi-Square    Pr > ChiSq
_____

Intercept            1       10.76        0.0010
BELT3                1       22.43        <.0001
BELT1                1       35.54        <.0001
SEVENTY              1        3.87        0.0492
BELT3*SEVENTY        1        1.59        0.2079
BELT3*BELT1          1        0.01        0.9129
BELT1*SEVENTY        1        1.90        0.1676


Likelihood Ratio     1        0.11        0.7439
```

Again, the BELT1*SEVENTY term is non-significant ($\chi^2 = 1.90$); the observed difference in effectiveness (51% for the younger drivers and 38% for the drivers age 70+) is not statistically significant. In the last CATMOD analysis, estimated effectiveness is 51 percent for the drivers age < 50 and -16 percent for the drivers 80+.

```
         Maximum Likelihood Analysis of Variance


Source              DF    Chi-Square    Pr > ChiSq
_____

Intercept            1        3.70        0.0543
BELT3                1        2.27        0.1319
BELT1                1        5.46        0.0194
EIGHTY               1        1.39        0.2391
BELT3*EIGHTY         1        2.50        0.1135
BELT3*BELT1          1        0.11        0.7427
BELT1*EIGHTY         1        2.79        0.0950


Likelihood Ratio     1        2.33        0.1268
```

Here, the observed difference is much larger, but the number of older-occupant cases is considerably smaller. Chi-square for the BELT1*EIGHTY interaction is 2.79, which is borderline-significant at the one-sided .05 level. The three CATMOD analyses have directionally similar results, namely, that 3-point belts are less effective for the older than for the younger groups of drivers in MY 1974-1982 cars. But in two of the three the observed difference is fairly small and not statistically significant, while the last one is only borderline-significant (at the one-sided .05 level). These three analyses, by themselves, are not sufficient to reach a conclusion that belts are less effective for older occupants in MY 1974-1982 cars. They need to be seen in the context of results for belts in other vehicles and at other seat positions (RF passengers). If the other analyses were to consistently show lower effectiveness for older occupants, the preceding results could be viewed as additional support for that conclusion.

Similar analyses may be performed to study belt effectiveness for the RF passengers in these vehicles – by exchanging the RF passenger and driver fatality cases and focusing on the interaction terms BELT3*MALE, BELT3*FIVEFIVE, etc. Similar analyses also work for drivers and RF passengers in other groups of vehicles; also for back-seat passengers – by double-pair comparison analyses of vehicles where there is a driver and a back-seat passenger and at least one of them is a fatality. A fourth independent variable needs to be included for the analysis of 3-point belts in the back seat – namely, whether or not the driver's seat is equipped with a frontal air bag (BAG1). This variable is unnecessary in the other analyses because (1) the front-seat analyses are performed separately for vehicles with dual air bags and for vehicles with no air bags; (2) the analyses of cars with lap belts only in the back seat are limited to vehicles without frontal air bags. However, notwithstanding the additional variable, CATMOD focuses on the same interaction term as in the other analyses.

FARS data from 1986 and later adds a complication when it is used for double-pair comparison analyses of belt effectiveness for drivers and RF passengers. The observed effectiveness is higher than with earlier FARS data, even for vehicles of the same model years. That is when large numbers of States began to pass buckle-up laws for drivers and RF passengers.

> "Specifically, New York was the first [S]tate to enact a belt use law, effective December 1, 1984. After a brief 'wait and see,' 21 [S]tates, including 9 of the 10 most populous [S]tates had belt laws effective by August 1986 for front-seat occupants of passenger cars. For the first time, unbelted people had a tangible incentive - avoidance of a fine - to report that they were belted. NHTSA hypothesized that uninjured or slightly injured occupants are often up and about before police arrive at the crash scene. Since the investigating officer is not an eye-witness to their belt use, they have an opportunity – and now also a motive – to say they wore belts, even if they hadn't. Mortally injured occupants may be in their original post-crash location when police arrive, often allowing direct observation of belt use. Thus, NHTSA believes belt use of fatalities is reported without net biases on FARS before and after belt laws. However, after the laws, belt use of survivors is over-reported. A bias has apparently been introduced in the reporting of this one data element, for survivors, as a consequence of belt use laws."[64]

NHTSA empirically derived a universal exaggeration factor (UEF) to adjust estimates based on FARS data after 1985 and make them comparable to estimates from earlier data.[65] The UEF is 1.369 and it is applied as follows to the observed effectiveness (percent fatality reduction) E* to estimate the actual effectiveness E:

$$E = 100 - [1.369 \times (100 - E^*)]$$

For example, if the observed effectiveness is 60 percent, the actual effectiveness is 45.24 percent. In the tables of this chapter, all effectiveness estimates for front-seat occupants are based either on FARS data up to CY 1985 or from CY 1986 and later, never a mixture of the two. All point estimates in the tables based on CY 1986 and later data have already been adjusted downward with the UEF. (Estimates of belt effectiveness for back-seat occupants are not adjusted with the UEF, because the buckle-up laws of the 1980s and 1990s usually did not apply to adult back-seat occupants.) However, even for the estimates based on CY 1986 and later data, the CATMOD analyses are based on the actual, observed data and are identical to the preceding example. The UEF is merely an adjustment factor that is applied in a uniform way to all estimates subsequent to the basic double-pair comparison analysis.

---

[64] Kahane (2000), pp. 2-3.
[65] Ibid., pp. 10-19.

9.4 Belt effectiveness by the occupant's gender

Table 9-1 compares fatality reduction by belts for men and women – for drivers and RF passengers of passenger cars – for six types/generations of seat belts. For example, use of a lap belt alone (in cars of the 1960s and early 1970s that were not yet equipped with 3-point belts) reduces a male driver's fatality risk by 26 percent and a female driver's risk by 41 percent. The results of the CATMOD analyses are color-coded in Table 9-1 and all subsequent tables. When the effectiveness estimate for females is printed **bold on a blue background with a frame**, the fatality reduction is significantly greater (at the two-sided .05 level) for women than for men. In regular type with a pale-blue background, the reduction for women is borderline-significantly greater (at the one-sided .05 level) than for men. In regular type with a pale-yellow background (not shown in Table 9-1, but will appear later), the observed fatality reduction is lower for women than for men and the difference is borderline-significant. **Bold type with an orange background and a frame** indicates the fatality reduction is significantly lower for women than for men.

The six types/generations of belts considered in Table 9-1 are:

- Lap belts only – includes cars of MY 1960 to 1967 equipped only with lap belts and MY 1968-1973 cars with separate lap belts and shoulder harnesses, as long as FARS does not explicitly state that the shoulder harness was in use (i.e., if MY $\geq$ 1968, include if MAN_REST is 2 or 8, exclude if it is 1 or 3). Based on double-pair comparison analyses of the available FARS data, lap belts reduce the overall fatality risk of drivers by an estimated 30 percent; they reduce the overall fatality risk of RF passengers by 23 percent.
- Two-point automatic belts – includes the motorized and non-motorized types furnished on some cars during MY 1975-1996. Included regardless whether the accompanying manual lap belt is worn or not. The estimated overall fatality-reducing effectiveness of 2-point belts is 42 percent for drivers and 27 percent for RF passengers: higher than the point estimates for lap-belt-only, but lower than any of the 3-point belts.
- Three-point belts, MY 1974-1982 – regardless of whether used correctly or incorrectly; estimates based on CY 1975-1985 FARS data (point estimates not adjusted with the UEF). Estimated overall fatality reduction: 50 percent for drivers, 39 percent for RF passengers.
- Three-point belts in cars without air bags, MY 1983-1996 – includes manual and automatic 3-point belts; based on CY 1986-2010 FARS data, (point estimates adjusted with the UEF). Estimated overall fatality reduction: 48 percent for drivers, 41 percent for RF passengers.
- Three-point belts in cars with air bags but neither pretensioners nor load limiters, MY 1985-2006. Estimated overall fatality reduction: 45 percent for drivers, 39 percent for RF passengers. These are the estimates of the incremental effectiveness of belts – i.e., the fatality reduction for a belted occupant in a car equipped with air bags relative to an unbelted person in a car equipped with air bags – not the combined effectiveness of belts and air bags.

214

TABLE 9-1: FATALITY REDUCTION (%) BY SEAT BELTS IN PASSENGER CARS, BY OCCUPANT'S GENDER
(Drivers and RF Passengers 13 and Older)

Gender	Lap Belt Only[66]	2-Point Automatic Belt	MY 1974-1982	MY 1983+ Without Air Bags	------- With Air Bags -------	
					No Pretensioner or Load Limiter	With Pretensioner & Load Limiter
DRIVERS						
Male	26	41	48	46	41	49
Female	41	45	55	51	56	52
RF PASSENGERS						
Male	33	36	56	49	41	44
Female	13	23	24	36	40	55

(header spanning: 3 – Point Belts)

---

[66] Includes MY 1968-1973 cars with separate lap and shoulder belts, as long as FARS does not specify shoulder belt was used.

- Three-point belts in cars with air bags, pretensioners, and load limiters, MY 1995-2011. Belt pretensioners retract the belt almost instantly in a crash to remove excess slack. By pulling in slack, they reduce occupants' impacts with interior surfaces and help the belt engage with the occupant to expedite "ride-down." Peak forces are reduced because force is applied over a longer period of time. Load limiters allow belts to yield in a crash, preventing the shoulder belt from exerting too much force on the chest of an occupant.[67] By MY 2007, all new cars were being equipped with pretensioners and load limiters for the driver's and RF belts. (Cars with pretensioners but not load limiters or *vice versa* are not included in the analyses.) Estimated fatality reduction: 50 percent for drivers and 48 percent for RF passengers.[68]

The salient feature of Table 9-1 is that every type of seat belt is more effective for female than for male <u>drivers</u>. The difference in observed effectiveness ranges from 3 to 15 percentage points; it is statistically significant in cars with air bags but without pretensioners and load limiters; it is borderline-significant in MY 1983+ cars without air bags. But for <u>RF passengers</u> the reverse is true in cars without air bags: belt effectiveness is 13 to 32 percentage points higher for males, reaching statistical significance for 3-point belts in MY 1974-1982 and MY 1983+ cars without air bags. Only in the next-to-last column is belt effectiveness about equal for males and females and only in the last column – cars with air bags, pretensioners, and load limiters – is the observed point estimate of effectiveness higher for females (55%) than for males (44%), although this difference is not statistically significant ($\chi^2 = 1.95$), as it is based on a somewhat limited number of FARS cases.

Overall, seat belts are more effective in LTVs than in cars, because "(1) ejection is substantially more frequent for unbelted occupants of LTVs than cars; (2) belts are much more effective in side impacts of LTVs…because intrusion is much less of a problem in LTVs; and (3) [historically, before ESC,] LTVs have [had] relatively more rollovers, where belts are most effective."[69] Overall, based on the latest FARS data, the estimated fatality reduction is 65 percent for drivers and 63 percent for RF passengers in LTVs without air bags; it is 74 percent for drivers and 67 percent for RF passengers in LTVs with air bags. Table 9-2 shows the same trend as Table 9-1 of belt effectiveness being higher for females at the driver's seat but higher for males at the RF seat.

For drivers, belt effectiveness is significantly higher for women in LTVs without air bags and in LTVs with air bags. For RF passengers, belt effectiveness is significantly lower for women than for men in LTVs without air bags. Even in LTVs with air bags the observed effectiveness is lower for women than men, a borderline-significant difference.

---

[67] Walz, M. C. (2003, March). *NCAP Test Improvements with Pretensioners and Load Limiters*. (Report No. DOT HS 809 562). Washington, DC: National Highway Traffic Safety Administration. Available at www-nrd.nhtsa.dot.gov/Pubs/809562.PDF

[68] In other words, the point estimates of effectiveness for belts with pretensioners and load limiters are higher than the corresponding estimates for belts without pretensioners and load limiters in cars with air bags. NHTSA plans to issue a separate report analyzing the incremental benefit of pretensioners and load limiters.

[69] Kahane (2000), p. 33.

TABLE 9-2: FATALITY REDUCTION (%) BY 3-POINT BELTS IN LTVs
BY OCCUPANT'S GENDER (Drivers and RF Passengers 13 and Older)

Gender	LTVs Without Air Bags (MY 1983-)	LTVs With Air Bags
	DRIVERS	
Male	64	71
Female	71	80
	RF PASSENGERS	
Male	68	70
Female	58	63

For back-seat occupants of passenger cars, based on the latest FARS data, lap belts reduce fatality risk by an estimated 41 percent and 3-point belts by 53 percent. However, Table 9-3 indicates that belt effectiveness is significantly lower for females than males: especially the lap belt but also the 3-point belt.

TABLE 9-3: FATALITY REDUCTION (%) BY SEAT BELTS FOR BACK-SEAT
PASSENGERS OF CARS, BY OCCUPANT'S GENDER (13 and Older)

Gender	Lap Belt Only	3-Point Belt
Male	52	60
Female	31	47

The historical trend of belts being more effective for females at the driver's seat and for males at the passenger seats is strong. The only observed exception is the latest cars with air bags, pretensioners, and load limiters, where the observed effectiveness is higher for females at both the driver and RF seats; however, there are not yet enough cases of these vehicles on FARS for firm conclusions. The principal differences between the driver and passenger seats are: (1) the steering assembly protrudes from the instrument panel toward the driver and (2) drivers with short legs (usually women) must move the seat forward to reach the pedals, placing them even closer to the steering assembly as well as other frontal components. At first glance, that would not seem to help female drivers, because the belts will not prevent their contacts with other frontal components. However, that may actually turn out to be an advantage if the other components absorb some of the occupant's kinetic energy, thereby reducing belt forces on the thorax. The contacts might also restrain excessive head motion relative to the torso, preventing neck injuries (a vulnerability of females, especially if belted without air bags – see Table 4-4d). Similarly, leg contacts, although sometimes resulting in nonfatal leg injuries (Table 5-2d), might reduce the possibility of the driver submarining under the belt and sustaining life-threatening abdominal injuries, another vulnerability of females (Table 4-1d). The latest results showing high belt effectiveness for female RF passengers in cars equipped with air bags, pretensioners, and load limiters offer hope that the combination of reduced belt loads and reduced head excursion are making the RF seat somewhat more like the driver seat for a belted female – but these results will need to be confirmed with additional data.

In addition to the preceding hypotheses based on vehicle design and human physiology, some covariates or artifacts in the data could also influence the results. For example, female drivers might be involved in less severe crashes, on the average, than male drivers – and belts are somewhat more effective in the less severe crashes (but female passengers are not necessarily in less severe crashes than male passengers, because in either case, the driver might be a male). Perhaps, surviving female passengers are less likely than other groups to misreport their belt use, resulting in lower observed effectiveness estimates.

9.5   Belt effectiveness by the occupant's age

Table 9-4 gives point estimates of the fatality reduction by belts for five age groups of occupants – for drivers and RF passengers of passenger cars – for six types/generations of seat belts. For example, use of a lap belt alone is estimated to reduce a driver's fatality risk by 31 percent for drivers 13 to 29, 28 percent at 30 to 54, 34 percent at 55 to 69, and 27 percent at 70 to 79, but to increase risk by 15 percent for drivers 80 and older. There is no color-coding in Table 9-4 because there are no CATMOD analyses, just point estimates; the CATMOD analyses are addressed in Table 9-5.

In general, belts are less effective for the older occupants. Sometimes (e.g., RF passengers MY 1983+ cars without air bags) estimated effectiveness decreases with each increase in age. In other cases (e.g., drivers' lap belts) effectiveness remains fairly constant and then drops off sharply in the oldest age group. Sometimes there is no clear pattern, especially when the point estimates have high sampling error due to limited data. However, the only two negative point estimates in Table 9-4 are for drivers 80+ in cars with a lap belt only (pre-1968) or with the earliest 3-point belts (MY 1974-1982); the estimates are based on few FARS cases, because belt use was low at the time, and those cars have mostly been retired. The exception to the general pattern is drivers' belts in cars with air bags, pretensioners, and load limiters: effectiveness is close to 50 percent in all five age groups, including a 54 percent fatality reduction for drivers 80+. The effect of this generation of belts for RF passengers is unclear; the observed fatality reduction is highest (63%) at 70 to 79 but lowest (27%) at 80+.

Table 9-5 documents the results of CATMOD analyses comparing belt effectiveness for a group of younger occupants (13 to 49) to one of three alternative, nested groups of older occupants (55+, 70+, and 80+), as discussed in Section 9.3. The distinctive feature of Table 9-5 is the numerous instances where effectiveness is significantly lower for the older occupants than for people 13 to 49, as evidenced by bold type with an orange background and a frame, or at least borderline-significantly lower, as evidenced by a pale-yellow background.

TABLE 9-4: FATALITY REDUCTION (%) BY SEAT BELTS IN PASSENGER CARS, BY OCCUPANT'S AGE GROUP
(Drivers and RF Passengers 13 and Older)

Age Group	Lap Belt Only[70]	2-Point Automatic Belt	MY 1974–1982	MY 1983+ Without Air Bags	No Pretensioner or Load Limiter	With Pretensioner & Load Limiter
				3 – P o i n t B e l t s	------- With Air Bags -------	
			DRIVERS			
13 – 29	31	44	52	50	46	50
30 – 54	28	44	50	48	45	50
55 – 69	34	31	48	41	35	43
70 – 79	27	17	51	42	44	50
80 & older	−15	30	−16	30	31	54
			RF PASSENGERS			
13 – 29	38	39	54	48	45	55
30 – 54	28	34	23	47	43	50
55 – 69	none	19	21	39	41	46
70 – 79	21	8	36	37	39	63
80 & older	15	7	53	29	27	27

[70] Includes MY 1968-1973 cars with separate lap and shoulder belts, as long as FARS does not specify shoulder belt was used.

TABLE 9-5: FATALITY REDUCTION (%) BY SEAT BELTS IN PASSENGER CARS, BY OCCUPANT'S AGE GROUP
(CATMOD comparisons of effectiveness at 13 to 49 to effectiveness in the older age group)

Age Group	Lap Belt Only[71]	2-Point Automatic Belt	3 – P o i n t B e l t s		——— With Air Bags ———	
			MY 1974-1982	MY 1983+ Without Air Bags	No Pretensioner or Load Limiter	With Pretensioner & Load Limiter
DRIVERS						
13 – 49	28	46	51	50	46	49
55 & older	28	25	44	40	38	48
70 & older	21	18	38	38	39	52
80 & older	–15	30	–16	30	31	54
RF PASSENGERS						
13 – 49	36	37	45	48	44	54
55 & older	9	12	32	36	35	46
70 & older	18	7	42	35	33	48
80 & older	15	7	53	29	27	27

---

[71] Includes MY 1968-1973 cars with separate lap and shoulder belts, as long as FARS does not specify shoulder belt was used.

Completely absent here (unlike the statistics for female drivers in Tables 9-1 and 9-2) are blue or pale-blue backgrounds that would have indicated significantly greater effectiveness for the older occupants. For example, in MY 1983+ cars without air bags (where the data samples are large), the observed 40-percent fatality reduction for drivers 55+, the 38-percent reduction for drivers 70+, and the 30-percent reduction for drivers 80+ are <u>all</u> significantly less than the observed 50-percent effectiveness for drivers 13 to 49; likewise, for the RF passengers in these cars, all three observed reductions for the older groups are significantly lower than the 48-percent effectiveness for RF passengers 13 to 49. The only columns without any orange or yellow squares are: lap belts only (limited data); MY 1974-1982 passengers (limited data and no clear pattern); and drivers with air bags, pretensioners, and load limiters (almost constant effectiveness at all ages).

Table 9-6 likewise shows consistently decreasing belt effectiveness for older occupants in LTVs.

TABLE 9-6: FATALITY REDUCTION (%) BY 3-POINT BELTS IN LTVs
BY OCCUPANT'S AGE GROUP (Drivers and RF Passengers 13 and Older)

Age Group	LTVs Without Air Bags (MY 1983-)	LTVs With Air Bags
DRIVERS		
13-29	69	76
30-54	68	76
55-69	61	71
70-79	27	53
80 & older	− 1	58
RF PASSENGERS		
13-29	67	73
30-54	68	70
55-69	58	65
70-79	48	52
80 & older	35	63

CATMOD analyses find belt effectiveness to be significantly lower for the older-occupant group than for people 13 to 49 in 10 out of 12 tests, as shown in Table 9-7. The downward trend in effectiveness is especially strong in the LTVs without air bags.

TABLE 9-7: FATALITY REDUCTION (%) BY 3-POINT BELTS IN LTVs
BY OCCUPANT'S AGE GROUP
(CATMOD comparisons of effectiveness at 13 to 49 to effectiveness in the older age group)

Age Group	LTVs Without Air Bags (MY 1983-)	LTVs With Air Bags
DRIVERS		
13-49	69	76
55 & older	53	66
70 & older	19	54
80 & older	– 1	58
RF PASSENGERS		
13-49	67	72
55 & older	53	60
70 & older	44	53
80 & older	35	63

For back-seat occupants of passenger cars, Table 9-8 shows that estimated belt effectiveness decreases with each increment in the passenger's age – both for lap belts and 3-point belts. Based on CATMOD, Table 9-9 demonstrates that belt effectiveness is significantly lower for the older-occupant group than for people 13 to 49 in five out of the six tests and borderline-significant in the sixth.

TABLE 9-8. FATALITY REDUCTION (%) BY SEAT BELTS FOR BACK-SEAT
PASSENGERS OF CARS, BY OCCUPANT'S AGE GROUP (13 and Older)

Age Group	Lap Belt Only	3-Point Belt
13 – 29	53	61
30 – 54	39	58
55 – 69	36	47
70 – 79	17	24
80 & older	15	7

TABLE 9-9: FATALITY REDUCTION (%) BY SEAT BELTS FOR BACK-SEAT
PASSENGERS OF CARS, BY OCCUPANT'S AGE GROUP
(CATMOD comparisons of effectiveness at 13 to 49 to effectiveness in the older age group)

Age Group	Lap Belt Only	3-Point Belt
13 – 49	51	61
55 & older	25	30
70 & older	16	17
80 & older	15	7

The results are no surprise. They are consistent with NHTSA's previous evaluations of seat-belt effectiveness, which already showed diminished effectiveness for older occupants: drivers, RF passengers, and back-seat passengers.[72] They are likewise consistent with all the findings of Chapters 2 to 5, which repeatedly showed a higher increase in fatality and injury risk per year of aging for belted occupants than for unrestrained occupants (see, for example, Table 5-1a). As Zhou, Rouhana, and Melvin explain, the effect of aging is more severe in belt loading than in blunt impact force: belt loading is a more static, less dynamic load than blunt impact, and thus has a less linear response. Also, belt force is concentrated on bone, rather than soft tissue. Bone deteriorates more rapidly with age than soft tissue.[73]

But there are two important caveats to the general rule that belts are less effective for older occupants: (1) though they are less effective, they are not ineffective, let alone harmful; all of the 3-point belt systems at seat positions equipped with air bags show substantial fatality reduction for older occupants in Tables 9-4 and 9-6, ranging from 27 percent for RF passengers 80+ in cars to 71 percent for drivers 55 to 69 in LTVs; (2) the near-constant effectiveness across all ages for belts in cars equipped with air bags, pretensioners, and load limiters, at least for drivers, offers hope that the latest belt systems depart from the historical trend by limiting belt force on the occupant – but more data is needed to confirm these results.

9.6   Effectiveness of non-belt technologies by the occupant's gender

Table 9-10 compares fatality reduction for males and females for three non-belt technologies designed to protect occupants of passenger cars in frontal crashes: energy-absorbing steering assemblies, barrier-certified frontal air bags, and sled-certified frontal air bags. For example, EA steering assemblies reduce fatality risk in frontal crashes by 11 percent for male drivers and by 16 percent for women. It is evident from the absence of any color-coding in Table 9-10 that, in all cases, there is no significant difference between the observed effect for males and the observed effect for females. In cars, air bags are almost equally effective for men and women at both the driver and the RF seats.

---

[72] Kahane (2000), pp. 35-38; Morgan (1999), pp. 47-48.
[73] Zhou, Rouhana, & Melvin (1996).

223

TABLE 9-10: FATALITY REDUCTION (%) IN FRONTAL CRASHES[74] OF **CARS**
FOR NON-BELT TECHNOLOGIES TO PROTECT OCCUPANTS IN **FRONTAL** IMPACTS,
BY OCCUPANT'S GENDER (Drivers and RF Passengers 13 and Older)

Gender	EA Steering Assemblies	-------------- Frontal Air Bags[75] --------------	
		Barrier-Certified	Sled-Certified
		DRIVERS	
Male	11	21	19
Female	16	22	18
		RF PASSENGERS	
Male		26	29
Female		25	28

EA steering assemblies were discussed in Section 9.2. They replaced rigid steering assemblies in passenger cars in MY 1967-1968. Their purpose is to limit force levels when a driver contacts the steering assembly in a frontal impact. The basic analysis of CY 1975-1985 FARS compares fatalities in frontal impacts (IMPACT2 = 11, 12, or 1) to a control group of side impact, rear impacts, and first-event rollovers. The overall fatality reduction for drivers in frontal impacts is a statistically significant 12 percent (χ^2 = 18.64). Section 9.2 explains how to derive separate effectiveness estimates for male drivers (11%) and female drivers (16%) and how testing with CATMOD indicates that these estimates are not significantly different.

Frontal air bags were standard equipment in some car models in MY 1985 and were standard on all new cars at the driver and RF seats by MY 1997. Air bags work in tandem with the EA steering assembly (or, on the RF side, with EA materials in the instrument panel) to provide much more cushioning and ride-down than the EA steering assembly alone and to reduce risk of head, neck, and torso injuries. "Barrier-certified" air bags include all air bags through MY 1997, some in MY 1998, and a few in MY 1999 certifying to the original requirements for air bags in FMVSS No. 208, which included a barrier-impact test with unrestrained dummies.[76] One of the basic effectiveness analyses compared fatalities in frontal crashes to a control group of side impact, rear impacts, and first-event rollovers (similar to the analysis of EA steering assemblies) – using data for cars of the first three MY with air bags versus cars of the last three MY without

---

[74] IMPACT2 = 11, 12, or 1
[75] For drivers: estimates the fatality reduction incremental over the EA steering assembly
[76] Kahane (2006), Chapter 1, especially p. 13.

air bags.[77] When the FARS data for these analyses is extended to CY 2010, the overall fatality reduction in frontal impacts is a statistically significant 21 percent for drivers ($\chi^2 = 69.08$) and a statistically significant 25 percent for RF passengers age 13+ ($\chi^2 = 33.60$). (For the driver, this estimate actually measures the incremental effect of the air bag plus EA steering assembly over just the EA steering assembly.) As shown in Table 9-10, effectiveness is almost the same for males and females at the driver's seat (21% versus 22%) and at the RF seat (26% versus 25%). The CATMOD analyses confirm these are not statistically significant differences.

"Sled-certified" air bags, starting in MY 1998, certified to an amended version of FMVSS No. 208 that NHTSA issued on March 19, 1997, which substituted a sled test with unrestrained dummies for the barrier-impact test. The sled test has a softer crash pulse than the typical barrier impact for cars of the 1990s, allowing air bags that deployed less forcefully. The objective was to reduce risks to children or people sitting close to the air bag, but for the average adult in most crashes, no large difference in effectiveness was expected between barrier- and sled-certified air bags. The basic effectiveness analyses actually compared sled-certified air bags to barrier-certified air bags – i.e., they compared fatalities in frontal crashes to the non-frontal control group – using data for cars of the first three model years with sled-certified air bags versus cars of the last three model years with barrier-certified air bags.[78] When the FARS data for these analyses is extended to CY 2010, there are no significant differences between barrier- and sled-certified air bags at either seat. However, Table 9-10 combines these results with the previous analyses of barrier-certified air bags to obtain point estimates of the fatality reduction for sled-certified air bags relative to no air bags at all. Again, effectiveness is almost the same for males and females at the driver's seat (19% versus 18%) and at the RF seat (29% versus 28%). Similarly, the CATMOD analyses for barrier-certified versus no air bags and for sled- versus barrier-certified air bags are combined to test if the point estimates in Table 9-10 (for sled-certified versus no air bags) differ significantly for males and females. It is accomplished as follows. The first CATMOD analysis (barrier-certified versus no air bag) generates the following regression coefficients:

Parameter		Estimate	Standard Error	Chi-Square	Pr > ChiSq
Intercept		0.4031	0.0153	691.79	<.0001
AIRBAG	DRIVER AIR BAG	-0.1208	0.0153	62.10	<.0001
MALE	MALE	0.1082	0.0153	49.85	<.0001
AIRBAG*MALE	DRIVER AIR BAG MALE	0.00189	0.0153	0.02	0.9021

The second CATMOD analysis (sled- versus barrier-certified) generates:

---

[77] Kahane, C. J. (1996). *Fatality Reduction by Air Bags: Analyses of Accident Data through Early 1996*. (Report No. DOT HS 808 470, Chapter 1). Washington, DC: National Highway Traffic Safety Administration Available at www-nrd.nhtsa.dot.gov/Pubs/808470.PDF

[78] Kahane, (2006), Chapters 3 and 4.

Parameter		Estimate	Standard Error	Chi-Square	Pr > ChiSq
Intercept		0.2415	0.0133	331.12	<.0001
SLED	SLED-CERTIFIED	0.0182	0.0133	1.89	0.1695
MALE	MALE	0.0929	0.0133	48.95	<.0001
SLED*MALE	SLED-CERTIFIED MALE	-0.00442	0.0133	0.11	0.7391

The sum of the regression coefficients for the interaction terms is .00189 - .00442 = -.00253. The standard error of the sum of these two estimates (which are statistically independent because they are based on entirely separate data) is $\sqrt{(.0153^2 + .0133^2)} = .0203$. The chi-square for the combined effect is $(-.00253/.0203)^2 = 0.02$, indicating that the difference in observed effectiveness for males and females is not statistically significant.

In LTVs, based on the latest FARS data, barrier-certified air bags are estimated to reduce the overall fatality risk of drivers by 24 percent and RF passengers by 22 percent in frontal crashes. Both reductions are statistically significant ($\chi^2 = 78.58$ and 12.36, respectively). The overall effectiveness of sled-certified air bags is not significantly different from barrier-certified air bags. Table 9-11 compares fatality reduction by frontal air bags for male and female occupants of LTVs.

TABLE 9-11: FATALITY REDUCTION (%) BY AIR BAGS IN FRONTAL CRASHES[79] OF **LTVs**, BY OCCUPANT'S GENDER (Drivers and RF Passengers 13 and Older)

Gender	Barrier-Certified	Sled-Certified
	DRIVERS	
Male	19	21
Female	31	33
	RF PASSENGERS	
Male	24	25
Female	20	27

Barrier-certified air bags are significantly more effective for female drivers of LTVs (31%) than for males (19%). The observed effectiveness of sled-certified air bags is also higher for female than for male drivers; the difference is borderline-significant. Air bags for RF passengers in LTVs are almost equally effective for men and women.

---

[79] IMPACT2 = 11, 12, or 1

226

Table 9-12 compares fatality reduction for four non-belt technologies: two designed to protect occupants of passenger cars in side-impact crashes – side door beams and the FMVSS No. 214 dynamic test/TTI(d) improvements; one to protect occupants of cars and LTVs from head impacts with the vehicle interior – the non-air bag technologies meeting the 1999-2003 upgrade of FMVSS No. 201; and a technology that protects occupants in side impacts and head impacts, namely side air bags with head protection (such as separate torso bags and curtains, or combination head-torso bags). The observed fatality reduction is higher for females than for males with all four technologies, but it is significantly higher only for the head-impact upgrade.

TABLE 9-12: FATALITY REDUCTION (%) FOR NON-BELT TECHNOLOGIES TO
PROTECT OCCUPANTS IN SIDE IMPACTS AND HEAD IMPACTS,
BY OCCUPANT'S GENDER (Occupants 13 and Older)

Gender	Side Door Beams[80]	FMVSS No. 214 TTI(d) Improvements[81]	Head Impact Upgrade[82]	Head + Torso Air Bags[83]
Male	7	8	0	32
Female	17	17	16	34

Side door beams were first installed in passenger cars in MY 1969-1973. Their purpose is to reduce intrusion in a side impact by strengthening the door structure of the struck vehicle and transmitting some of the impact force away from the door and into the pillars and frame of the struck vehicle. The original FMVSS No. 214 – Side Impact Protection – was a static test to confirm the strength of the beams. However, the primary fatality-reducing effect is in side impacts with fixed objects such as trees, where forces are partially frontal. rather than merely absorbing energy, the beam acts like an internal "guard rail" to deflect the vehicle and allow it to slide past the tree, with a longer, shallower crush pattern. NHTSA's original analysis of CY 1975-1981 FARS compared overall fatality risk of occupants of outboard seats (i.e., next to a door: drivers; right-front, left-rear and right-rear passengers) in side impacts with fixed objects, before and after the installation of side door beams, relative to a control group of fatalities in frontal impacts with fixed objects.[84] When this analysis is extended to CY 1975-1990 FARS, the overall fatality reduction is a statistically significant 9 percent ($\chi^2 = 10.20$). Separate point estimates for male occupants (7%) are lower than for females (17%); however testing with CATMOD indicates that these estimates are not significantly different.

---

[80] Fatality reduction in side impacts with fixed objects, nearside or far-side: drivers + RF passengers + back-seat outboard passengers of passenger cars

[81] Fatality reduction in nearside impacts by another vehicle: drivers + RF passengers of passenger cars

[82] Head injury reduction in FARS MCOD for vehicles meeting the FMVSS No. 201 head-impact upgrade without curtain or side air bags: car and LTV occupants at any seat position

[83] Including combination bags or separate torso and curtain bags; fatality reduction in nearside impacts, incremental over the side door beams and TTI(d) improvements: drivers + RF passengers of passenger cars

[84] Kahane, C. J. (1982). *An Evaluation of Side Structure Improvements in Response to Federal Motor Vehicle Safety Standard 214*. (Report No. DOT HS 806 314, Chapter 6, pp. 143-157). Washington, DC: National Highway Traffic Safety Administration. Available at www-nrd.nhtsa.dot.gov/Pubs/806314.PDF

During the early 1990s, NHTSA upgraded FMVSS No. 214 with a dynamic test, where a moving deformable barrier strikes a vehicle in the side and a thoracic trauma index called TTI(d) is measured on side-impact dummies designed for this test. During the MY 1994-1997 phase-in, cars usually certified to the new requirement by a combination of energy-absorbing materials in the interior of the door and improved side door beams and other structural modifications. These technologies succeeded in reducing fatality risk to outboard occupants in nearside impacts by other vehicles. NHTSA's original analysis of CY 1993-2005 FARS focused on 15 groups of make-models whose TTI(d) performance on the dynamic test was substantially better after certifying to FMVSS No. 214 than in the model years before the standard.[85] It compared overall fatality risk of occupants of front-outboard seats (i.e., drivers and RF passengers) in nearside impacts by other vehicles, before and after certification to FMVSS No. 214, relative to a control group of drivers and RF passengers of cars of these makes and models involved in crashes that were fatal to a pedestrian or another non-occupant, but were not fatal to the driver or RF passenger. When this analysis is extended to CY 1975-1990 FARS, the overall fatality reduction is a statistically significant 12 percent ($\chi^2 = 6.10$). Separate point estimates for male occupants (8%) are lower than for females (17%); however testing with CATMOD indicates that these estimates are not significantly different.

Later in the 1990s, NHTSA upgraded FMVSS No. 201 – Occupant Protection in Interior Impact – to reduce occupants' risk of head injury from contact with a vehicle's upper interior, including its pillars, roof headers and side rails, and the upper roof. Initially, during the MY 1999-2003 phase-in, energy-absorbing materials alone were used to meet the standard. These materials succeeded in reducing the risk of fatal and life-threatening head injuries to occupants of cars and LTVs at all interior seat positions. NHTSA's analysis of CY 1999-2007 FARS-MCOD data compares, for vehicles produced before and after certification to FMVSS No. 201, occupants' risk of head injuries that "contributed to death" relative to a control group of injuries to other body regions.[86] (Injuries to the neck or to unknown body regions are excluded from the analysis.) The head-injury reduction for all occupants, by this analysis, is a statistically significant 6 percent ($\chi^2 = 4.84$).

The point estimate for males is close to zero, whereas the observed fatality reduction for females is 16 percent. Effectiveness is significantly higher for women than for men (χ^2 for the FMVSS201*MALE interaction in the CATMOD analysis is 9.27). Although statistically significant, the difference should perhaps be viewed with caution because: (1) it is based on still relatively limited FARS-MCOD data, which is only available through CY 2007 as of October 2012; (2) females have only moderately higher risk of fatal head injury than males and the injury patterns are similar (see Tables 4-2d and 6-3); thus, it is surprising for a safety technology to be much more effective for females; (3) an analysis of CDS data in NHTSA's 2011 evaluation report only partially confirms this result – it shows higher effectiveness for females with the unweighted data, but higher effectiveness for males with weighted data.[87] At any rate, though, the data make it clear that the head-impact upgrade is no less effective for females than for males.

---

[85] Kahane (2007), Chapter 2, especially pp. 37-41 and Table 2-1a; these 15 groups of models reduced their TTI(d) score by an average of 23 units after FMVSS No. 214 certification; other make-models were able to certify to FMVSS No. 214 with little or no modification, or their pre-standard performance is unknown.
[86] Kahane (2011), Chapter 3, especially pp. 37-42 and Table 3-1.
[87] Ibid., Chapter 2, especially pp. 29-32 and Table 2-3.

Side air bags designed to protect an occupant's torso in a nearside impact first appeared on some MY 1996 car models. Head-protection air bags – such as curtains, inflatable tubes, or side air bags that extend upwards for head protection – began to appear on some MY 1998 models; they protect the occupant's head in nearside impacts and possibly, depending on their design, in other types of crashes such as far-side impacts, rollovers, and oblique frontal impacts. Side air bags with head protection, such as separate torso and curtain bags or combination head-torso bags are quite effective in nearside impacts. NHTSA's 2007 evaluation estimated a 24 percent fatality reduction for front-outboard occupants of cars, based on CY 1993-2005 FARS data. That estimate is not based on a single analysis; it is the average of results from the three analyses considered most likely to produce accurate results and it is also close to the median of the results from 15 analyses discussed in the report.[88]

This report will use a slightly different effectiveness estimate, because it is somewhat simpler and it can be based on a large number of FARS cases. It includes all MY 1994-2009 makes and models of passenger cars that switched at some point from no side air bags to some type of side air bags (limited to the MY in which they certified to the FMVSS No. 214 dynamic test and were also equipped with dual frontal air bags). It compares fatality risk of drivers and RF passengers in nearside impacts (IMPACT2 = 8-10 for the driver or 2-4 for the RF passenger), without any side air bags and with side air bags that offer head protection, relative to a control group of drivers and RF passengers involved in directly frontal or directly rear impacts (IMPACT2 = 12 or 6). Overall, with CY 1993-2010 FARS data, this analysis finds a statistically significant 34 percent fatality reduction for side air bags with head protection ($\chi^2 = 214.61$). This estimate is somewhat higher than the 24 percent in NHTSA's 2007 evaluation; it is possibly biased upward by the inclusion of such a wide range of model years. However, the objective of this report is not to estimate overall effectiveness, but to compare effectiveness by occupants' gender or age. These comparisons would still be valid, because any biases in the estimation process would tend to be directionally similar for males and females, or for older and younger occupants.

Table 9-12 shows that the air bags are at least as effective for females (34%) as for males (32%); the CATMOD analysis indicates these estimates are not significantly different. The finding is not inconsistent with Table 3-4d, which suggested that the added fatality risk for females is possibly lower or at least not higher with the air bags (2.6 ± 21.5% increase) than for belted occupants in nearside impacts without the air bags (15.5 ± 10.8%); likewise with Table 8-5, which showed females vulnerable in nearside impacts to head and torso injuries from the side interior surface, torso injuries from the armrest, and head injuries from the A-pillar – thus, suggesting high potential benefits for both side and head-protection air bags.

Tables 9-10, 9-11, and 9-12 address a total of 13 non-belt technologies to protect occupants from injury. The observed fatality reduction is higher for females than males for nine of these 13 technologies; in two cases the difference is significant at the two-sided .05 level and in one, borderline-significant. The observed fatality reduction is higher for males than females for just four of the 13 technologies; never significantly higher, in one case 4 percentage points higher, and in the remaining three cases just 1 percentage point higher. In summary, these non-belt technologies are helping females just as much and quite possibly more than they protect males.

---

[88] Kahane (2007), Chapter 3, especially pp. 82-90 and 123-124.

They are contributing to shrinking the historical risk increase for females relative to males of the same age.

9.7   Effectiveness of non-belt technologies by the occupant's age

Table 9-13 gives point estimates of the fatality reduction by belts for five age groups of occupants – for drivers and RF passengers of passenger cars – for three non-belt technologies designed to protect occupants in frontal crashes: EA steering assemblies and two types of frontal air bags. For example, EA steering assemblies are estimated to reduce a driver's fatality risk in frontal crashes by 16 percent for drivers 13 to 29, 9 percent at 30 to 54, 13 percent at 55 to 69, and 12 percent at 70 to 79, but to increase risk by 20 percent for drivers 80 and older. There is no color-coding in Table 9-13 because there are no CATMOD analyses, just point estimates; the CATMOD analyses are addressed in Table 9-14.

TABLE 9-13: FATALITY REDUCTION (%) IN FRONTAL CRASHES[89] OF **CARS** FOR NON-BELT TECHNOLOGIES TO PROTECT OCCUPANTS IN **FRONTAL** IMPACTS, BY OCCUPANT'S AGE GROUP (Drivers and RF Passengers 13 and Older)

Age Group	EA Steering Assemblies	-------------- Frontal Air Bags[90] --------------	
		Barrier-Certified	Sled-Certified
DRIVERS			
13-29	16	23	18
30-54	9	19	15
55-69	13	21	26
70-79	12	27	29
80 & older	- 20	14	14
RF PASSENGERS			
13-29		23	24
30-54		23	32
55-69		10	35
70-79		38	32
80 & older		37	31

---

[89] IMPACT2 = 11, 12, or 1

[90] For drivers: estimates the fatality reduction incremental over the EA steering assembly

In fact, the negative effectiveness for EA steering assemblies for drivers 80 and older is the only clear-cut indication of low effectiveness for older occupants in Table 9-13. The fatality reduction for EA steering assemblies is fairly steady up through age 79. The two types of driver air bags have slightly lower effectiveness at 80+, but there is no overall downward trend because the highest observed effectiveness is at 70 to 79. Barrier-certified air bags reach the highest effectiveness for the two oldest groups of RF passengers. The fatality reduction for sled-certified passenger air bags is almost constant across the five age groups.

Table 9-14 presents CATMOD analyses comparing effectiveness for a group of younger occupants (13 to 49) to one of three alternative, nested groups of older occupants (55+, 70+, and 80+), as discussed in Section 9.2.

TABLE 9-14: FATALITY REDUCTION (%) IN FRONTAL CRASHES[91] OF **CARS** FOR NON-BELT TECHNOLOGIES, BY OCCUPANT'S AGE GROUP
(CATMOD comparisons of effectiveness at 13 to 49 to effectiveness in the older age group)

Age Group	EA Steering Assemblies	-------------- Frontal Air Bags[92] --------------	
		Barrier-Certified	Sled-Certified
DRIVERS			
13-49	16	22	18
55 & older	5	22	24
70 & older	2	22	23
80 & older	-20	14	14
RF PASSENGERS			
13-49		24	25
55 & older		30	33
70 & older		38	31
80 & older		37	31

EA steering assemblies in all three cases have a lower point estimate of effectiveness for the older than for the younger group of drivers. The difference is statistically significant (at the two-sided .05 level), as evidenced by the orange background, for the 80+ versus 13 to 49 comparison; it is borderline-significant (at the one-sided .05 level), as evidenced by the yellow background,

---

[91] IMPACT2 = 11, 12, or 1
[92] For drivers: estimates the fatality reduction incremental over the EA steering assembly

231

for the 55+ versus 13-49 comparison. The results are moderately strong but not unequivocal evidence that the EA steering assemblies are less effective for older than for young occupants. They are also consistent with Table 3-2a, which shows a perhaps slightly larger effect of aging one year ($2.81 \pm .15\%$ increase in fatality risk) for unrestrained occupants of MY 1969-1982 cars with EA steering assemblies than in pre-1968 cars with rigid steering assemblies ($2.58 \pm .46\%$). (FARS and FARS-MCOD do not have enough cases of these early vehicles for statistically meaningful analyses of frontal crashes or of chest injuries as a separate group.)

By contrast, the four types of frontal air bags are usually as effective, if not more so, for the older occupants than for young drivers and passengers. The only exceptions are the two types of air bags for drivers 80+, where the point estimate is slightly, but not significantly lower than for 13 to 49. For RF passengers 70 and older, the observed fatality reduction is 38 percent, a borderline-significant difference from the 24 percent effectiveness at 13 to 49, as evidenced by the light-blue background.

Likewise, in LTVs, air bags are about equally effective at all ages, as shown in Table 9-15.

TABLE 9-15: FATALITY REDUCTION (%) BY AIR BAGS IN FRONTAL CRASHES[93] OF **LTVs**, BY OCCUPANT'S AGE GROUP (Drivers and RF Passengers 13 and Older)

Age Group	Barrier-Certified	Sled-Certified
	DRIVERS	
13-29	30	33
30-54	19	23
55-69	23	22
70-79	28	24
80 & older	33	17
	RF PASSENGERS	
13-29	27	38
30-54	20	23
55-69	23	17
70-79	23	1
80 & older	25	37

---

[93] IMPACT2 = 11, 12, or 1

Barrier-certified air bags for drivers have effectiveness fairly close to 25 percent in all five age groups, with the highest point estimate (33%) for the oldest drivers. Sled-certified air bags for drivers have the closest thing to a visible trend, with somewhat higher effectiveness for the youngest group, slightly lower for the oldest, and nearly 23 percent for the three middle groups. Fatality reduction is always close to 25 percent for the barrier-certified air bags at the RF seat. As for sled-certified air bags at this seat (where data are more limited and the estimates have more sampling error), the estimated 1-percent effectiveness at 70 to 79 is low, but it is offset by one of the highest estimates, 37 percent, at 80 and older.

The CATMOD analyses in Table 9-16 confirm that there are no significant differences in air bag effectiveness between the group of younger occupants (13 to 49) and any of the three alternative groups of older occupants (55+, 70+, and 80+): as evidenced by the absence of any color-shading in the table. The somewhat lower observed effectiveness estimates for drivers 80+ and for RF passengers age 55+ and 70+ are not significantly lower, given the number of cases available for the analyses.

TABLE 9-16: FATALITY REDUCTION (%) BY AIR BAGS
IN FRONTAL CRASHES[94] OF **LTVs**, BY OCCUPANT'S AGE GROUP
(CATMOD comparisons of effectiveness at 13 to 49 to effectiveness in the older age group)

Age Group	Barrier-Certified	Sled-Certified
DRIVERS		
13-49	25	29
55 & older	25	22
70 & older	30	22
80 & older	33	17
RF PASSENGERS		
13-49	28	37
55 & older	22	15
70 & older	22	15
80 & older	25	37

Table 9-17 presents point estimates of fatality reduction by age group for four non-belt technologies designed to protect occupants in side impacts and/or from head impacts with the vehicle interior. Side air bags with head protection – i.e., separate torso bags and head curtains or head-torso combination bags – show a trend of increasing effectiveness for older occupants. The

---

[94] IMPACT2 = 11, 12, or 1

other three technologies do not have clear trends. Side door beams have a negative point estimate for 70 to 79 but return to positive at 80+. The TTI(d) improvements with the dynamic test for FMVSS No. 214 have negative results at 70 to 79 and especially 80+, but the highest effectiveness at 55 to 69. Similarly, the head-impact upgrade of FMVSS No. 201 in MY 1999-2003 shows a negative result for 80+ but its most positive estimate at 70 to 79. The various observed negative estimates for the FMVSS Nos. 214 and 201 upgrades are themselves puzzling because they primarily involve static, padded materials that would not likely exert more force on an occupant than the unpadded surfaces they superseded.[95] The observed negative effects might just be indicators of close-to-zero actual effectiveness.

TABLE 9-17: FATALITY REDUCTION (%) FOR NON-BELT TECHNOLOGIES TO PROTECT OCCUPANTS IN SIDE IMPACTS AND HEAD IMPACTS, BY OCCUPANT'S AGE GROUP (Occupants 13 and Older)

Age Group	Side Door Beams[96]	FMVSS No. 214 TTI(d) Improvements[97]	Head Impact Upgrade[98]	Head + Torso Air Bags[99]
13-29	9	12	3	31
30-54	11	10	10	27
55-69	1	24	3	38
70-79	− 26	− 1	15	44
80 & older	10	− 63	− 16	47

Table 9-18 presents the corresponding CATMOD analyses. The CATMOD tests confirm that side air bags with head protection are especially effective for older occupants. Of course, they are quite effective in reducing injury severity at all ages; however, in a crash that a young person survives even without air bags (where the bags merely reduce injury severity to a lower level), these air bags could make the difference between life and death for an older occupant. The finding is consistent with Table 3-4a, where the effect of aging one year is possibly lower with the air bags ($2.28 \pm 1.22\%$ increase) than for belted occupants in nearside impacts without the air bags ($3.19 \pm .32\%$); also with Table 8-2, which suggested potential benefits for side air bags

---

[95] An actual negative effect for side door beams is conceivable, if not particularly likely, because the beams stiffen the door structure and could increase force on the occupant in some situations. To the extent that the FMVSS Nos. 214 and 201 upgrades include structural reinforcement as well as energy-absorbing padding, it is even imaginable that forces on occupants could increase in vehicles where the effect of the reinforcement is greater than the effect of the padding.

[96] Fatality reduction in side impacts with fixed objects, nearside or far-side: drivers + RF passengers + back-seat outboard passengers of passenger cars

[97] Fatality reduction in nearside impacts by another vehicle: drivers + RF passengers of passenger cars

[98] Head injury reduction in FARS MCOD for vehicles meeting the FMVSS No. 201 head-impact upgrade without curtain or side air bags: car and LTV occupants at any seat position

[99] Including combination bags or separate torso and curtain bags; fatality reduction in nearside impacts, incremental over the side door beams and TTI(d) improvements: drivers + RF passengers of passenger cars

because older occupants are vulnerable in nearside impacts to torso injuries from contacting the side interior surface or the armrest.

TABLE 9-18: FATALITY REDUCTION (%) FOR NON-BELT TECHNOLOGIES TO PROTECT OCCUPANTS IN SIDE/HEAD IMPACTS, BY OCCUPANT'S AGE GROUP (CATMOD comparisons of effectiveness at 13 to 49 to effectiveness in the older age group)

Age Group	Side Door Beams	FMVSS No. 214 TTI(d) Improvements	Head Impact Upgrade	Head + Torso Air Bags
13-49	10	11	6	30
55 & older	− 4	7	3	43
70 & older	− 13	− 19	3	45
80 & older	10	−63	− 16	47

The three earlier technologies without air bags do not show increasing effectiveness for older occupants but, on the contrary, one case of significantly lower effectiveness for an older-occupant group and two cases of borderline significance.

Tables 9-14, 9-16, and 9-18 display the results of CATMOD tests for 13 non-belt technologies to protect occupants for three alternative older-occupant groups: a total of 39 comparisons with the effect for younger occupants. The analyses show a contrast between the technologies without air bags and the inflatable crash protection. In the 12 tests of technologies without air bags, effectiveness is almost always lower for the older occupants – significantly lower in two cases and borderline-significant in another three. Frontal air bags are almost equally effective across all ages: in 24 tests, effectiveness is higher for the younger occupant in 10, for the older occupant in 10, and four ties (when effectiveness is rounded to the nearest percentage point). For side air bags with head protection, effectiveness is significantly higher for the older occupants on all three tests. In summary, the early, non-inflatable non-belt technologies did not help older people as much as young people and may have contributed to augmenting the historical risk increase associated with aging. Frontal air bags are helping people of all ages about equally, neither augmenting nor easing the risk increase. Side air bags with head protection may be especially helpful for older occupants and may be helping to shrink the excess risk of an older person relative to a young person, given the same crash situation.

REFERENCES

Austin, R. A., and Faigin, B. M. (2003). "Effect of Vehicle and Crash Factors on Older Occupants," *Journal of Safety Research*, Vol. 34, pp. 441-452.

Bean, J.D., Kahane, C. J., Mynatt, M., Rudd, R.W., Rush, C.J., and Wiacek, C. (2009). *Fatalities in Frontal Crashes Despite Seat Belts and Air Bags*, NHTSA Technical Report. DOT HS 811 202. Washington, DC: National Highway Traffic Safety Administration, http://www-nrd.nhtsa.dot.gov/pubs/811102.pdf.

Bose, D., Segui-Gomez, M., and Crandall, J. F. (2011). "Vulnerability of Female Drivers Involved in Motor Vehicle Crashes: An Analysis of US Population at Risk," *American Journal of Public Health*, Vol. 101, pp. 2368-2373.

CDC (a) *Classification of Diseases and Injuries – ICD-9-CM Tabular List of Diseases (FY07)*. Dtab07.zip at ftp://ftp.cdc.gov/pub/Health_Statistics/NCHS/Publications/ICD9-CM/2006/.

CDC (b). *Multiple Cause of Death, 1999-2006*. http://wonder.cdc.gov/wonder/help/mcd.html.

Dischinger, P. C., Kerns, T. J., and Kufera, J. A. (1995). "Lower Extremity Fractures in Motor Vehicle Collisions: The Role of Driver Gender and Height," *Accident Analysis and Prevention*, Vol. 27, pp. 601-606.

Evans, L. (1986a). "Double Pair Comparison – A New Method to Determine How Occupant Characteristics Affect Fatality Risk in Traffic Crashes," *Accident Analysis and Prevention*, Vol. 18, pp. 217-227.

Evans, L. (1986b). "The Effectiveness of Safety Belts in Preventing Fatalities," *Accident Analysis and Prevention*, Vol. 18, pp. 229-241.

Evans, L. (1988). "Risk of Fatality from Physical Trauma Versus Sex and Age," *Journal of Trauma*, Vol. 28, pp. 368-378.

Evans, L. (1991). *Traffic Safety and the Driver*. New York: Van Nostrand Reinhold.

Gennarelli, T. A., and Wodzin, E. (2006). "AIS 2005: A Contemporary Injury Scale," *Injury*, Vol. 37, pp. 1083-1091.

Hanna, R., and Hershman, L. (2009). *Evaluation of Thoracic Injuries Among Older Motor Vehicle Occupants*, NHTSA Report No. DOT HS 811 101. Washington, DC: National Highway Traffic Safety Administration.

Kahane, C. J. (1981). *An Evaluation of Federal Motor Vehicle Safety Standards for Passenger Car Steering Assemblies*, NHTSA Technical Report No. DOT HS 805 705. Washington, DC: National Highway Traffic Safety Administration, http://www-nrd.nhtsa.dot.gov/Pubs/805705.PDF.

Kahane, C. J. (1982). *An Evaluation of Side Structure Improvements in Response to Federal Motor Vehicle Safety Standard 214*, NHTSA Technical Report No. DOT HS 806 314. Washington, DC: National Highway Traffic Safety Administration, http://www-nrd.nhtsa.dot.gov/Pubs/806314.PDF.

Kahane, C. J. (1986). *An Evaluation of Child Passenger Safety: The Effectiveness and Benefits of Safety Seats*, NHTSA Technical Report No. DOT HS 806 890. Washington, DC: National Highway Traffic Safety Administration, http://www-nrd.nhtsa.dot.gov/Pubs/806890.PDF.

Kahane, C. J. (1988). *An Evaluation of Occupant Protection in Frontal Interior Impact for Unrestrained Front Seat Occupants of Cars and Light Trucks*, NHTSA Technical Report No. DOT HS 807 203. Washington, DC: National Highway Traffic Safety Administration, http://www-nrd.nhtsa.dot.gov/pubs/807203.pdf.

Kahane, C. J. (1994). *Correlation of NCAP Performance with Fatality Risk in Actual Head-On Collisions*, NHTSA Technical Report No. DOT HS 808 061. Washington, DC: National Highway Traffic Safety Administration, http://www-nrd.nhtsa.dot.gov/Pubs/808061.PDF.

Kahane, C. J. (1996). *Fatality Reduction by Air Bags: Analyses of Accident Data through Early 1996*, NHTSA Technical Report No. DOT HS 808 470. Washington, DC: National Highway Traffic Safety Administration, http://www-nrd.nhtsa.dot.gov/Pubs/808470.PDF.

Kahane, C. J. (2000). *Fatality Reduction by Safety Belts for Front-Seat Occupants of Cars and Light Trucks: Updated and Expanded Estimates Based on 1986-99 FARS Data*, NHTSA Technical Report No. DOT HS 809 199. Washington, DC: National Highway Traffic Safety Administration, http://www-nrd.nhtsa.dot.gov/Pubs/809199.PDF.

Kahane, C. J. (2004). *Lives Saved by the Federal Motor Vehicle Safety Standards and Other Vehicle Safety Technologies, 1960-2002*, NHTSA Technical Report No. DOT HS 809 833. Washington, DC: National Highway Traffic Safety Administration, http://www-nrd.nhtsa.dot.gov/Pubs/809833.PDF.

Kahane, C. J. (2006). *An Evaluation of the 1998-1999 Redesign of Frontal Air Bags*, NHTSA Technical Report No. DOT HS 810 685. Washington, DC: National Highway Traffic Safety Administration, http://www-nrd.nhtsa.dot.gov/Pubs/810685.PDF.

Kahane, C. J. (2007). *An Evaluation of Side Impact Protection – FMVSS 214 TTI(d) Improvements and Side Air Bags*, NHTSA Technical Report No. DOT HS 810 748. Washington, DC: National Highway Traffic Safety Administration, http://www-nrd.nhtsa.dot.gov/Pubs/810748.PDF.

Kahane, C. J. (2009). *The Long-Term Effect of ABS in Passenger Cars and LTVs*, NHTSA Technical Report No. DOT HS 811 182. Washington, DC: National Highway Traffic Safety Administration, http://www-nrd.nhtsa.dot.gov/Pubs/811182.PDF.

Kahane, C. J. (2011). *Evaluation of the 1999-2003 Head Impact Upgrade of FMVSS No. 201 – Upper-Interior Components: Effectiveness of Energy-Absorbing Materials Without Head-Protection Air Bags*, NHTSA Technical Report No. DOT HS 811 538. Washington, DC: National Highway Traffic Safety Administration, http://www-nrd.nhtsa.dot.gov/Pubs/811538.PDF.

Kahane, C. J. (2012). *Relationships Between Fatality Risk, Mass, and Footprint in Model Year 2000-2007 Passenger Cars and LTVs – Final Report*, NHTSA Technical Report No. DOT HS 811 665. Washington, DC: National Highway Traffic Safety Administration, http://www-nrd.nhtsa.dot.gov/Pubs/811665.PDF.

Karnath, B. (2004). "Subdural Hematoma, Presentation and Management in Older Adults," *Geriatrics*, Vol. 58, pp. 18-23. https://ssl-w03dnn0374.websiteseguro.com/sbn-neurocirurgia/site/download/artigos/article.pdf

Kent, R., and Patrie, J. (2005). "Chest Deflection Tolerance to Blunt Anterior Loading Is Sensitive to Age but Not Load Distribution," *Forensic Science International*, Vol. 149, pp. 121-128.

Kent, R., Trowbridge, M., Lopez-Valdes, F. J., Heredero Ordoyo, R., & Segui-Gomez, M. (2009). "How Many People Are Injured and Killed as a Result of Aging? Frailty, Fragility and the Elderly Risk-Exposure Tradeoff Assessed via a Risk Saturation Model," *Annals of Advances in Automotive Medicine,* Vol. 53, pp. 41-50.

Morgan, C. (1999). *Effectiveness of Lap/Shoulder Belts in the Back Outboard Seating Positions*, NHTSA Technical Report No. DOT HS 808 945. Washington, DC: National Highway Traffic Safety Administration, http://www-nrd.nhtsa.dot.gov/Pubs/808945.PDF.

NCHS (2006). *Multiple Causes of Mortality, 2003; Documentation of the Mortality Tape File for 2003 Data*. Hyattsville, MD: National Center for Health Statistics. http://wonder.cdc.gov/wonder/sci_data/mort/mcmort/type_txt/mcmort03/mcmort03.asp.

Partyka, S. C. (1984). *Restraint Use and Fatality Risk for Infants and Toddlers*. Washington, DC: National Highway Traffic Safety Administration.

Partyka, S. C. (1988). "Belt Effectiveness in Pickup Trucks and Passenger Cars by Crash Direction and Accident Year," *Papers on Adult Seat Belts – Effectiveness and Use*, NHTSA Report No. DOT HS 807 285. Washington, DC: National Highway Traffic Safety Administration.

Ridella, S. A., Rupp, J. D., and Poland, K. (2012). "Age-Related Differences in AIS 3+ Crash Injury Risk, Types, Causation and Mechanisms," *International IRCOBI Conference on the Biomechanics of Impact*.

Sivinski, R. (2010). *Booster Seat Effectiveness Estimates Based on CDS and State Data*, NHTSA Technical Report No. DOT HS 811 338. Washington, DC: National Highway Traffic Safety Administration. http://www-nrd.nhtsa.dot.gov/Pubs/811338.pdf.

Walz, M. C. (2003). *NCAP Test Improvements with Pretensioners and Load Limiters*, NHTSA Technical Report No. DOT HS 809 562. Washington, DC: National Highway Traffic Safety Administration. http://www-nrd.nhtsa.dot.gov/Pubs/809562.PDF.

Wang, S., and Rupp, J. (2006). *Alterations in Injury Patterns and Body Composition With Aging*. http://www.nhtsa.gov/DOT/NHTSA/NVS/CIREN/2006%20Presentations/MI_0306b.pdf.

Zhou, Q., Rouhana, S. W., and Melvin, J. W. (1996). "Age Effects on Thoracic Injury Tolerance," *40th Stapp Car Crash Conference Proceedings*, Paper No. 962421. (Publication No. P-305). Warrendale, PA: Society of Automotive Engineers.

APPENDIX A

CLASSIFICATION OF ICD-9 INJURY CODES

Red =   torso injuries, excluding minor injuries (includes chest, abdomen, thoracic or lumbar
         spine, pelvis, clavicle, and scapula)
                (C): chest injury (skeletal or soft tissue)
                (A): abdominal injury (soft tissue)
Blue =   head injuries, excluding minor injuries (includes brain, skull, and face)
Green = neck injuries, excluding minor injuries (includes cervical spine and throat)
Black = arm or leg injuries, multiple or unspecified body regions, minor injuries to any body
         region, burns to any body region, poisoning, complications after an injury
Bold italics (for red, blue, or green only) = evidently severe injury
Regular print (for red, blue, or green only) = injury of unknown severity

800 FRACTURE OF VAULT OF SKULL
801 FRACTURE OF BASE OF SKULL
802 FRACTURE OF FACIAL BONES
803 OTHER/UNSPECIFIED SKULL FRACTURES
805 FRACTURE OF VERTEBRAL COLUMN
806 FX VERTEBRA WITH CORD, LOCATION UNSPEC
807 (C) FRACTURE OF RIBS, STERNUM, LARYNX OR TRACHEA
808 FRACTURE OF PELVIS
809 UNKNOWN FRACTURE OF BONES OF TRUNK
810 FRACTURE OF CLAVICLE
811 FRACTURE OF SCAPULA
812 FRACTURE OF HUMERUS
813 FRACTURE OF RADIUS AND/OR ULNA
814 FRACTURE OF HAND
815 FX HAND
816 FX FINGER
817 FX MULTIPLE HAND BONES
818 FRACTURE OF UPPER LIMB
819 MULT ARM FRACTURES, POSSIBLY WITH RIBS/STERNUM
820 FRACTURE OF NECK OF FEMUR
821 FRACTURE, OTHER PART OF FEMUR
822 FRACTURE OF PATELLA
823 FRACTURE OF TIBIA AND FIBULA
824 FRACTURE OF ANKLE
825 FRACTURE OF FOOT
826 FRACTURE OF TOE(S)
827 OTHER/MULT/UNSPEC FRACTURES OF LOWER LIMBS
828 MULT LEG FRACTURES, POSSIBLY WITH RIBS/STERNUM
829 FRACTURE OF UNSPECIFIED BONES
830 DISLOCATION OF JAW
831 DISLOCATION OF SHOULDER
832 DISLOCATION OF ELBOW
833 DISLOCATION OF WRIST
834 DISLOCATION OF FINGER
835 DISLOCATION OF HIP
836 DISLOCATION OF KNEE
837 DISLOCATION OF ANKLE
838 DISLOCATION OF FOOT

839 OTHER/UNSPECIFIED DISLOCATIONS
843 SPRAIN OF HIP
844 SPRAIN OF LEG
845 SPRAIN OF FOOT
846 SPRAIN OF BACK
847 UNKNOWN BACK SPRAINS
848 UNKNOWN SPRAINS
850 CONCUSSION
851 CEREBRAL LACERATION OR CONTUSION
852 SUBARACHNOID, SUBDURAL OR EXTRADURAL HEMORRHAGE
853 OTHER/UNSPECIFIED INTRACRANIAL HEMORRHAGE
854 OTHER/UNSPECIFIED INTRACRANIAL INJURY
860 (C) PNEUMOTHORAX AND/OR HEMOTHORAX
861 (C) INJURY TO HEART AND/OR LUNG
862 (C) INJURY TO OTHER/UNSPECIFIED THORACIC ORGANS
863 (A) INJURY OF GASTRO-INTESTINAL TRACT
864 (A) INJURY TO LIVER
865 (A) INJURY TO SPLEEN
866 (A) INJURY TO KIDNEY
867 (A) INJURY TO PELVIC ORGANS
868 (A) INJURY TO INTRA-ABDOMINAL ORGANS
869 INJURY TO UNSPECIFIED ORGANS
870 OPEN WOUND OF OCULAR ADNEXA
871 OPEN WOUND OF EYEBALL
872 OPEN WOUND OF EAR
873 OPEN HEAD WOUND
874 OPEN NECK WOUND
875 (C) OPEN WOUND OF CHEST WALL
876 OPEN WOUND OF BACK
877 OPEN WOUND OF BUTTOCK
878 (A) OPEN WOUND OF GENITALIA
879 OTHER/UNSPECIFIED OPEN WOUND
880-883 OPEN WOUND OF ARM
884 MULTIPLE/UNSPECIFIED OPEN WOUNDS OF UPPER LIMB
886 AMPUTATION OF FINGER
887 ARM OR HAND AMPUTATION
890 OPEN WOUND OF HIP OR THIGH
891-894 OPEN WOUND OF LEG
895 AMPUTATION OF TOE
896 AMPUTATION OF FOOT
897 AMPUTATION OF LEG
900 INJURY TO BLOOD VESSELS OF HEAD & NECK
901 (C) INJURY TO BLOOD VESSELS OF THORAX
902 (A) INJURY TO BLOOD VESSELS OF ABDOMEN OR PELVIS
903 INJURY TO BLOOD VESSELS OF ARM
904 INJURY TO BLOOD VESSELS OF LEG
905.0 LATE EFFECT OF SKULL FRACTURE
905.1 LATE EFFECT OF SPINE FRACTURE (UNK LOCATION)
905.2 LATE EFFECT OF ARM FRACTURE
905.3 LATE EFFECT OF FEMUR-NECK FRACTURE
905.9 LATE EFFECT OF AMPUTATION
907.0 LATE EFFECT OF INTRACRANIAL INJURY
907.2 LATE EFFECT OF SPINAL CORD INJURY
908.0 (C) LATE EFFECT OF CHEST INJURY
908.2 LATE EFFECT OF UNK INTERNAL INJURY
908.4 LATE EFFECT OF INJURY TO TORSO BLOOD VESSELS
908.9 LATE EFFECT OF UNK INJURY

909.3 LATE EFFECT OF SURGICAL COMPLICATIONS
910 SUPERFICIAL HEAD/FACE INJURY
911 SUPERFICIAL TRUNK INJURY
912-915 SUPERFICIAL ARM INJURY
916-917 SUPERFICIAL LEG INJURY
918 SUPERFICIAL EYE INJURY
919 OTHER/UNSPEC/MULT SUPERFICIAL INJURIES
920 CONTUSION OF FACE/HEAD
921 CONTUSION OF EYE
922 CONTUSION OF TRUNK
923 CONTUSION OF ARM
924 CONTUSION OF LEG
925 CRUSHING OF FACE, SCALP OR NECK
926 CRUSHING OF TRUNK
927 CRUSHED ARM
928 CRUSHED LEG
929 CRUSHING OF MULTIPLE/UNSPECIFIED SITES
932 FOREIGN BODY IN NOSE
933 FOREIGN BODY IN PHARYNX OR LARYNX
934 FOREIGN BODY IN TRACHEA, BRONCHUS OR LUNG
935-937 F.B. IN ALIMENTARY TRACT
939 (A) F.B. IN GENITO-URINARY TRACT
941.0-949.9 BURNS
950 INJURY TO OPTIC NERVE
951 INJURY TO OTHER CRANIAL NERVES
952 SPINAL CORD INJURY
953 INJURY TO NERVE ROOTS OR PLEXUS
954 INJURY TO OTHER NERVES OF TRUNK
955 INJURY TO ARM NERVE
956 INJURY TO LEG NERVE
957 INJURIES TO OTHER/UNSPECIFIED NERVES
958.0-958.9 COMPLICATIONS OF TRAUMA (NOT NECESSARILY IN ORIGINAL BODY REGION)
959.0 UNKNOWN HEAD INJURIES
959.1 UNKNOWN TRUNK INJURIES
959.2 UNKNOWN SHOULDER OR UPPER ARM INJURY
959.3-959.5 UNKNOWN ARM INJURY
959.6 UNKNOWN HIP & THIGH INJURY
959.7 UNKNOWN LEG INJURY
959.8 OTHER OR MULTIPLE UNSPECIFIED INJURIES
959.9 UNKNOWN INJURIES
962.0-989.9 POISONING, ADVERSE EFFECTS OF DRUGS, TOXINS
990 EFFECTS OF RADIATION
991 EFFECTS OF COLD
992 EFFECTS OF HEAT
993 EFFECTS OF AIR PRESSURE
994.0 STRUCK BY LIGHTNING
994.1 DROWNING
994.2 EFFECT OF HUNGER
994.5 OVEREXERTION
994.6 MOTION SICKNSS
994.7 ASPHYXIATION
994.8 EFFECT OF ELECTRIC CURRENT
994.9 OTHER EFFECTS OF EXTERNAL CAUSES
995.0 ANAPHALACTIC SHOCK
995.1 ANGIONEUROTIC EDEMA
995.2 OTHER DRUG REACTION
995.4 REACTION TO ANESTHESIA

995.8 OTHER ADVERSE EFFECTS
996.0-999.9 COMPLICATIONS OF MEDICAL PROCEDURES

APPENDIX B

CLASSIFICATION OF ICD-10 INJURY CODES

Red =   torso injuries, excluding minor injuries (includes chest, abdomen, thoracic or lumbar spine, pelvis, clavicle, and scapula)
        (C): chest injury (skeletal or soft tissue)
        (A): abdominal injury (soft tissue)
Blue =   head injuries, excluding minor injuries (includes brain, skull, and face)
Green =   neck injuries, excluding minor injuries (includes cervical spine and throat)
Black =   arm or leg injuries, multiple or unspecified body regions, minor injuries to any body region, burns to any body region, poisoning, complications after an injury
Bold italics (for red, blue, or green only) = evidently severe injury
Regular print (for red, blue, or green only) = injury of unknown severity

```
S00.0   Superficial injury of scalp
S00.3   Superficial injury of nose
S00.5   Superficial injury of lip and oral cavity
S00.7   Multiple superficial injuries of head
S00.8   Superficial injury of other parts of head
S00.9   Superficial injury of head, part unspecified
S01.0   Open wound of scalp
S01.1   Open wound of eyelid and periocular area
S01.2   Open wound of nose
S01.3   Open wound of ear
S01.5   Open wound of lip and oral cavity
S01.7   Multiple open wounds of head
S01.8   Open wound of other parts of head
S01.9   Open wound of head, part unspecified
S02.0   Fracture of vault of skull
S02.1   Fracture of base of skull
S02.2   Fracture of nasal bones
S02.3   Fracture of orbital floor
S02.4   Fracture of malar and maxillary bones
S02.5   Fracture of tooth
S02.6   Fracture of mandible
S02.7   Multiple fractures involving skull and facial bones
S02.8   Fractures of other skull and facial bones
S02.9   Fracture of skull and facial bones, part unspecified
S03.1   Dislocation of septal cartilage of nose
S03.3   Dislocation of other and unspecified parts of head
S04.8   Injury of other cranial nerves
S04.9   Injury of unspecified cranial nerve
S05.1   Contusion of eyeball and orbital tissues
S05.3   Ocular laceration without prolapse or loss of intraocular tissue
S05.4   Penetrating wound of orbit with or without foreign body
S05.7   Avulsion of eye
S05.8   Other injuries of eye and orbit
S05.9   Injury of eye and orbit, unspecified
S06.0   Concussion
S06.1   Traumatic cerebral edema
S06.2   Diffuse brain injury
S06.3   Focal brain injury
```

S06.4   Epidural haemorrhage
S06.5   Traumatic subdural hemorrhage
S06.6   Traumatic subarachnoid hemorrhage
S06.7   Intracranial injury with prolonged coma
S06.8   Other intracranial injuries
S06.9   Intracranial injury, unspecified
S07.0   Crushing injury of face
S07.1   Crushing injury of skull
S07.8   Crushing injury of other parts of head
S07.9   Crushing injury of head, part unspecified
S08.0   Avulsion of scalp
S08.8   Traumatic amputation of other parts of head
S08.9   Traumatic amputation of unspecified part of head
S09.0   Injury of blood vessels of head, not elsewhere classified
S09.1   Injury of muscle and tendon of head
S09.2   Traumatic rupture of ear drum
S09.7   Multiple injuries of head
S09.8   Other specified injuries of head
S09.9   Unspecified injury of head
S10.0   Contusion of throat
S10.1   Other and unspecified superficial injuries of throat
S10.8   Superficial injury of other parts of neck
S10.9   Superficial injury of neck, part unspecified
S11.0   Open wound involving larynx and trachea
S11.7   Multiple open wounds of neck
S11.8   Open wound of other parts of neck
S11.9   Open wound of neck, part unspecified
S12.0   Fracture of first cervical vertebra (atlas)
S12.1   Fracture of second cervical vertebra (axis)
S12.2   Fracture of other specified cervical vertebra
S12.7   Multiple fractures of cervical spine
S12.8   Fracture of other parts of neck
S12.9   Fracture of neck, part unspecified
S13.0   Traumatic rupture of cervical intervertebral disc
S13.1   Dislocation of cervical vertebra
S13.3   Multiple dislocations of neck
S13.4   Sprain and strain of cervical spine
S13.6   Sprain and strain of joints and ligaments of o. & u. parts of neck
S14.0   Concussion and edema of cervical spinal cord
S14.1   Other and unspecified injuries of cervical spinal cord
S14.2   Injury of nerve root of cervical spine
S14.3   Injury of brachial plexus
S14.6   Injury of other and unspecified nerves of neck
S15.0   Injury of carotid artery
S15.1   Injury of vertebral artery
S15.2   Injury of external jugular vein
S15.3   Injury of internal jugular vein
S15.7   Injury of multiple blood vessels at neck level
S15.9   Injury of unspecified blood vessel at neck level
S16   Injury of muscle and tendon at neck level
S17.0   Crushing injury of larynx and trachea
S17.8   Crushing injury of other parts of neck
S17.9   Crushing injury of neck, part unspecified
S18   Other and unspecified injuries of neck
S19.7   Multiple injuries of neck
S19.8   Other specified injuries of neck
S19.9   Unspecified injury of neck

S20.1 Other and unspecified superficial injuries of breast
S20.2 Contusion of thorax
S20.3 Other superficial injuries of front wall of thorax
S20.7 Multiple superficial injuries of thorax
S20.8 Superficial injury of other and unspecified parts of thorax
S21.0 Open wound of breast
S21.1 Open wound of front wall of thorax
S21.2 Open wound of back wall of thorax
S21.7 Multiple open wounds of thoracic wall
S21.8 Open wound of other parts of thorax
S21.9 Open wound of thorax, part unspecified
S22.0 Fracture of thoracic vertebra
S22.1 Multiple fractures of thoracic spine
S22.2 (C) Fracture of sternum
S22.3 (C) Fracture of rib
S22.4 (C) Multiple fractures of ribs
S22.5 (C) Flail chest
S22.8 (C) Fracture of other parts of bony thorax
S22.9 (C) Fracture of bony thorax, part unspecified
S23.1 Dislocation of thoracic vertebra
S23.2 Dislocation of other and unspecified parts of thorax
S24.0 Concussion and edema of thoracic spinal cord
S24.1 Other and unspecified injuries of thoracic spinal cord
S24.2 Injury of nerve root of thoracic spine
S24.4 Injury of thoracic sympathetic nerves
S24.6 Injury of unspecified nerve of thorax
S25.0 (C) Injury of thoracic aorta
S25.1 (C) Injury of innominate or subclavian artery
S25.2 (C) Injury of superior vena cava
S25.3 (C) Injury of innominate or subclavian vein
S25.4 (C) Injury of pulmonary blood vessels
S25.5 (C) Injury of intercostal blood vessels
S25.7 I(C) njury of multiple blood vessels of thorax
S25.8 (C) Injury of other blood vessels of thorax
S25.9 (C) Injury of unspecified blood vessel of thorax
S26.0 (C) Injury of heart with hemopericardium
S26.8 (C) Other injuries of heart
S26.9 (C) Injury of heart, unspecified
S27.0 (C) Traumatic pneumothorax
S27.1 (C) Traumatic hemothorax
S27.2 (C) Traumatic hemopneumothorax
S27.3 (C) Other injuries of lung
S27.4 (C) Injury of bronchus
S27.6 (C) Injury of pleura
S27.7 (C) Multiple injuries of intrathoracic organs
S27.8 (C) Injury of other specified intrathoracic organs
S27.9 (C) Injury of unspecified intrathoracic organ
S28.0 (C) Crushed chest
S28.1 (C) Traumatic amputation of part of thorax
S29.0 Injury of muscle and tendon at thorax level
S29.7 (C) Multiple injuries of thorax
S29.8 (C) Other specified injuries of thorax
S29.9 (C) Unspecified injury of thorax
S30.0 Contusion of lower back and pelvis
S30.1 Contusion of abdominal wall
S30.2 Contusion of external genital organs
S30.7 Multiple superficial injuries of abdomen, lower back and pelvis

S30.8 Other superficial injuries of abdomen, lower back and pelvis
S30.9 Superficial injury of abdomen, lower back and pelvis, part unspecified
S31.0 Open wound of lower back and pelvis
S31.1 Open wound of abdominal wall
S31.3 (A) Open wound of scrotum and testes
S31.4 (A) Open wound of vagina and vulva
S31.7 Multiple open wounds of abdomen, lower back and pelvis
S31.8 Open wound of other and unspecified parts of abdomen
S32.0 Fracture of lumbar vertebra
S32.1 Fracture of sacrum
S32.2 Fracture of coccyx
S32.3 Fracture of ilium
S32.4 Fracture of acetabulum
S32.5 Fracture of pubis
S32.7 Multiple fractures of lumbar spine and pelvis
S32.8 Fracture of other and unspecified parts of lumbar spine and pelvis
S33.1 Dislocation of lumbar vertebra
S34.1 Other injury of lumbar spinal cord
S34.8 Injury of o & u nerves at abdomen, lower back and pelvis level
S35.0 (A) Injury of abdominal aorta
S35.1 (A) Injury of inferior vena cava
S35.2 (A) Injury of coeliac or mesenteric artery
S35.3 (A) Injury of portal or splenic vein
S35.4 (A) Injury of renal blood vessels
S35.5 (A) Injury of iliac blood vessels
S35.8 (A) Injury of other blood vessels at abdomen, lower back and pelvis level
S35.9 (A) Injury of u blood vessel at abdomen, lower back and pelvis level
S36.0 (A) Injury of spleen
S36.1 (A) Injury of liver or gall bladder
S36.2 (A) Injury of pancreas
S36.3 (A) Injury of stomach
S36.4 (A) Injury of small intestine
S36.5 (A) Injury of colon
S36.6 (A) Injury of rectum
S36.7 (A) Injury of multiple intra-abdominal organs
S36.8 (A) Injury of other intra-abdominal organs
S36.9 (A) Injury of unspecified intra-abdominal organ
S37.0 (A) Injury of kidney
S37.2 (A) Injury of bladder
S37.3 (A) Injury of urethra
S37.6 (A) Injury of uterus
S37.7 (A) Injury of multiple pelvic organs
S37.8 (A) Injury of other pelvic organs
S37.9 (A) Injury of unspecified pelvic organ
S38.0 (A) Crushing injury of external genital organs
S38.1 (A) Crushing of o. & u. parts of abdomen, lower back and pelvis
S38.2 (A) Traumatic amputation of external genital organs
S38.3 (A) Traumatic amputation of o. & u. parts of abdomen, lower back and pelvis
S39.0 Injury of muscle and tendon of abdomen, lower back and pelvis
S39.6 (A) Injury of intra-abdominal organ(s) with pelvic organ(s)
S39.7 (A) Other multiple injuries of abdomen, lower back and pelvis
S39.8 (A) Other specified injuries of abdomen, lower back and pelvis
S39.9 (A) Unspecified injury of abdomen, lower back and pelvis
S40.0 Contusion of shoulder and upper arm
S40.7 Multiple superficial injuries of shoulder and upper arm
S40.9 Superficial injury of shoulder and upper arm, unspecified
S41.0 Open wound of shoulder

S41.1 Open wound of upper arm
S41.8 Open wound of other and unspecified parts of shoulder girdle
S42.0 Fracture of clavicle
S42.1 Fracture of scapula
S42.2 Fracture of upper end of humerus
S42.3 Fracture of shaft of humerus
S42.4 Fracture of lower end of humerus
S42.7 Multiple fractures of clavicle, scapula and humerus
S42.8 Fracture of other parts of shoulder and upper arm
S42.9 Fracture of shoulder girdle, part unspecified
S43.0 Dislocation of shoulder joint
S43.1 Dislocation of acromioclavicular joint
S43.2 Dislocation of sternoclavicular joint
S45.0 Injury of axillary artery
S45.1 Injury of brachial artery
S45.2 Injury of axillary or brachial vein
S45.9 Injury of unspecified blood vessel at shoulder and upper arm level
S47   Crushing injury of shoulder and upper arm
S48.0 Traumatic amputation at shoulder joint
S48.9 Traumatic amputation of shoulder and upper arm, level unspecified
S49.7 Multiple injuries of shoulder and upper arm
S49.8 Other specified injuries of shoulder and upper arm
S49.9 Unspecified injury of shoulder and upper arm
S50.7 Multiple superficial injuries of forearm
S50.9 Superficial injury of forearm, unspecified
S51.0 Open wound of elbow
S51.9 Open wound of forearm, part unspecified
S52.0 Fracture of upper end of ulna
S52.1 Fracture of upper end of radius
S52.2 Fracture of shaft of ulna
S52.4 Fracture of shafts of both ulna and radius
S52.5 Fracture of lower end of radius
S52.7 Multiple fractures of forearm
S52.8 Fracture of other parts of forearm
S52.9 Fracture of forearm, part unspecified
S53.1 Dislocation of elbow, unspecified
S55.1 Injury of radial artery at forearm level
S58.9 Traumatic amputation of forearm, level unspecified
S59.7 Multiple injuries of forearm
S59.9 Unspecified injury of forearm
S60.8 Other superficial injuries of wrist and hand
S60.9 Superficial injury of wrist and hand, unspecified
S61.9 Open wound of wrist and hand part, part unspecified
S62.1-S62.8 Arm fractures
S63.0 Dislocation of wrist
S68.1 Traumatic amputation of other single finger
S68.4 Traumatic amputation of hand at wrist level
S68.9 Traumatic amputation of wrist and hand, level unspecified
S69.7 Multiple injuries of wrist and hand
S69.9 Unspecified injury of wrist and hand
S70.0 Contusion of hip
S70.1 Contusion of thigh
S70.7 Multiple superficial injuries of hip and thigh
S70.8 Other superficial injuries of hip and thigh
S70.9 Superficial injury of hip and thigh, unspecified
S71.0 Open wound of hip
S71.1 Open wound of thigh

S72.0 Fracture of hip and/or fracture of neck of femur
S72.1 Trochanteric fracture of femur
S72.2 Subtrochanteric fracture of femur
S72.3 Fracture of shaft of femur
S72.4 Fracture of lower end of femur
S72.7 Multiple fractures of femur
S72.8 Fractures of other parts of femur
S72.9 Fracture of femur, part unspecified
S73.0 Dislocation of hip
S75.0 Injury of femoral artery
S75.1 Injury of femoral vein at hip and thigh level
S75.9 Injury of unspecified blood vessel at hip and thigh level
S76.0 Injury of muscle and tendon of hip
S77.1 Crushing injury of thigh
S78.1 Traumatic amputation of thigh at level between hip and knee
S78.9 Traumatic amputation of hip and thigh, level unspecified
S79.7 Multiple injuries of hip and thigh
S79.8 Other specified injuries of hip and thigh
S79.9 Unspecified injury of hip and thigh
S80.1 Contusion of other and unspecified parts of lower leg
S80.7 Multiple superficial injuries of lower leg
S80.8 Other superficial injuries of lower leg
S80.9 Superficial injury of lower leg, unspecified
S81.0 Open wound of knee
S81.8 Open wound of other parts of lower leg
S81.9 Open wound of lower leg, part unspecified
S82.0 Fracture of patella
S82.1 Fracture of upper end of tibia
S82.2 Fracture of shaft of tibia
S82.3 Fracture of lower end of tibia
S82.4 Fracture of fibula alone
S82.7 Multiple fractures of lower leg
S82.8 Fractures of other parts of lower leg
S82.9 Fracture of lower leg, part unspecified
S83.0 Dislocation of patella
S83.1 Dislocation of knee
S83.2 Tear of meniscus of knee, current
S83.7 Injury to multiple structures of knee
S84.7 Injury of multiple nerves at lower leg level
S85.0 Injury of popliteal artery
S85.1 Injury of (anterior)(posterior) tibial artery
S85.2 Injury of peroneal artery
S85.5 Injury of popliteal vein
S85.9 Injury of unspecified blood vessel at lower leg level
S86.0 Injury of Achilles tendon
S87.8 Crushing injury of other and unspecified parts of lower leg
S88.0 Traumatic amputation of leg at knee level
S88.1 Traumatic amputation of leg at level between knee and ankle
S88.9 Traumatic amputation of lower leg, level unspecified
S89.7 Multiple injuries of lower leg
S89.9 Unspecified injury of lower leg
S90.0 Contusion of ankle
S90.9 Superficial injury of ankle and foot, unspecified
S91.0 Open wound of ankle
S91.3 Open wound of other parts of foot
S92.0-S92.9 Foot fractures
S93.0 Dislocation of ankle joint

```
S93.3  Dislocation of other and unspecified parts of foot
S96.1  Injury of long extensor muscle of toe at ankle and foot level
S96.8  Injury of other muscles and tendons at ankle and foot level
S97.0  Crushing injury of ankle
S98.0  Traumatic amputation of foot at ankle level
S98.2  Traumatic amputation of two or more toes
S98.4  Traumatic amputation of foot, level unspecified
S99.9  Unspecified injury of ankle and foot
T00.2  Superficial injuries involving multiple regions of upper limb(s)
T00.3  Superficial injuries involving multiple regions of lower limb(s)
T00.8  Superficial injuries involving other combinations of body regions
T00.9  Multiple superficial injuries, unspecified
T01.2  Open wounds involving multiple regions of upper limb(s)
T01.3  Open wounds involving multiple regions of lower limb(s)
T01.9  Multiple open wounds, unspecified
T02.1  Fractures involving thorax with lower back and pelvis
T02.2  Fractures involving multiple regions of one upper limb
T02.3  Fractures involving multiple regions of one lower limb
T02.4  Fractures involving multiple regions of both upper limbs
T02.5  Fractures involving multiple regions of both lower limbs
T02.8  Fractures involving other combinations of body regions
T02.9  Multiple fractures, unspecified
T03.8  Dislocations, sprains and strains involving combinations of body regions
T03.9  Multiple dislocations, sprains and strains, unspecified
T04.1  Crushing injuries involving thorax with abdomen, lower back and pelvis
T04.2  Crushing injuries involving multiple regions of upper limb(s)
T04.3  Crushing injuries involving multiple regions of lower limb(s)
T04.8  Crushing injuries involving other combinations of body regions
T04.9  Multiple crushing injuries, unspecified
T05.5  Traumatic amputation of both legs [any level]
T05.8  Traumatic amputations involving other combinations of body regions
T05.9  Multiple traumatic amputations, unspecified
T06.2  Injuries of nerves involving multiple body regions
T06.3  Injuries of blood vessels involving multiple body regions
T06.4  Injuries of muscles and tendons involving multiple body regions
T06.5  Injuries of intrathoracic, intra-abdominal & pelvic organs
T07    Unspecified multiple injuries
T08    Fracture of spine, level unspecified
T09.0  Superficial injury of trunk, level unspecified
T09.1  Open wound of trunk, level unspecified
T09.2  Dislocation, sprain and strain of unknown joint or ligament of trunk
T09.3  Injury of spinal cord, level unspecified
T09.4  Injury of unspecified nerve, spinal nerve root and plexus of trunk
T09.5  Injury of unspecified muscle and tendon of trunk
T09.6  Traumatic amputation of trunk, level unspecified
T09.8  Other specified injuries of trunk, level unspecified
T09.9  Unspecified injury of trunk, level unspecified
T10    Fracture of upper limb, level unspecified
T11.0  Superficial injury of upper limb, level unspecified
T11.1  Open wound of upper limb, level unspecified
T11.2  Dislocation, sprain, strain of u. joint/ligament of upper limb, level u.
T11.4  Injury of unspecified blood vessel of upper limb, level unspecified
T11.5  Injury of u. muscle and tendon of upper limb, level unspecified
T11.6  Traumatic amputation of upper limb, level unspecified
T11.8  Other specified injuries of upper limb, level unspecified
T11.9  Unspecified injury of upper limb, level unspecified
T12    Fracture of lower limb, level unspecified
```

```
T13.0  Superficial injury of lower limb, level unspecified
T13.1  Open wound of lower limb, level unspecified
T13.2  Dislocation, sprain, strain of u. joint/ligament of lower limb, level u.
T13.4  Injury of unspecified blood vessel of lower limb, level unspecified
T13.5  Injury of u. muscle and tendon of lower limb, level unspecified
T13.6  Traumatic amputation of lower limb, level unspecified
T13.8  Other specified injuries of lower limb, level unspecified
T13.9  Unspecified injury of lower limb, level unspecified
T14.0  Superficial injury of unspecified body region
T14.1  Open wound of unspecified body region
T14.2  Fracture of unspecified body region
T14.3  Dislocation, sprain and strain of unspecified body region
T14.4  Injury of nerve(s) of unspecified body region
T14.5  Injury of blood vessel(s) of unspecified body region
T14.6  Injury of muscles and tendons of unspecified body region
T14.7  Crushing injury and traumatic amputation of unspecified body region
T14.8  Other injuries of unspecified body region
T14.9  Injury, unspecified
T15.1  Foreign body in conjunctival sac of eye
T15.9  Foreign body on external eye, part unspecified
T17.1  Foreign body in nostril
T17.3  Foreign body in larynx
T17.4  Foreign body in trachea
T17.5  (C) Foreign body in bronchus
T17.8  Foreign body in other and multiple parts of respiratory tract
T17.9  Foreign body in respiratory tract, part unspecified
T18.9  Foreign body in alimentary tract, part unspecified
T20.0-T31.9  Burns
T35.5  Unspecified frostbite of lower limb
T36.8-T65.9  Poisoning, adverse effects of drugs, toxins
T67.0-T67.9  Effects of heat
T68.0-T69.9  Effects of cold
T70.9  Effect of air pressure and water pressure, unspecified
T71    Asphyxiation
T73.8  Other effects of deprivation (exposure, overexertion)
T75.1  Drowning
T75.4  Effects of electric current
T75.8  Other specified effects of external causes
T78.2  Anaphylactic shock
T78.4  Allergy, unspecified
T79.0  Air embolism (traumatic)
T79.1  Fat embolism (traumatic)
T79.2  Traumatic secondary and recurrent hemorrhage
T79.3  Post-traumatic wound infection
T79.4  Traumatic shock
T79.5  Traumatic anuria
T79.6  Traumatic ischaemia of muscle
T79.7  Traumatic subcutaneous emphysema
T79.8  Other early complications of trauma
T79.9  Unspecified early complication of trauma
T80.2-T88.9  Complications of medical procedures
T90.2-T98.3  Sequelae of injuries, toxic effects, or procedures
```

APPENDIX C

CLASSIFICATION OF CDS 3-LETTER INJURY CODES

Red = torso injuries (includes chest, abdomen, thoracic or lumbar spine, pelvis, clavicle, and scapula)

Blue = head injuries (includes brain, skull, and face)

Green = neck injuries (includes cervical spine and throat)

Black = arm or leg injuries, all burns

CDS Body Region codes:

A	ARM (UPPER)
B	BACK
C	CHEST
E	ELBOW
F	FACE
H	HEAD - SKULL
K	KNEE
L	LEG/LOWER
M	ABDOMEN
N	NECK
O	WHOLE BODY
P	PELVIC/HIP
Q	ANKLE/FOOT
R	FOREARM
S	SHOULDER
T	THIGH
U	INJURED/UNK REG
W	WRIST/HAND
X	UPPER LIMBS
Y	LOWER LIMBS

CDS Lesion codes:

A	ABRASION
B	BURN
C	CONTUSION
D	DISLOCATION
E	TOTAL SEVERANCE
F	FRACTURE
G	DETACHMENT
K	CONCUSSION
L	LACERATION
M	AMPUTATION
N	CRUSH
O	OTHER
P	PERFORATION
R	RUPTURE
S	SPRAIN
T	STRAIN
U	INJURED/UNK LES
V	AVULSION
Z	FRACTURE/DISLOC

CDS System/Organ codes:

```
A    ARTERIES/VEINS
B    BRAIN
C    SPINAL CORD
D    DIGESTIVE
E    EARS
G    UROGENITAL
H    HEART
I    INTEGUMENTARY
J    JOINTS
K    KIDNEYS
L    LIVER
M    MUSCLES
N    NERVOUS SYSTEM
O    EYE
P    PULMONARY-LUNGS
Q    SPLEEN
R    RESPIRATORY
S    SKELETAL
T    THYROID
U    INJURED/UNK SYS
V    VERTEBRAE
W    ALL IN REGION
```

CDS 3-letter combinations included in the analyses:

```
ADJ = ARM DISLOCATED JOINT (also includes AZJ)
ALA = ARM ARTERY LACERATION
ALI = ARM SUPERFICIAL INJURY (also includes AAI, ACI, ACN, ALM, ALN, API, ARM, ASJ, AVI, AVM)
ALJ = ARM JOINT LACERATION (also includes ARJ)
AMW = ARM AMPUTATION
AFS = ARM FRACTURE
ANW = ARM CRUSHED
NOTE: ADJ also includes EDJ, RDJ, WDJ, and XDJ, etc. (and similarly for the other A__ codes)

BCC = THORACOLUMBAR CORD CONTUSION (also includes BUC)
BEC = THORACOLUMBAR CORD TRANSECTION
BDV = THORACOLUMBAR DISLOCATION
BFS = THORACOLUMBAR FRACTURE (also includes BZV)
BLC = THORACOLUMBAR CORD LACERATION

CCH = HEART CONTUSION
CCP = LUNG CONTUSION (also includes CCR)
CCS = CHEST BONE CONTUSION
CEA = TORSO ARTERY TRANSECTION/RUPTURE (also includes CRA, MEA, MRA)
CFS = CHEST FRACTURES (also includes CFR, CCS)
CLA = CHEST LACERATED ARTERY
CLH = HEART LACERATION (also includes CPH, CVH)
CLI = TORSO SUPERFICIAL INJURY (also includes BAI, BCI, BLI, BLM, BTM, CAI, CCI, CLM, CPI, CUN,
CVM, CVI, MAI, MCI, MCM, MLI, MLM, MVI, PAI, PCI, PVI, SAI, SCI, SLI, SLM, SPM, SVI, SVM)
CLP = LUNG LACERATION (also includes CLR, CPP)
CNW = CHEST CRUSHED (also includes BNW)
CRH = HEART RUPTURE
```

253

```
FDJ = FACIAL DISLOCATION (also includes FZJ)
FLI = FACIAL LACERATION (also includes FAD, FAI, FCI, FCM, FCN, FCO, FLD, FLN, FVD, FVI, FVM)
FFS = FACIAL FRACTURE
FVO = EYE AVULSION/DETACHMENT (also includes FGO)


HCB = BRAIN CONTUSION
HEB = BRAIN STEM TRANSECTION
HFS = SKULL FRACTURE
HKB = CONCUSSION
HLA = HEAD ARTERY LACERATION (also includes HEA, HOA, HUA)
HLB = BRAIN LACERATION (also includes HRB, HVB)
HLE = EAR EXTERNAL INJURY (also includes HAE, HRE, HVE)
HLI = HEAD SUPERFICIAL INJURY (also includes HAI, HCI, HCM, HCN, HCO, HLM, HLN, HVI, HVM)
HNW = HEAD CRUSHED
HUB = BRAIN UNK INJURY


LDJ = LEG DISLOCATED JOINT (also includes LZJ)
LFS = LOWER LEG FRACTURE
LLI = LEG SUPERFICIAL INJURY (also includes LAI, LCI, LCN, LLM, LPI, LPM, LVI)
LLA = LEG ARTERY LACERATION (also includes LEA)
LLJ = LEG JOINT LACERATION (also includes LRJ, LVJ)
LMW = LEG AMPUTATION
LNW = LEG CRUSHED
LSJ = KNEE OR ANKLE SPRAIN
NOTE: LDJ also includes KDJ, QDJ, TDJ, and YDJ, etc. (and similarly for the other L__ codes)
except TFS is listed separately (see below)

MCD = INTESTINAL CONTUSION (also includes MUD)
MCG = CONTUSION OF GENITALS
MCK = KIDNEY CONTUSION (also includes MUK)
MCL = LIVER CONTUSION (also includes MUL)
MCQ = SPLEEN CONTUSION (also includes MUQ)
MCR = DIAPHRAGM CONTUSION (also includes MUR)
MLD = INTESTINAL LACERATION (also includes CLD, MPD, MVD)
MLA = ABDOMEN ARTERY LACERATION (also includes MVA)
MLG = LACERATION OF GENITALS (also includes MVG)
MLK = KIDNEY LACERATION (also includes MVK)
MLL = LIVER LACERATION (also includes MVL)
MLQ = SPLEEN LACERATION (also includes MVQ)
MLR = DIAPHRAGM LACERATION (also includes MVR)
MRD = INTESTINAL RUPTURE
MRG = RUPTURE OF GENITALS (also includes MGG)
MRK = KIDNEY RUPTURE
MRL = LIVER RUPTURE
MRQ = SPLEEN RUPTURE
MRR = DIAPHRAGM RUPTURE


NCC = CERVICAL CORD CONTUSION (also includes NOC, NUC)
NCR = TRACHEA CONTUSION
NCT = THYROID CONTUSION
NDV = NECK DISLOCATED VERTEBRA (also includes NZV)
NEC = CERVICAL CORD TRANSECTION
NFS = NECK FRACTURE
NLA = NECK ARTERY LACERATION (also includes NEA)
NLC = CERVICAL CORD LACERATION (also includes NVC)
NLI = NECK SUPERFICIAL INJURY (also includes NAI, NCI, NLM, NTM, NVI)
NLR = TRACHEA LACERATION
```

254

```
NLT = THYROID LACERATION
NMW = NECK AMPUTATION
NRV = NECK RUPTURED VERTEBRA
```

OBI = BURNS (includes all burns: LESION = B, regardless of BODYREG and SYSORG)

```
PDJ = HIP DISLOCATION (also includes PZJ)
PFS = PELVIS FRACTURE
PGJ = HIP JOINT SEPARATION
PNW = PELVIS CRUSHED

SDJ = SHOULDER DISLOCATED (also includes SRJ, SZJ)
SFS = SHOULDER FRACTURE
```

TFS = FEMUR FRACTURE

CDS 3-letter combinations excluded because it is unclear what the injury is:

```
AOJ , AUA , AUI , AUN , AUS , AUU , BLI , BVI , BRU , BUN , BUV , CLU , COR , CUA , CUH , CUP ,
CUR , CUU , FAO , FLO , FRO , FUI , FUN , HCS , HOE , HUE , HUN , LUA , LUI , LUN , LUU , MUA ,
MUG , MUT , MUU , MUW , NUA , NUD , NUN , NUR , NUU , NUV , OCI , OLI , OVI , PVM , UOU
```

APPENDIX D

CLASSIFICATION OF CDS INJURY SOURCES

CDS injury source codes have been classified into 18 groups. Here are the names of the groups and the specific INJSOU codes included for CY 1988-1992, CY 1993-1994, and CY 1995-2010:

AIR BAG includes:
    CY 1988-1992
```
45  AIR BAG
93  AIR BAG EXHAUST GAS
```
    CY 1993-1994
```
16  BAGCOVER DR SIDE
17  BAGCOVER PS SIDE
45  AIR BAG
93  AIR BAG EXHAUST GAS
```
    CY 1995-2010
```
170-195  AIR BAG OR AIR BAG COVER
604  AIR BAG EXH GAS
```

BELT SYSTEM includes:
    CY 1988-1994
```
41  BELT WEBB/BUCKLE
42  BELT B PILL ATCH
43  OTHER COMPONENT
```
    CY 1995-2010
```
152  BELT RESTRAINT WEBB/BUCKLE
153  BELT B PILL ATCH
154  OTH RESTR SYS COMPON
```

STEERING ASSEMBLY includes:
    CY 1988-1994
```
4  STEERING RIM
5  STEERING HUB
6  STEERING COMBO OF RIM AND HUB
7  STEERING COLUMN
```
    CY 1995-2010
```
4  STEERING WHEEL RIM
5  STEERING WHEEL HUB/SPOKE
6  STEERING WHEEL COMBO OF RIM AND HUB/SPOKE
7  STEERING COLUMN, TRANSMISSION SELECTOR LEVER, OTH ATTACH
```

INSTRUMENT PANEL includes:
CY 1988-1994
```
 9   LEFT PANEL
10   CENTER PANEL
11   RIGHT PANEL
12   GLOVE DOOR
13   KNEE BOLSTER
```
CY 1995-2010
```
10   LEFT PANEL
11   CENTER PANEL
12   RIGHT PANEL
13   GLOVE DOOR
14   KNEE BOLSTER
21   LEFT INSTR PANEL
22   CENTER INS PANEL
23   RIGHT INS PANEL
24   LOW LT INSTRU PANEL
25   CE LOW INSTRU PANEL
26   RL INSTRU PANEL
```

OTHER FRONTAL COMPONENT includes:
CY 1988-1992
```
 1   WINDSHIELD
 2   MIRROR
 3   SUNVISOR
14   WINDSHLD DR SIDE
15   WINDSHLD PS SIDE
16   OTHER FRONT OBJ
```
CY 1993-1994
```
 1   WINDSHIELD
 2   MIRROR
 3   SUNVISOR
14   WINDSHLD DR SIDE
15   WINDSHLD PS SIDE
19   OTHER FRONT OBJ
```
CY 1995-2010
```
 1   WINDSHIELD
 2   MIRROR
 3   SUNVISOR
15   WINDSHLD DR SIDE ONLY
16   WINDSHLD PS SIDE ONLY
19   OTHER FRONT OBJ
```

A-PILLAR includes:
CY 1988-1994
```
 22   LEFT A PILLAR
 32   RIGHT A PILLAR
```
CY 1995-2010
```
 53   LEFT A (A1/A2)PILLAR
103   RIGHT A (A1/A2) PILLAR
```

B-PILLAR includes:
    CY 1988-1994
        23 LEFT B PILLAR
        33 RIGHT B PILLAR
    CY 1995-2010
        54 LEFT B PILLAR
       104 RIGHT B PILLAR

SIDE HARDWARE OR ARMREST includes:
    CY 1988-1994
        21 LEFT HARDWARE
        31 RIGHT HARDWARE
    CY 1995-2010
        52 LEFT HARDWARE
      77-80 Left armrest/hardware
       102 RIGHT HARDWARE
    125-128 Right armrest/hardware

SIDE INTERIOR SURFACE includes:
    CY 1988-1994
        20 LEFT INTERIOR
        30 RIGHT INTERIOR
    CY 1995-2010
        51 LEFT INTERIOR
        62 Left side panel forward of A1/A2 pillar
        63 Left side panel rear of the B-pillar
        73 Left forward upper quadrant
        74 Left forward lower quadrant
        75 Left rear upper quadrant
        76 Left rear lower quadrant
       101 RIGHT INTERIOR
       112 Right side panel forward of A1/A2 pillar
       113 Right side panel rear of the B-pillar
       121 Right forward upper quadrant
       122 Right forward lower quadrant
       123 Right rear upper quadrant
       124 Right rear lower quadrant

OTHER SIDE COMPONENT includes:
CY 1988-1994
```
24   OTH LEFT PILLAR
25   LEFT WINDOW
26   LEFT WINDOW+OTH
27   OTHER LEFT OBJ
28   LEFT WINDOW SILL
34   OTH RIGHT PILLAR
35   RIGHT WINDOW
36   RIGHT WINDOW+OTH
37   OTH RIGHT OBJ
38   RIGHT WINDOW SILL
```
CY 1995-2010
```
55   OTH LEFT PILLAR (SPECIFY)
56   LEFT WINDOW GLAS
57   LEFT WINDOW FRAME
58   LEFT WINDOW SILL
59   LEFT WINDOW+OTH
60   LEFT SIDE GLASS REINFORCED BY EXT OBJ (SPECIFY)
106  RIGHT SIDE WIND GLASS
107  RIGHT SIDE WIND FRAME
108  RIGHT SIDE WINDOW SILL
109  RIGHT SIDE WINDOW+OTH
110  RIGHT SIDE GLASS REINFORCED BY EXT OBJ (SPECIFY)
```

ROOF & ITS COMPONENTS includes:
CY 1988-1994
```
50   FRONT HEADER
51   REAR HEADER
52   ROOF LEFT RAIL
53   ROOF RIGHT RAIL
54   ROOF
```
CY 1995-2010
```
20   SUNVISOR REINFORCED BY EXT OBJ
201  FRONT HEADER
202  REAR HEADER
203  ROOF LEFT RAIL
204  ROOF RIGHT RAIL
205  ROOF/CONVERTIBLE TOP
206  ROOF MAPLIGHT/CONSOLE
207  SUNROOF/COMP
208  ROLL-BAR
```

FLOOR & ITS COMPONENTS includes:
CY 1988-1994
```
56   FLOOR
57   TRANSMISS LEVER
58   BRAKE HANDLE
59   FOOT CONTROLS
```
CY 1995-2010
```
251  FLOOR (INCLUDING TOE PAN)
252  FLOOR OR CONSOLE TRANSMISS LEVER
253  PARKING BRAKE HANDLE
254  FOOT CONTROLS INCLUD PARKING BRAKE
```

SEAT & ITS COMPONENTS includes:
CY 1988-1994
  40 SEAT, BACK
  44 HEAD RESTRAINT
CY 1995-2010
  151 SEAT, BACK SUPPORT
  155 HEAD RESTRAINT SYS
  575 SEATBACK TRAYS

EXTERIOR includes:
CY 1988-1992
  65 HOOD
  66 OUTSIDE HARDWARE
  67 OTHER EXTERIOR
  68 UNK EXTERIOR OBJ
  70-83 OTHER MOTOR VEHICLE
  84 GROUND
  85 OTHER VEH OR OBJ
  86 UNK VEH OR OBJ
CY 1993-1994
  18 WINDSHIELD REINFORCED BY EXTERIOR OBJECT
  65 HOOD
  66 OUTSIDE HARDWARE
  67 OTHER EXTERIOR
  68 UNK EXTERIOR OBJ
  70-83 OTHER MOTOR VEHICLE
  84 GROUND
  85 OTHER VEH OR OBJ
  86 UNK VEH OR OBJ
CY 1995-2010
  17 REINFORCED WNDSH BY EXT OBJ
  451 HOOD
  452 OUTSIDE HARDWARE
  453 OTHER EXTERIOR SURFACE OR TIRES
  454 UNK EXTERIOR OBJ
  501-514 OTHER MOTOR VEHICLE
  551 GROUND
  574 ENGINE SHROUD

FIRE includes:
CY 1988-1994
  90 FIRE IN VEHICLE
CY 1995-2010
  601 FIRE IN VEHICLE

OTHER OCCUPANTS, SELF, CARGO includes:
CY 1988-1994
  46 OTHER OCCUPANTS
  47 INT LOOSE OBJ
CY 1995-2010
  160 OTHER OCCUPANTS (SPECIFY)
  161 INT LOOSE OBJ (SPECIFY)
  570 SAME OCC CONTACT
  571 INT LOOSE OBJ

NON-CONTACT includes:
  CY 1988-1994
      91 FLYING GLASS
      92 OTHER NONCONTACT
  CY 1995-2010
      602 FLYING GLASS
      603 OTHER NONCONTACT INJ SOURCE (SPECIFY)

OTHER includes:
  CY 1988-1994
      8   ADD ON EQUIPMENT
      48 CHILD SEAT
      49 OTH INTERIOR OBJ
      60 BACKLIGHT
      61 BACK STORAGE
      62 OTHER REAR OBJ
  CY 1995-2010
      8   CELLTELP/CBRADIO
      9   ADD ON EQUIPMENT
      162 CHILD SEAT
      163 OTH INTERIOR OBJ (SPECIFY)
      271-273 CHILD SAFETY SEAT COMPONENTS
      301 BACKLIGHT (REAR WINDOW)
      302 BACKLIGHT STORAGE RACK
      303 OTHER REAR OBJ
      401-412 ADAPTIVE DEVICES
      598 OTH OBJ (SPECIFY)

APPENDIX E

INJURY SOURCES RANKED BY AVERAGE AGE OF THE VICTIMS
OR BY PERCENT FEMALE VICTIMS

TABLE E-2: AVERAGE AGE OF DRIVERS AGE 18-96 WITH AIS 3+ INJURIES, CDS 1988-2010

IMPACT TYPE	OCCUPANT PROTECTION	BODY REGION	AVERAGE AGE	PERCENT FEMALE	N	INJURY SOURCE
FRONTAL	BELTED, AIR BAG	HEAD	54.91	45.45	11	AIR BAG
			50.47	60.00	15	INSTRUMENT PANEL
			46.87	46.67	30	NON-CONTACT
			44.45	48.73	2048	average injured occupant
			43.57	28.57	14	OTHER FRONTAL COMPONENT
			43.43	33.33	30	EXTERIOR
			43.30	50.91	110	STEERING ASSEMBLY
			40.96	31.65	79	ROOF & ITS COMPONENTS
			39.23	27.27	22	B-PILLAR
			39.04	54.35	46	A-PILLAR
			36.89	15.79	19	OTHER
FRONTAL	BELTED, AIR BAG	NECK	* 53.33	76.19	21	NON-CONTACT
			50.85	40.00	20	STEERING ASSEMBLY
			50.42	41.67	24	ROOF & ITS COMPONENTS
			44.45	48.73	2048	average injured occupant
FRONTAL	BELTED, AIR BAG	TORSO	** 64.45	69.70	33	AIR BAG
			52.23	38.46	13	NON-CONTACT
			47.66	47.73	44	SIDE HARDWARE OR ARMREST
			* 47.17	50.20	245	BELT SYSTEM
			46.26	42.57	343	STEERING ASSEMBLY
			44.97	46.36	110	INSTRUMENT PANEL
			44.76	47.44	78	SIDE INTERIOR SURFACE
			44.45	48.73	2048	average injured occupant
			44.22	36.11	36	SEAT & ITS COMPONENTS
			39.57	43.48	23	FLOOR & ITS COMPONENTS
FRONTAL	BELTED, AIR BAG	ARM	* 49.25	86.84	76	AIR BAG
			47.64	60.00	25	OTHER FRONTAL COMPONENT
			45.97	53.44	131	INSTRUMENT PANEL
			45.33	33.33	15	EXTERIOR
			44.45	48.73	2048	average injured occupant
			44.32	71.59	88	STEERING ASSEMBLY
			43.68	67.57	37	A-PILLAR
			43.60	57.14	35	ROOF & ITS COMPONENTS
			41.05	46.58	73	SIDE INTERIOR SURFACE
FRONTAL	BELTED, AIR BAG	LEG	45.37	64.45	256	FLOOR & ITS COMPONENTS
			44.45	48.73	2048	average injured occupant
			43.20	45.00	20	SIDE HARDWARE OR ARMREST
			42.07	45.16	589	INSTRUMENT PANEL
			39.32	23.68	38	SIDE INTERIOR SURFACE
			33.40	32.00	50	STEERING ASSEMBLY
FRONTAL	BELTED, NO AIR BAG	HEAD	43.81	41.97	1413	average injured occupant
			41.66	42.13	216	STEERING ASSEMBLY
			41.48	38.46	52	OTHER FRONTAL COMPONENT
			41.11	42.55	47	A-PILLAR
			39.69	28.13	32	ROOF & ITS COMPONENTS
			39.50	15.00	20	INSTRUMENT PANEL
			38.39	38.89	18	EXTERIOR
FRONTAL	BELTED, NO AIR BAG	NECK	50.22	48.15	27	STEERING ASSEMBLY
			43.81	41.97	1413	average injured occupant
FRONTAL	BELTED, NO AIR BAG	TORSO	* 48.14	43.75	144	BELT SYSTEM
			** 47.44	38.07	415	STEERING ASSEMBLY
			44.74	25.71	35	SIDE INTERIOR SURFACE
			43.81	41.97	1413	average injured occupant
			42.58	29.17	48	INSTRUMENT PANEL
			39.80	30.00	10	SEAT & ITS COMPONENTS
			39.25	50.00	12	SIDE HARDWARE OR ARMREST
FRONTAL	BELTED, NO AIR BAG	ARM	46.57	54.22	83	STEERING ASSEMBLY
			43.81	41.97	1413	average injured occupant
			43.26	42.11	19	SIDE INTERIOR SURFACE
			41.67	47.13	87	INSTRUMENT PANEL
FRONTAL	BELTED, NO AIR BAG	LEG	43.81	41.97	1413	average injured occupant
			42.50	59.09	132	FLOOR & ITS COMPONENTS
			39.78	39.62	318	INSTRUMENT PANEL
			37.48	30.43	23	SIDE INTERIOR SURFACE
			36.77	28.21	39	STEERING ASSEMBLY

**3+ standard deviations higher than average injured occupant (p < .002).
*1.96-3 standard deviations higher than average injured occupant (p < .05).

IMPACT TYPE	OCCUPANT PROTECTION	BODY REGION	AVERAGE AGE	PERCENT FEMALE	N	INJURY SOURCE
FRONTAL	NO BELT, AIR BAG	HEAD	* 49.55	63.64	11	AIR BAG
			41.11	15.87	63	A-PILLAR
			39.44	18.52	81	OTHER FRONTAL COMPONENT
			38.15	29.71	1518	average injured occupant
			37.85	23.73	59	STEERING ASSEMBLY
			37.25	25.00	24	INSTRUMENT PANEL
			36.37	18.56	97	ROOF & ITS COMPONENTS
			31.99	16.05	81	EXTERIOR
			31.52	33.33	21	OTHER
FRONTAL	NO BELT, AIR BAG	NECK	41.00	20.00	20	OTHER FRONTAL COMPONENT
			40.50	50.00	16	STEERING ASSEMBLY
			39.71	14.29	14	A-PILLAR
			38.15	29.71	1518	average injured occupant
			37.68	10.53	19	ROOF & ITS COMPONENTS
			33.61	17.39	23	EXTERIOR
FRONTAL	NO BELT, AIR BAG	TORSO	* 47.11	52.63	19	AIR BAG
			41.60	30.00	10	FLOOR & ITS COMPONENTS
			* 40.58	32.06	496	STEERING ASSEMBLY
			38.97	26.36	129	INSTRUMENT PANEL
			38.15	29.71	1518	average injured occupant
			37.40	20.00	15	SIDE HARDWARE OR ARMREST
			37.25	16.67	12	SEAT & ITS COMPONENTS
			36.98	28.33	60	SIDE INTERIOR SURFACE
			33.38	23.21	56	EXTERIOR
FRONTAL	NO BELT, AIR BAG	ARM	42.06	55.56	18	SIDE INTERIOR SURFACE
			41.09	63.64	11	OTHER FRONTAL COMPONENT
			40.08	22.97	74	INSTRUMENT PANEL
			38.50	30.00	10	OTHER
			38.15	29.71	1518	average injured occupant
			36.71	42.86	14	A-PILLAR
			34.22	36.11	36	STEERING ASSEMBLY
			33.00	21.43	14	EXTERIOR
FRONTAL	NO BELT, AIR BAG	LEG	40.55	47.86	117	FLOOR & ITS COMPONENTS
			38.15	29.71	1518	average injured occupant
			36.75	28.76	372	INSTRUMENT PANEL
			36.38	27.72	101	STEERING ASSEMBLY
			32.33	16.67	12	EXTERIOR
			31.13	12.50	16	SIDE INTERIOR SURFACE
FRONTAL	NO BELT, NO AIR BAG	HEAD	40.41	19.78	91	ROOF & ITS COMPONENTS
			39.72	42.19	64	INSTRUMENT PANEL
			39.66	28.26	2788	average injured occupant
			39.58	27.74	310	OTHER FRONTAL COMPONENT
			39.22	29.07	172	STEERING ASSEMBLY
			38.13	25.00	16	OTHER SIDE COMPONENT
			36.47	18.64	118	EXTERIOR
			35.04	27.62	105	A-PILLAR
			30.07	13.33	15	OTHER
FRONTAL	NO BELT, NO AIR BAG	NECK	* 46.52	37.50	48	OTHER FRONTAL COMPONENT
			41.22	14.81	27	ROOF & ITS COMPONENTS
			39.67	33.33	30	STEERING ASSEMBLY
			39.66	28.26	2788	average injured occupant
			37.92	25.00	12	EXTERIOR
			31.30	30.00	10	A-PILLAR
FRONTAL	NO BELT, NO AIR BAG	TORSO	* 48.10	35.00	20	FLOOR & ITS COMPONENTS
			43.13	29.51	61	SIDE INTERIOR SURFACE
			** 42.83	25.29	1103	STEERING ASSEMBLY
			40.83	21.88	192	INSTRUMENT PANEL
			39.66	28.26	2788	average injured occupant
			38.43	21.43	14	SEAT & ITS COMPONENTS
			37.55	18.18	11	OTHER FRONTAL COMPONENT
			33.10	34.62	52	EXTERIOR
			32.30	0.00	10	ROOF & ITS COMPONENTS
FRONTAL	NO BELT, NO AIR BAG	ARM	44.27	36.36	11	A-PILLAR
			41.05	42.27	97	STEERING ASSEMBLY
			40.82	37.40	131	INSTRUMENT PANEL
			39.66	28.26	2788	average injured occupant
			37.58	26.92	26	SIDE INTERIOR SURFACE
			30.93	14.29	14	EXTERIOR

**3+ standard deviations higher than average injured occupant (p < .002).
*1.96-3 standard deviations higher than average injured occupant (p < .05).

IMPACT TYPE	OCCUPANT PROTECTION	BODY REGION	AVERAGE AGE	PERCENT FEMALE	N	INJURY SOURCE
FRONTAL	NO BELT, NO AIR BAG	LEG	41.06	9.68	31	SIDE INTERIOR SURFACE
			39.66	28.26	2788	average injured occupant
			37.66	47.19	178	FLOOR & ITS COMPONENTS
			37.04	28.41	542	INSTRUMENT PANEL
			36.97	31.97	122	STEERING ASSEMBLY
			31.07	40.00	15	EXTERIOR
FRONTAL	NO BELT, NO AIR BAG	BURN	39.66	28.26	2788	average injured occupant
			30.70	20.00	10	FIRE
NEARSIDE IMPACT	BELTED	HEAD	* 56.00	50.00	14	STEERING ASSEMBLY
			48.41	43.59	39	OTHER SIDE COMPONENT
			45.08	57.38	61	A-PILLAR
			45.04	53.85	26	NON-CONTACT
			44.77	46.34	1830	average injured occupant
			44.51	46.55	174	B-PILLAR
			43.64	52.31	130	EXTERIOR
			43.61	63.64	33	SIDE INTERIOR SURFACE
			38.64	31.52	92	ROOF & ITS COMPONENTS
			34.77	28.33	60	OTHER
NEARSIDE IMPACT	BELTED	NECK	* 53.04	50.00	26	B-PILLAR
			49.25	50.00	16	NON-CONTACT
			44.88	31.25	16	ROOF & ITS COMPONENTS
			44.77	46.34	1830	average injured occupant
			40.92	33.33	12	EXTERIOR
NEARSIDE IMPACT	BELTED	TORSO	* 46.46	45.94	849	SIDE INTERIOR SURFACE
			45.84	52.66	395	SIDE HARDWARE OR ARMREST
			45.79	41.05	95	B-PILLAR
			45.66	46.15	65	STEERING ASSEMBLY
			44.77	46.34	1830	average injured occupant
			43.89	35.56	45	BELT SYSTEM
			43.15	61.54	13	INSTRUMENT PANEL
			43.03	55.88	34	FLOOR & ITS COMPONENTS
			37.27	57.69	26	SEAT & ITS COMPONENTS
NEARSIDE IMPACT	BELTED	ARM	47.44	52.00	25	INSTRUMENT PANEL
			46.27	54.55	22	STEERING ASSEMBLY
			44.77	46.34	1830	average injured occupant
			42.79	42.86	14	SIDE HARDWARE OR ARMREST
			40.43	40.54	74	SIDE INTERIOR SURFACE
NEARSIDE IMPACT	BELTED	LEG	44.77	46.34	1830	average injured occupant
			41.97	50.82	61	INSTRUMENT PANEL
			41.53	35.29	17	FLOOR & ITS COMPONENTS
			38.53	06.11	94	SIDE INTERIOR SURFACE
			38.47	49.33	75	SIDE HARDWARE OR ARMREST
NEARSIDE IMPACT	NO BELT	HEAD	44.31	75.00	16	SIDE INTERIOR SURFACE
			40.33	41.67	24	OTHER SIDE COMPONENT
			39.80	37.50	56	B-PILLAR
			38.39	31.95	1280	average injured occupant
			36.81	29.63	27	OTHER FRONTAL COMPONENT
			36.58	28.38	74	A-PILLAR
			36.00	37.01	154	EXTERIOR
			32.74	27.17	92	ROOF & ITS COMPONENTS
			30.09	16.98	53	OTHER
NEARSIDE IMPACT	NO BELT	NECK	41.44	11.11	18	ROOF & ITS COMPONENTS
			40.44	22.22	18	EXTERIOR
			38.39	31.95	1280	average injured occupant
NEARSIDE IMPACT	NO BELT	TORSO	42.53	34.88	43	B-PILLAR
			* 40.79	36.05	516	SIDE INTERIOR SURFACE
			40.53	39.88	173	SIDE HARDWARE OR ARMREST
			39.16	47.37	19	FLOOR & ITS COMPONENTS
			38.50	0.00	10	INSTRUMENT PANEL
			38.39	31.95	1280	average injured occupant
			36.54	29.23	65	EXTERIOR
			34.63	29.79	94	STEERING ASSEMBLY
NEARSIDE IMPACT	NO BELT	ARM	43.35	58.82	17	STEERING ASSEMBLY
			38.44	31.11	45	SIDE INTERIOR SURFACE
			38.39	31.95	1280	average injured occupant
			30.00	40.00	15	EXTERIOR

3+ standard deviations higher than average injured occupant (p < .002).
*1.96-3 standard deviations higher than average injured occupant (p < .05).

IMPACT TYPE	OCCUPANT PROTECTION	BODY REGION	AVERAGE AGE	PERCENT FEMALE	N	INJURY SOURCE
NEARSIDE IMPACT	NO BELT	LEG	41.53	21.05	19	FLOOR & ITS COMPONENTS
			38.39	31.95	1280	average injured occupant
			37.06	18.75	48	INSTRUMENT PANEL
			35.52	15.91	44	SIDE HARDWARE OR ARMREST
			34.31	15.38	26	EXTERIOR
			32.96	33.33	90	SIDE INTERIOR SURFACE
FAR-SIDE IMPACT	BELTED	HEAD	53.64	35.71	14	EXTERIOR
			48.13	33.33	15	INSTRUMENT PANEL
			46.60	30.00	10	OTHER OCCS, SELF, CARGO
			44.05	42.75	648	average injured occupant
			41.23	41.94	31	B-PILLAR
			40.45	48.72	78	SIDE INTERIOR SURFACE
			39.60	60.00	10	OTHER SIDE COMPONENT
			37.52	28.00	50	ROOF & ITS COMPONENTS
			33.70	30.00	10	NON-CONTACT
FAR-SIDE IMPACT	BELTED	NECK	44.05	42.75	648	average injured occupant
			34.09	18.18	11	ROOF & ITS COMPONENTS
FAR-SIDE IMPACT	BELTED	TORSO	** 55.05	44.87	78	BELT SYSTEM
			51.11	55.56	27	STEERING ASSEMBLY
			51.00	38.10	21	OTHER
			46.00	47.06	68	FLOOR & ITS COMPONENTS
			44.80	50.00	10	INSTRUMENT PANEL
			44.05	42.75	648	average injured occupant
			40.95	42.11	19	SIDE HARDWARE OR ARMREST
			39.38	47.67	86	SIDE INTERIOR SURFACE
			37.75	28.57	28	OTHER OCCS, SELF, CARGO
			36.56	34.55	55	SEAT & ITS COMPONENTS
FAR-SIDE IMPACT	BELTED	ARM	* 54.06	83.33	18	STEERING ASSEMBLY
			47.38	38.46	13	INSTRUMENT PANEL
			44.05	42.75	648	average injured occupant
			38.75	75.00	16	SIDE INTERIOR SURFACE
FAR-SIDE IMPACT	BELTED	LEG	53.33	66.67	18	FLOOR & ITS COMPONENTS
			50.54	42.31	26	INSTRUMENT PANEL
			44.05	42.75	648	average injured occupant
FAR-SIDE IMPACT	NO BELT	HEAD	* 51.14	22.73	22	INSTRUMENT PANEL
			41.83	28.81	59	OTHER FRONTAL COMPONENT
			41.76	29.03	62	A-PILLAR
			40.15	37.08	89	SIDE INTERIOR SURFACE
			39.74	29.25	988	average injured occupant
			37.16	28.00	25	OTHER SIDE COMPONENT
			36.61	21.11	90	ROOF & ITS COMPONENTS
			35.30	33.33	27	B-PILLAR
			34.81	22.45	98	EXTERIOR
			32.59	13.64	22	OTHER
			30.09	18.18	11	STEERING ASSEMBLY
FAR-SIDE IMPACT	NO BELT	NECK	46.41	31.82	22	OTHER FRONTAL COMPONENT
			45.42	25.00	12	A-PILLAR
			43.61	27.78	18	SIDE INTERIOR SURFACE
			39.74	29.25	988	average injured occupant
			36.25	25.00	20	EXTERIOR
			35.05	18.18	22	ROOF & ITS COMPONENTS
FAR-SIDE IMPACT	NO BELT	TORSO	48.23	30.77	13	OTHER
			* 44.74	27.03	74	INSTRUMENT PANEL
			42.07	35.71	14	OTHER OCCS, SELF, CARGO
			41.12	29.90	204	SIDE INTERIOR SURFACE
			40.34	24.14	29	FLOOR & ITS COMPONENTS
			39.74	29.25	988	average injured occupant
			39.12	32.65	49	STEERING ASSEMBLY
			38.83	24.14	29	SIDE HARDWARE OR ARMREST
			37.97	25.71	35	SEAT & ITS COMPONENTS
			35.09	35.94	64	EXTERIOR
FAR-SIDE IMPACT	NO BELT	ARM	43.67	40.00	15	INSTRUMENT PANEL
			39.95	42.86	21	SIDE INTERIOR SURFACE
			39.74	29.25	988	average injured occupant
			39.13	33.33	15	EXTERIOR

**3+ standard deviations higher than average injured occupant (p < .002).
*1.96 3 standard deviations higher than average injured occupant (p < .05).

IMPACT TYPE	OCCUPANT PROTECTION	BODY REGION	AVERAGE AGE	PERCENT FEMALE	N	INJURY SOURCE
FAR-SIDE IMPACT	NO BELT	LEG	* 52.33	41.67	12	STEERING ASSEMBLY
			43.66	42.86	35	INSTRUMENT PANEL
			43.53	26.67	15	FLOOR & ITS COMPONENTS
			39.74	29.25	988	average injured occupant
			33.50	50.00	12	EXTERIOR
			32.22	16.67	18	SIDE INTERIOR SURFACE
ROLLOVER FIRST/WORST EVENT	BELTED	HEAD	39.55	45.45	11	B-PILLAR
			39.13	36.32	669	average injured occupant
			37.32	37.64	178	ROOF & ITS COMPONENTS
			36.10	36.67	30	EXTERIOR
ROLLOVER FIRST/WORST EVENT	BELTED	NECK	41.94	29.27	82	ROOF & ITS COMPONENTS
			39.13	36.32	669	average injured occupant
ROLLOVER FIRST/WORST EVENT	BELTED	TORSO	48.55	27.27	11	B-PILLAR
			44.60	40.00	10	FLOOR & ITS COMPONENTS
			41.62	42.86	21	SEAT & ITS COMPONENTS
			41.59	37.04	27	SIDE HARDWARE OR ARMREST
			39.13	36.32	669	average injured occupant
			38.56	41.76	91	SIDE INTERIOR SURFACE
			35.83	29.17	24	ROOF & ITS COMPONENTS
			35.45	22.45	49	STEERING ASSEMBLY
			35.07	37.04	27	BELT SYSTEM
ROLLOVER FIRST/WORST EVENT	BELTED	ARM	* 51.00	50.00	12	INSTRUMENT PANEL
			44.36	54.55	22	SIDE INTERIOR SURFACE
			41.79	42.42	33	EXTERIOR
			40.97	48.28	29	ROOF & ITS COMPONENTS
			39.13	36.32	669	average injured occupant
ROLLOVER FIRST/WORST EVENT	BELTED	LEG	39.13	36.32	669	average injured occupant
			38.10	33.33	21	INSTRUMENT PANEL
			35.50	30.00	10	FLOOR & ITS COMPONENTS
ROLLOVER FIRST/WORST EVENT	NO BELT	HEAD	34.00	20.00	15	A-PILLAR
			33.20	23.67	1293	average injured occupant
			33.18	18.98	137	ROOF & ITS COMPONENTS
			32.37	23.99	296	EXTERIOR
			30.14	36.36	22	OTHER FRONTAL COMPONENT
ROLLOVER FIRST/WORST EVENT	NO BELT	NECK	35.53	13.95	43	ROOF & ITS COMPONENTS
			33.20	23.67	1293	average injured occupant
			33.16	26.79	56	EXTERIOR
ROLLOVER FIRST/WORST EVENT	NO BELT	TORSO	* 41.11	0.00	18	SEAT & ITS COMPONENTS
			35.61	16.25	80	STEERING ASSEMBLY
			35.18	9.09	11	INSTRUMENT PANEL
			34.65	17.50	40	ROOF & ITS COMPONENTS
			33.20	23.67	1293	average injured occupant
			32.43	28.53	354	EXTERIOR
			32.18	30.26	76	SIDE INTERIOR SURFACE
ROLLOVER FIRST/WORST EVENT	NO BELT	ARM	33.20	23.67	1293	average injured occupant
			33.17	31.91	47	EXTERIOR
			31.80	13.33	15	ROOF & ITS COMPONENTS
ROLLOVER FIRST/WORST EVENT	NO BELT	LEG	34.92	16.67	24	INSTRUMENT PANEL
			33.20	23.67	1293	average injured occupant
			33.07	36.00	75	EXTERIOR
			26.80	10.00	10	STEERING ASSEMBLY
REAR & OTHER	BELTED	HEAD	* 56.00	54.55	11	B-PILLAR
			45.44	46.88	32	SEAT & ITS COMPONENTS
			44.24	45.74	188	average injured occupant
			39.07	40.00	15	ROOF & ITS COMPONENTS
REAR & OTHER	BELTED	NECK	* 54.62	30.77	13	SEAT & ITS COMPONENTS
			44.24	45.74	188	average injured occupant
REAR & OTHER	BELTED	TORSO	46.58	57.69	52	SEAT & ITS COMPONENTS
			44.24	45.74	188	average injured occupant
			43.33	38.89	18	STEERING ASSEMBLY
REAR & OTHER	NO BELT	HEAD	40.89	26.32	19	SEAT & ITS COMPONENTS
			40.10	20.00	10	B-PILLAR
			36.65	17.48	143	average injured occupant
			31.42	8.33	12	EXTERIOR
			28.69	6.25	16	ROOF & ITS COMPONENTS

**3+ standard deviations higher than average injured occupant (p < .002).
*1.96-3 standard deviations higher than average injured occupant (p < .05).

IMPACT TYPE	OCCUPANT PROTECTION	BODY REGION	AVERAGE AGE	PERCENT FEMALE	N	INJURY SOURCE
REAR & OTHER	NO BELT	NECK	36.65	17.48	143	average injured occupant
			36.62	23.08	13	SEAT & ITS COMPONENTS
REAR & OTHER	NO BELT	TORSO	43.14	21.43	14	EXTERIOR
			39.45	13.64	22	SEAT & ITS COMPONENTS
			38.00	20.00	10	SIDE INTERIOR SURFACE
			36.65	17.48	143	average injured occupant
			35.86	9.52	21	STEERING ASSEMBLY

**3+ standard deviations higher than average injured occupant (p < .002).
*1.96-3 standard deviations higher than average injured occupant (p < .05).

268

TABLE E-3: AVERAGE AGE OF DRIVERS AGE 18-96 WITH AIS 4+ INJURIES, CDS 1988-2010

IMPACT TYPE	OCCUPANT PROTECTION	BODY REGION	AVERAGE AGE	PERCENT FEMALE	N	INJURY SOURCE
FRONTAL	BELTED, AIR BAG	HEAD	59.57	42.86	7	AIR BAG
			49.47	47.37	19	NON-CONTACT
			49.00	44.44	9	INSTRUMENT PANEL
			47.79	40.68	649	average injured occupant
			47.20	20.00	5	OTHER OCCS, SELF, CARGO
			46.60	40.00	5	SEAT & ITS COMPONENTS
			45.63	37.50	8	OTHER FRONTAL COMPONENT
			44.00	28.00	25	EXTERIOR
			43.41	57.63	59	STEERING ASSEMBLY
			40.93	26.67	15	B-PILLAR
			39.87	29.63	54	ROOF & ITS COMPONENTS
			36.81	62.96	27	A-PILLAR
			34.46	7.69	13	OTHER
FRONTAL	BELTED, AIR BAG	NECK	54.56	33.33	9	STEERING ASSEMBLY
			47.79	40.68	649	average injured occupant
			47.29	42.86	7	ROOF & ITS COMPONENTS
FRONTAL	BELTED, AIR BAG	TORSO	** 71.00	68.75	16	AIR BAG
			55.71	35.71	14	SIDE HARDWARE OR ARMREST
			50.40	50.68	73	BELT SYSTEM
			47.79	40.68	649	average injured occupant
			47.11	43.01	186	STEERING ASSEMBLY
			44.00	50.00	28	SIDE INTERIOR SURFACE
			40.80	20.00	5	NON-CONTACT
			37.00	50.00	6	INSTRUMENT PANEL
FRONTAL	BELTED, AIR BAG	BURN	47.79	40.68	649	average injured occupant
			45.86	0.00	7	FIRE
FRONTAL	BELTED, NO AIR BAG	HEAD	46.05	37.73	546	average injured occupant
			42.95	33.33	21	OTHER FRONTAL COMPONENT
			42.62	41.18	34	A-PILLAR
			42.60	40.00	5	B-PILLAR
			42.54	35.29	85	STEERING ASSEMBLY
			42.40	25.00	20	ROOF & ITS COMPONENTS
			42.36	9.09	11	INSTRUMENT PANEL
			38.50	33.33	6	OTHER
			37.92	38.46	13	EXTERIOR
FRONTAL	BELTED, NO AIR BAG	NECK	61.50	50.00	6	STEERING ASSEMBLY
			46.05	37.73	546	average injured occupant
FRONTAL	BELTED, NO AIR BAG	TORSO	49.86	50.00	42	BELT SYSTEM
			48.47	38.66	194	STEERING ASSEMBLY
			46.05	37.73	546	average injured occupant
			43.20	10.00	10	SIDE INTERIOR SURFACE
FRONTAL	NO BELT, AIR BAG	HEAD	42.43	20.00	35	A-PILLAR
			38.94	25.20	742	average injured occupant
			38.36	33.33	36	STEERING ASSEMBLY
			38.09	15.56	45	OTHER FRONTAL COMPONENT
			37.00	16.67	54	ROOF & ITS COMPONENTS
			35.86	28.57	14	INSTRUMENT PANEL
			31.93	28.57	14	OTHER
			31.01	16.42	67	EXTERIOR
FRONTAL	NO BELT, AIR BAG	NECK	43.20	80.00	5	STEERING ASSEMBLY
			38.94	25.20	742	average injured occupant
			37.71	0.00	7	EXTERIOR
			36.88	0.00	8	ROOF & ITS COMPONENTS
FRONTAL	NO BELT, AIR BAG	TORSO	49.86	42.86	7	AIR BAG
			41.23	29.69	256	STEERING ASSEMBLY
			38.94	25.20	742	average injured occupant
			37.29	28.57	7	SIDE HARDWARE OR ARMREST
			37.24	20.00	25	INSTRUMENT PANEL
			35.49	18.92	37	EXTERIOR
			34.84	20.00	25	SIDE INTERIOR SURFACE
			28.33	16.67	6	ROOF & ITS COMPONENTS
FRONTAL	NO BELT, AIR BAG	BURN	38.94	25.20	742	average injured occupant
			29.00	20.00	5	FIRE

3+ standard deviations higher than average injured occupant (p < .002).
*1.96-3 standard deviations higher than average injured occupant (p < .05).

TABLE E-3 (CONTINUED): AVERAGE AGE OF DRIVERS AGE 18-96 WITH AIS 4+ INJURIES, CDS 1988-2010

IMPACT TYPE	OCCUPANT PROTECTION	BODY REGION	AVERAGE AGE	PERCENT FEMALE	N	INJURY SOURCE
FRONTAL	NO BELT, NO AIR BAG	HEAD	41.83	33.33	6	SIDE INTERIOR SURFACE
			41.46	25.02	1431	average injured occupant
			41.03	32.50	80	STEERING ASSEMBLY
			40.94	28.37	141	OTHER FRONTAL COMPONENT
			40.08	15.25	59	ROOF & ITS COMPONENTS
			37.00	20.00	5	B-PILLAR
			36.44	22.22	9	OTHER SIDE COMPONENT
			35.93	29.85	67	A-PILLAR
			35.81	17.86	84	EXTERIOR
			35.05	43.24	37	INSTRUMENT PANEL
			29.78	11.11	9	OTHER
FRONTAL	NO BELT, NO AIR BAG	NECK	50.50	33.33	6	ROOF & ITS COMPONENTS
			45.63	31.58	19	OTHER FRONTAL COMPONENT
			45.55	45.45	11	STEERING ASSEMBLY
			41.46	25.02	1431	average injured occupant
FRONTAL	NO BELT, NO AIR BAG	TORSO	* 43.78	25.04	603	STEERING ASSEMBLY
			41.88	36.00	25	SIDE INTERIOR SURFACE
			41.46	25.02	1431	average injured occupant
			38.60	20.00	30	INSTRUMENT PANEL
			33.69	40.63	32	EXTERIOR
FRONTAL	NO BELT, NO AIR BAG	BURN	41.46	25.02	1431	average injured occupant
			32.00	22.22	9	FIRE
NEARSIDE IMPACT	BELTED	HEAD	55.00	33.33	6	STEERING ASSEMBLY
			49.81	47.62	21	OTHER SIDE COMPONENT
			45.92	43.34	1006	average injured occupant
			45.91	54.55	22	NON-CONTACT
			43.90	46.96	115	B-PILLAR
			43.44	61.11	36	A-PILLAR
			43.05	70.00	20	SIDE INTERIOR SURFACE
			42.90	51.04	96	EXTERIOR
			40.65	30.91	55	ROOF & ITS COMPONENTS
			36.20	29.41	51	OTHER
NEARSIDE IMPACT	BELTED	NECK	45.92	43.34	1006	average injured occupant
			44.80	20.00	5	NON-CONTACT
			42.50	62.50	8	B-PILLAR
			41.29	28.57	7	EXTERIOR
NEARSIDE IMPACT	BELTED	TORSO	48.80	38.79	116	SIDE HARDWARE OR ARMREST
			47.87	51.61	31	STEERING ASSEMBLY
			47.18	42.00	400	SIDE INTERIOR SURFACE
			45.92	43.34	1006	average injured occupant
			45.81	39.62	53	B-PILLAR
			45.20	40.00	10	BELT SYSTEM
			43.30	40.00	10	FLOOR & ITS COMPONENTS
			36.40	20.00	10	SEAT & ITS COMPONENTS
			33.80	0.00	5	EXTERIOR
NEARSIDE IMPACT	NO BELT	HEAD	51.33	66.67	9	SIDE INTERIOR SURFACE
			41.46	37.84	37	B-PILLAR
			39.36	35.71	14	OTHER SIDE COMPONENT
			38.00	30.52	842	average injured occupant
			36.88	37.50	8	NON-CONTACT
			36.35	32.38	105	EXTERIOR
			36.24	30.61	49	A-PILLAR
			33.12	29.41	51	ROOF & ITS COMPONENTS
			29.50	40.00	10	OTHER FRONTAL COMPONENT
			29.26	16.28	43	OTHER
NEARSIDE IMPACT	NO BELT	NECK	39.00	30.52	842	average injured occupant
			29.78	22.22	9	EXTERIOR
NEARSIDE IMPACT	NO BELT	TORSO	45.52	24.00	25	B-PILLAR
			42.40	0.00	5	ROOF & ITS COMPONENTS
			42.16	29.82	57	SIDE HARDWARE OR ARMREST
			* 42.02	34.53	278	SIDE INTERIOR SURFACE
			39.00	30.52	842	average injured occupant
			36.05	32.50	40	EXTERIOR
			34.18	29.55	44	STEERING ASSEMBLY
			31.83	50.00	6	FLOOR & ITS COMPONENTS

**3+ standard deviations higher than average injured occupant (p < .002).
*1.96-3 standard deviations higher than average injured occupant (p < .05).

270

IMPACT TYPE	OCCUPANT PROTECTION	BODY REGION	AVERAGE AGE	PERCENT FEMALE	N	INJURY SOURCE
FAR-SIDE IMPACT	BELTED	HEAD	54.33	33.33	12	EXTERIOR
			52.71	28.57	7	OTHER OCCS, SELF, CARGO
			50.18	36.36	11	INSTRUMENT PANEL
			42.83	16.67	6	SEAT & ITS COMPONENTS
			42.52	37.01	354	average injured occupant
			40.31	49.18	61	SIDE INTERIOR SURFACE
			40.04	42.31	26	B-PILLAR
			39.60	60.00	10	OTHER SIDE COMPONENT
			39.50	12.50	8	A-PILLAR
			38.22	27.03	37	ROOF & ITS COMPONENTS
			27.17	33.33	6	NON-CONTACT
FAR-SIDE IMPACT	BELTED	NECK	42.52	37.01	354	average injured occupant
			35.00	20.00	5	ROOF & ITS COMPONENTS
FAR-SIDE IMPACT	BELTED	TORSO	* 69.40	80.00	5	STEERING ASSEMBLY
			57.17	50.00	6	OTHER
			51.52	23.81	21	FLOOR & ITS COMPONENTS
			49.40	50.00	20	BELT SYSTEM
			44.58	50.00	12	SIDE HARDWARE OR ARMREST
			42.52	37.01	354	average injured occupant
			39.25	43.18	44	SIDE INTERIOR SURFACE
			38.73	33.33	15	OTHER OCCS, SELF, CARGO
			37.47	31.58	19	SEAT & ITS COMPONENTS
			34.80	40.00	5	EXTERIOR
FAR-SIDE IMPACT	NO BELT	HEAD	54.33	33.33	6	SIDE HARDWARE OR ARMREST
			44.33	0.00	12	INSTRUMENT PANEL
			44.00	24.14	29	OTHER FRONTAL COMPONENT
			41.84	27.91	43	A-PILLAR
			40.23	35.48	62	SIDE INTERIOR SURFACE
			39.34	25.69	654	average injured occupant
			38.06	22.22	18	OTHER SIDE COMPONENT
			36.34	21.43	70	ROOF & ITS COMPONENTS
			35.40	20.00	5	SEAT & ITS COMPONENTS
			34.67	27.78	18	B-PILLAR
			34.29	20.59	68	EXTERIOR
			32.28	16.67	18	OTHER
			29.40	0.00	5	STEERING ASSEMBLY
FAR-SIDE IMPACT	NO BELT	NECK	* 59.00	40.00	5	OTHER FRONTAL COMPONENT
			43.20	40.00	5	SIDE INTERIOR SURFACE
			40.80	40.00	5	A-PILLAR
			39.34	25.69	654	average injured occupant
			39.33	0.00	6	EXTERIOR
			38.00	0.00	5	ROOF & ITS COMPONENTS
FAR-SIDE IMPACT	NO BELT	TORSO	* 45.97	25.00	32	INSTRUMENT PANEL
			42.06	25.00	32	EXTERIOR
			41.75	25.00	8	OTHER
			41.57	7.14	14	SEAT & ITS COMPONENTS
			41.41	27.64	123	SIDE INTERIOR SURFACE
			40.14	57.14	7	OTHER OCCS, SELF, CARGO
			39.34	25.69	654	average injured occupant
			39.29	0.00	7	SIDE HARDWARE OR ARMREST
			36.71	21.43	14	FLOOR & ITS COMPONENTS
			35.57	34.78	23	STEERING ASSEMBLY
ROLLOVER FIRST/WORST EVENT	BELTED	HEAD	38.29	29.91	351	average injured occupant
			36.68	36.84	19	EXTERIOR
			35.76	31.40	121	ROOF & ITS COMPONENTS
			27.20	40.00	5	OTHER
ROLLOVER FIRST/WORST EVENT	BELTED	NECK	44.06	33.33	18	ROOF & ITS COMPONENTS
			38.29	29.91	351	average injured occupant
ROLLOVER FIRST/WORST EVENT	BELTED	TORSO	43.50	33.33	6	EXTERIOR
			41.50	33.33	18	STEERING ASSEMBLY
			40.10	40.48	42	SIDE INTERIOR SURFACE
			38.33	25.00	12	SIDE HARDWARE OR ARMREST
			38.29	29.91	351	average injured occupant
			35.36	27.27	11	BELT SYSTEM
			30.88	23.53	17	ROOF & ITS COMPONENTS
			30.50	33.33	6	SEAT & ITS COMPONENTS

**3+ standard deviations higher than average injured occupant (p < .002).
*1.96-3 standard deviations higher than average injured occupant (p < .05).

IMPACT TYPE	OCCUPANT PROTECTION	BODY REGION	AVERAGE AGE	PERCENT FEMALE	N	INJURY SOURCE
ROLLOVER FIRST/WORST EVENT	NO BELT	HEAD	38.57	0.00	7	A-PILLAR
			34.05	21.29	902	average injured occupant
			32.72	19.35	93	ROOF & ITS COMPONENTS
			32.55	27.27	11	OTHER FRONTAL COMPONENT
			32.34	21.92	219	EXTERIOR
			25.43	14.29	7	OTHER
ROLLOVER FIRST/WORST EVENT	NO BELT	NECK	41.15	15.38	13	EXTERIOR
			35.38	15.38	13	ROOF & ITS COMPONENTS
			34.05	21.29	902	average injured occupant
ROLLOVER FIRST/WORST EVENT	NO BELT	TORSO	37.57	17.14	35	STEERING ASSEMBLY
			35.73	0.00	11	SEAT & ITS COMPONENTS
			34.05	21.29	902	average injured occupant
			33.57	13.04	23	ROOF & ITS COMPONENTS
			33.10	25.77	194	EXTERIOR
			32.74	29.03	31	SIDE INTERIOR SURFACE
			28.83	0.00	6	OTHER SIDE COMPONENT
REAR & OTHER	BELTED	HEAD	55.50	50.00	6	B-PILLAR
			46.68	46.53	101	average injured occupant
			43.00	44.44	18	SEAT & ITS COMPONENTS
			39.42	41.67	12	ROOF & ITS COMPONENTS
REAR & OTHER	BELTED	NECK	53.80	40.00	5	SEAT & ITS COMPONENTS
			46.68	46.53	101	average injured occupant
REAR & OTHER	BELTED	TORSO	52.73	60.00	30	SEAT & ITS COMPONENTS
			46.68	46.53	101	average injured occupant
			46.18	27.27	11	STEERING ASSEMBLY
REAR & OTHER	NO BELT	HEAD	40.33	33.33	6	B-PILLAR
			40.33	25.00	12	SEAT & ITS COMPONENTS
			36.41	14.89	94	average injured occupant
			30.33	0.00	9	EXTERIOR
			28.83	8.33	12	ROOF & ITS COMPONENTS
REAR & OTHER	NO BELT	NECK	36.60	0.00	5	SEAT & ITS COMPONENTS
			36.41	14.89	94	average injured occupant
REAR & OTHER	NO BELT	TORSO	42.15	15.38	13	SEAT & ITS COMPONENTS
			39.20	10.00	10	STEERING ASSEMBLY
			36.41	14.89	94	average injured occupant

**3+ standard deviations higher than average injured occupant (p < .002).
*1.96-3 standard deviations higher than average injured occupant (p < .05).

TABLE E-4: AVERAGE AGE OF RIGHT-FRONT PASSENGERS AGE 18-96 WITH AIS 2+ INJURIES, CDS 1988-2010

IMPACT TYPE	OCCUPANT PROTECTION	BODY REGION	AVERAGE AGE	PERCENT FEMALE	N	INJURY SOURCE
FRONTAL	BELTED, AIR BAG	HEAD	52.41	75.86	29	AIR BAG
			48.10	70.00	20	B-PILLAR
			45.76	69.37	790	average injured occupant
			45.36	77.27	22	NON-CONTACT
			41.07	40.00	15	A-PILLAR
			40.48	51.52	33	ROOF & ITS COMPONENTS
			33.62	50.00	26	INSTRUMENT PANEL
FRONTAL	BELTED, AIR BAG	TORSO	* 55.75	70.00	20	AIR BAG
			** 51.75	76.37	237	BELT SYSTEM
			50.71	52.38	21	SIDE HARDWARE OR ARMREST
			49.56	77.78	36	SIDE INTERIOR SURFACE
			45.76	69.37	790	average injured occupant
			42.53	76.47	17	FLOOR & ITS COMPONENTS
			38.26	59.57	47	INSTRUMENT PANEL
			33.77	59.09	22	SEAT & ITS COMPONENTS
FRONTAL	BELTED, AIR BAG	ARM	* 57.13	87.50	16	OTHER FRONTAL COMPONENT
			51.46	78.57	28	SIDE INTERIOR SURFACE
			48.71	76.19	63	AIR BAG
			47.20	70.27	74	INSTRUMENT PANEL
			45.76	69.37	790	average injured occupant
FRONTAL	BELTED, AIR BAG	LEG	45.83	80.30	132	FLOOR & ITS COMPONENTS
			45.76	69.37	790	average injured occupant
			43.79	71.88	160	INSTRUMENT PANEL
FRONTAL	BELTED, NO AIR BAG	HEAD	45.93	70.46	765	average injured occupant
			41.36	64.00	25	ROOF & ITS COMPONENTS
			40.00	49.25	67	INSTRUMENT PANEL
			37.81	51.72	58	OTHER FRONTAL COMPONENT
			35.93	46.67	15	A-PILLAR
FRONTAL	BELTED, NO AIR BAG	TORSO	** 51.78	73.49	298	BELT SYSTEM
			51.20	70.77	65	INSTRUMENT PANEL
			46.16	50.00	32	SIDE INTERIOR SURFACE
			45.93	70.46	765	average injured occupant
			45.41	74.07	27	SEAT & ITS COMPONENTS
FRONTAL	BELTED, NO AIR BAG	ARM	* 51.46	73.95	119	INSTRUMENT PANEL
			45.93	70.46	765	average injured occupant
FRONTAL	BELTED, NO AIR BAG	LEG	47.31	72.99	137	INSTRUMENT PANEL
			45.93	70.46	765	average injured occupant
			40.89	70.10	97	FLOOR & ITS COMPONENTS
FRONTAL	NO BELT, AIR BAG	HEAD	36.19	52.07	484	average injured occupant
			34.98	44.64	56	OTHER FRONTAL COMPONENT
			34.18	58.82	34	INSTRUMENT PANEL
			32.00	37.14	35	ROOF & ITS COMPONENTS
			30.00	50.00	28	EXTERIOR
FRONTAL	NO BELT, AIR BAG	TORSO	40.45	59.09	22	SIDE INTERIOR SURFACE
			39.05	54.47	123	INSTRUMENT PANEL
			36.19	52.07	484	average injured occupant
			33.20	45.00	20	EXTERIOR
FRONTAL	NO BELT, AIR BAG	ARM	40.30	62.12	66	INSTRUMENT PANEL
			36.19	52.07	484	average injured occupant
FRONTAL	NO BELT, AIR BAG	LEG	38.63	56.03	116	INSTRUMENT PANEL
			36.19	52.07	484	average injured occupant
			35.24	65.93	91	FLOOR & ITS COMPONENTS
FRONTAL	NO BELT, NO AIR BAG	HEAD	36.76	55.48	146	INSTRUMENT PANEL
			36.53	51.34	1498	average injured occupant
			35.58	51.04	480	OTHER FRONTAL COMPONENT
			33.92	36.73	49	ROOF & ITS COMPONENTS
			31.81	34.88	43	EXTERIOR
			29.11	37.14	35	A-PILLAR
FRONTAL	NO BELT, NO AIR BAG	NECK	** 48.78	62.96	27	INSTRUMENT PANEL
			** 48.78	45.00	40	OTHER FRONTAL COMPONENT
			36.53	51.34	1498	average injured occupant

**3+ standard deviations higher than average injured occupant (p < .002).
*1.96-3 standard deviations higher than average injured occupant (p < .05).

IMPACT TYPE	OCCUPANT PROTECTION	BODY REGION	AVERAGE AGE	PERCENT FEMALE	N	INJURY SOURCE
FRONTAL	NO BELT, NO AIR BAG	TORSO	41.53	40.00	15	FLOOR & ITS COMPONENTS
			** 40.99	52.40	479	INSTRUMENT PANEL
			39.71	41.18	17	STEERING ASSEMBLY
			38.24	52.38	21	SIDE INTERIOR SURFACE
			36.96	41.67	24	SEAT & ITS COMPONENTS
			36.53	51.34	1498	average injured occupant
			30.80	10.00	20	EXTERIOR
FRONTAL	NO BELT, NO AIR BAG	ARM	41.86	57.14	21	OTHER FRONTAL COMPONENT
			* 39.83	61.42	254	INSTRUMENT PANEL
			36.53	51.34	1498	average injured occupant
FRONTAL	NO BELT, NO AIR BAG	LEG	37.81	54.67	300	INSTRUMENT PANEL
			36.87	59.52	168	FLOOR & ITS COMPONENTS
			36.53	51.34	1498	average injured occupant
NEARSIDE IMPACT	BELTED	HEAD	49.27	60.00	15	B-PILLAR
			47.45	68.73	339	average injured occupant
			46.77	83.33	30	OTHER OCCS, SELF, CARGO
			42.57	71.43	21	SEAT & ITS COMPONENTS
			32.40	35.00	20	ROOF & ITS COMPONENTS
NEARSIDE IMPACT	BELTED	TORSO	* 61.56	62.50	16	OTHER
			* 53.73	79.10	67	BELT SYSTEM
			53.55	79.31	29	OTHER OCCS, SELF, CARGO
			52.33	83.33	30	FLOOR & ITS COMPONENTS
			48.13	71.88	32	SEAT & ITS COMPONENTS
			47.45	68.73	339	average injured occupant
NEARSIDE IMPACT	BELTED	LEG	51.60	80.00	15	FLOOR & ITS COMPONENTS
			47.45	68.73	339	average injured occupant
			39.35	69.23	26	INSTRUMENT PANEL
NEARSIDE IMPACT	NO BELT	HEAD	38.33	42.86	21	INSTRUMENT PANEL
			34.49	44.48	353	average injured occupant
			32.94	38.24	34	ROOF & ITS COMPONENTS
			30.56	37.50	16	OTHER OCCS, SELF, CARGO
			29.50	33.33	30	OTHER FRONTAL COMPONENT
			28.94	46.88	32	EXTERIOR
NEARSIDE IMPACT	NO BELT	TORSO	41.32	53.57	28	STEERING ASSEMBLY
			* 41.15	41.03	39	INSTRUMENT PANEL
			34.49	44.48	353	average injured occupant
			31.67	38.10	21	SEAT & ITS COMPONENTS
			30.91	27.27	22	OTHER OCCS, SELF, CARGO
			30.19	25.00	16	SIDE INTERIOR SURFACE
			29.18	36.36	22	EXTERIOR
NEARSIDE IMPACT	NO BELT	ARM	41.50	68.18	22	INSTRUMENT PANEL
			34.49	44.48	353	average injured occupant
NEARSIDE IMPACT	NO BELT	LEG	39.48	52.00	25	INSTRUMENT PANEL
			34.49	44.48	353	average injured occupant
FAR-SIDE IMPACT	BELTED	HEAD	48.15	77.50	40	OTHER SIDE COMPONENT
			45.62	55.17	29	SIDE INTERIOR SURFACE
			45.55	66.49	749	average injured occupant
			45.47	71.29	101	B-PILLAR
			44.21	68.42	38	EXTERIOR
			40.58	45.83	24	A-PILLAR
			35.50	87.50	16	NON-CONTACT
			35.49	56.60	53	ROOF & ITS COMPONENTS
			28.93	20.00	15	OTHER
FAR-SIDE IMPACT	BELTED	NECK	48.56	80.00	25	B-PILLAR
			45.71	41.18	17	ROOF & ITS COMPONENTS
			45.55	66.49	749	average injured occupant
FAR-SIDE IMPACT	BELTED	TORSO	51.29	67.86	56	BELT SYSTEM
			49.34	64.15	53	B-PILLAR
			47.68	69.39	196	SIDE HARDWARE OR ARMREST
			47.65	60.87	23	FLOOR & ITS COMPONENTS
			47.34	65.92	355	SIDE INTERIOR SURFACE
			46.93	80.00	15	OTHER OCCS, SELF, CARGO
			45.55	66.49	749	average injured occupant
			38.18	88.24	17	SEAT & ITS COMPONENTS
FAR-SIDE IMPACT	BELTED	ARM	55.27	73.33	15	INSTRUMENT PANEL
			45.55	66.49	749	average injured occupant
			42.15	60.27	73	SIDE INTERIOR SURFACE

**3+ standard deviations higher than average injured occupant (p < .002).
*1.96-3 standard deviations higher than average injured occupant (p < .05).

274

IMPACT TYPE	OCCUPANT PROTECTION	BODY REGION	AVERAGE AGE	PERCENT FEMALE	N	INJURY SOURCE
FAR-SIDE IMPACT	BELTED	LEG	45.55	66.49	749	average injured occupant
			44.16	76.32	38	FLOOR & ITS COMPONENTS
			43.40	45.00	20	SIDE HARDWARE OR ARMREST
			39.96	61.76	68	SIDE INTERIOR SURFACE
			38.04	60.00	25	INSTRUMENT PANEL
FAR-SIDE IMPACT	NO BELT	HEAD	36.92	53.33	60	B-PILLAR
			35.61	50.00	74	EXTERIOR
			34.63	54.29	35	OTHER FRONTAL COMPONENT
			34.46	48.42	632	average injured occupant
			33.93	42.86	56	A-PILLAR
			32.24	50.00	34	OTHER SIDE COMPONENT
			28.95	28.21	39	ROOF & ITS COMPONENTS
			27.06	25.00	16	SIDE INTERIOR SURFACE
			25.39	16.67	18	OTHER
FAR-SIDE IMPACT	NO BELT	NECK	34.46	48.42	632	average injured occupant
			31.00	47.06	17	EXTERIOR
FAR-SIDE IMPACT	NO BELT	TORSO	* 37.61	50.00	234	SIDE INTERIOR SURFACE
			37.58	53.85	26	OTHER OCCS, SELF, CARGO
			37.29	50.00	34	B-PILLAR
			36.97	46.15	39	INSTRUMENT PANEL
			35.33	57.55	106	SIDE HARDWARE OR ARMREST
			34.59	38.64	44	EXTERIOR
			34.46	48.42	632	average injured occupant
FAR-SIDE IMPACT	NO BELT	ARM	36.40	52.00	25	EXTERIOR
			34.46	48.42	632	average injured occupant
			34.45	52.27	44	SIDE INTERIOR SURFACE
FAR-SIDE IMPACT	NO BELT	LEG	34.46	48.42	632	average injured occupant
			32.54	44.64	56	SIDE INTERIOR SURFACE
			31.17	56.10	41	INSTRUMENT PANEL
			25.35	43.48	23	FLOOR & ITS COMPONENTS
ROLLOVER FIRST/WORST EVENT	BELTED	HEAD	37.06	49.85	337	average injured occupant
			36.66	55.41	74	ROOF & ITS COMPONENTS
ROLLOVER FIRST/WORST EVENT	BELTED	NECK	38.94	50.00	48	ROOF & ITS COMPONENTS
			37.06	49.85	337	average injured occupant
ROLLOVER FIRST/WORST EVENT	BELTED	TORSO	39.16	31.58	19	ROOF & ITS COMPONENTS
			38.03	57.14	35	SIDE INTERIOR SURFACE
			37.06	49.85	337	average injured occupant
			36.81	56.25	16	SIDE HARDWARE OR ARMREST
			35.33	54.76	42	BELT SYSTEM
			35.09	40.91	22	SEAT & ITS COMPONENTS
ROLLOVER FIRST/WORST EVENT	BELTED	ARM	37.35	52.17	23	ROOF & ITS COMPONENTS
			37.06	49.85	337	average injured occupant
			32.00	50.00	16	SIDE INTERIOR SURFACE
			29.53	47.06	17	EXTERIOR
ROLLOVER FIRST/WORST EVENT	BELTED	LEG	39.60	65.00	20	FLOOR & ITS COMPONENTS
			37.90	65.00	20	INSTRUMENT PANEL
			37.06	49.85	337	average injured occupant
ROLLOVER FIRST/WORST EVENT	NO BELT	HEAD	31.06	40.92	501	average injured occupant
			30.44	34.67	75	ROOF & ITS COMPONENTS
			28.34	42.15	121	EXTERIOR
ROLLOVER FIRST/WORST EVENT	NO BELT	NECK	33.36	40.00	25	ROOF & ITS COMPONENTS
			31.06	40.92	501	average injured occupant
			29.34	37.50	32	EXTERIOR
ROLLOVER FIRST/WORST EVENT	NO BELT	TORSO	33.04	36.00	25	INSTRUMENT PANEL
			32.52	48.48	33	SIDE INTERIOR SURFACE
			32.16	36.84	38	ROOF & ITS COMPONENTS
			31.06	40.92	501	average injured occupant
			30.01	44.22	147	EXTERIOR
ROLLOVER FIRST/WORST EVENT	NO BELT	ARM	32.70	51.52	33	EXTERIOR
			31.06	40.92	501	average injured occupant
ROLLOVER FIRST/WORST EVENT	NO BELT	LEG	31.06	40.92	501	average injured occupant
			30.68	58.06	31	EXTERIOR

**3+ standard deviations higher than average injured occupant (p < .002).
*1.96-3 standard deviations higher than average injured occupant (p < .05).

IMPACT TYPE	OCCUPANT PROTECTION	BODY REGION	AVERAGE AGE	PERCENT FEMALE	N	INJURY SOURCE
REAR & OTHER	BELTED	TORSO	48.17	61.11	18	SEAT & ITS COMPONENTS
			41.54	60.00	120	average injured occupant
REAR & OTHER	NO BELT	TORSO	38.00	73.33	15	SEAT & ITS COMPONENTS
			36.53	57.14	77	average injured occupant

3+ standard deviations higher than average injured occupant (p < .002).
*1.96-3 standard deviations higher than average injured occupant (p < .05).

TABLE E-5: AVERAGE AGE OF RIGHT-FRONT PASSENGERS AGE 18-96 WITH AIS 3+ INJURIES, CDS 1988-2010

IMPACT TYPE	OCCUPANT PROTECTION	BODY REGION	AVERAGE AGE	PERCENT FEMALE	N	INJURY SOURCE
FRONTAL	BELTED, AIR BAG	HEAD	48.81	69.61	385	average injured occupant
			46.73	54.55	11	B-PILLAR
			45.07	60.00	15	ROOF & ITS COMPONENTS
			43.80	40.00	10	A-PILLAR
			38.50	50.00	12	INSTRUMENT PANEL
FRONTAL	BELTED, AIR BAG	TORSO	57.40	70.00	10	AIR BAG
			53.51	72.16	97	BELT SYSTEM
			49.81	76.19	21	SIDE INTERIOR SURFACE
			48.81	69.61	385	average injured occupant
			47.92	50.00	12	SIDE HARDWARE OR ARMREST
			39.83	68.97	29	INSTRUMENT PANEL
			32.64	45.45	11	SEAT & ITS COMPONENTS
FRONTAL	BELTED, AIR BAG	ARM	56.33	83.33	12	SIDE INTERIOR SURFACE
			55.86	81.82	22	AIR BAG
			49.94	69.44	36	INSTRUMENT PANEL
			48.81	69.61	385	average injured occupant
FRONTAL	BELTED, AIR BAG	LEG	49.12	80.00	25	FLOOR & ITS COMPONENTS
			48.81	69.61	385	average injured occupant
			45.97	71.11	90	INSTRUMENT PANEL
FRONTAL	BELTED, NO AIR BAG	HEAD	49.86	68.00	375	average injured occupant
			48.38	42.31	26	INSTRUMENT PANEL
			46.00	72.73	11	ROOF & ITS COMPONENTS
FRONTAL	BELTED, NO AIR BAG	TORSO	* 55.52	70.92	141	BELT SYSTEM
			50.17	69.05	42	INSTRUMENT PANEL
			49.86	68.00	375	average injured occupant
			45.24	38.10	21	SIDE INTERIOR SURFACE
FRONTAL	BELTED, NO AIR BAG	ARM	50.21	74.47	47	INSTRUMENT PANEL
			49.86	68.00	375	average injured occupant
FRONTAL	BELTED, NO AIR BAG	LEG	49.86	68.00	375	average injured occupant
			45.11	65.28	72	INSTRUMENT PANEL
			43.43	73.91	23	FLOOR & ITS COMPONENTS
FRONTAL	NO BELT, AIR BAG	HEAD	42.36	64.29	14	INSTRUMENT PANEL
			38.10	55.87	281	average injured occupant
			34.30	20.00	10	A-PILLAR
			33.50	37.50	16	OTHER FRONTAL COMPONENT
			32.41	36.36	22	ROOF & ITS COMPONENTS
			30.95	52.63	19	EXTERIOR
FRONTAL	NO BELT, AIR BAG	TORSO	39.19	62.50	16	SIDE INTERIOR SURFACE
			39.01	58.33	72	INSTRUMENT PANEL
			38.10	55.87	281	average injured occupant
			35.20	46.67	15	EXTERIOR
FRONTAL	NO BELT, AIR BAG	ARM	43.00	76.67	30	INSTRUMENT PANEL
			38.10	55.87	281	average injured occupant
FRONTAL	NO BELT, AIR BAG	LEG	38.14	53.09	81	INSTRUMENT PANEL
			38.10	55.87	281	average injured occupant
			32.95	52.63	19	FLOOR & ITS COMPONENTS
FRONTAL	NO BELT, NO AIR BAG	HEAD	40.82	52.85	123	OTHER FRONTAL COMPONENT
			38.82	50.65	768	average injured occupant
			38.76	48.15	54	INSTRUMENT PANEL
			36.52	44.83	29	ROOF & ITS COMPONENTS
			31.30	33.33	30	EXTERIOR
			30.57	28.57	14	A-PILLAR
FRONTAL	NO BELT, NO AIR BAG	NECK	48.73	72.73	11	INSTRUMENT PANEL
			47.20	46.67	15	OTHER FRONTAL COMPONENT
			38.82	50.65	768	average injured occupant
FRONTAL	NO BELT, NO AIR BAG	TORSO	** 42.74	52.07	290	INSTRUMENT PANEL
			40.93	57.14	14	SIDE INTERIOR SURFACE
			38.83	41.67	12	STEERING ASSEMBLY
			38.82	50.65	768	average injured occupant
			28.21	14.29	14	EXTERIOR
FRONTAL	NO BELT, NO AIR BAG	ARM	40.39	62.26	106	INSTRUMENT PANEL
			38.82	50.65	768	average injured occupant

3+ standard deviations higher than average injured occupant (p < .002).
*1.96-3 standard deviations higher than average injured occupant (p < .05).

IMPACT TYPE	OCCUPANT PROTECTION	BODY REGION	AVERAGE AGE	PERCENT FEMALE	N	INJURY SOURCE
FRONTAL	NO BELT, NO AIR BAG	LEG	39.59	59.26	27	FLOOR & ITS COMPONENTS
			38.82	50.65	768	average injured occupant
			38.39	52.00	200	INSTRUMENT PANEL
NEARSIDE IMPACT	BELTED	HEAD	53.92	84.62	13	OTHER OCCS, SELF, CARGO
			51.20	80.00	10	B-PILLAR
			49.34	67.48	163	average injured occupant
			32.00	38.46	13	ROOF & ITS COMPONENTS
NEARSIDE IMPACT	BELTED	TORSO	* 60.81	77.78	27	BELT SYSTEM
			52.44	81.25	16	FLOOR & ITS COMPONENTS
			50.82	76.47	17	OTHER OCCS, SELF, CARGO
			49.34	67.48	163	average injured occupant
			44.85	75.00	20	SEAT & ITS COMPONENTS
NEARSIDE IMPACT	NO BELT	HEAD	35.45	43.06	209	average injured occupant
			34.22	50.00	18	ROOF & ITS COMPONENTS
			32.00	45.00	20	EXTERIOR
NEARSIDE IMPACT	NO BELT	TORSO	43.21	64.29	14	STEERING ASSEMBLY
			43.10	45.00	20	INSTRUMENT PANEL
			35.45	43.06	209	average injured occupant
			31.86	50.00	14	SEAT & ITS COMPONENTS
			30.94	37.50	16	EXTERIOR
			30.67	8.33	12	OTHER OCCS, SELF, CARGO
			26.91	27.27	11	SIDE INTERIOR SURFACE
NEARSIDE IMPACT	NO BELT	ARM	45.33	83.33	12	INSTRUMENT PANEL
			35.45	43.06	209	average injured occupant
NEARSIDE IMPACT	NO BELT	LEG	36.90	30.00	10	INSTRUMENT PANEL
			35.45	43.06	209	average injured occupant
FAR-SIDE IMPACT	BELTED	HEAD	59.27	81.82	11	OTHER SIDE COMPONENT
			48.41	63.49	63	B-PILLAR
			48.00	57.14	21	SIDE INTERIOR SURFACE
			46.99	64.41	503	average injured occupant
			44.94	70.59	34	EXTERIOR
			42.92	25.00	12	A-PILLAR
			39.31	55.17	29	ROOF & ITS COMPONENTS
			28.08	16.67	12	OTHER
FAR-SIDE IMPACT	BELTED	NECK	54.50	85.71	14	B-PILLAR
			46.99	64.41	503	average injured occupant
			42.90	50.00	10	ROOF & ITS COMPONENTS
FAR-SIDE IMPACT	BELTED	TORSO	49.77	54.84	31	B-PILLAR
			49.01	64.31	255	SIDE INTERIOR SURFACE
			48.72	66.98	106	SIDE HARDWARE OR ARMREST
			46.99	64.41	503	average injured occupant
			44.92	61.54	13	BELT SYSTEM
FAR-SIDE IMPACT	BELTED	ARM	46.99	64.41	503	average injured occupant
			42.10	75.00	20	SIDE INTERIOR SURFACE
FAR-SIDE IMPACT	BELTED	LEG	46.99	64.41	503	average injured occupant
			44.21	47.37	19	SIDE HARDWARE OR ARMREST
			38.19	57.45	47	SIDE INTERIOR SURFACE
FAR-SIDE IMPACT	NO BELT	HEAD	37.76	56.76	37	B-PILLAR
			35.43	48.52	439	average injured occupant
			35.09	45.45	55	EXTERIOR
			33.50	35.71	28	A-PILLAR
			28.36	18.18	11	SIDE INTERIOR SURFACE
			27.92	16.67	24	ROOF & ITS COMPONENTS
			26.60	20.00	15	OTHER
FAR-SIDE IMPACT	NO BELT	TORSO	41.06	44.44	18	B-PILLAR
			* 39.25	51.18	170	SIDE INTERIOR SURFACE
			36.30	55.56	54	SIDE HARDWARE OR ARMREST
			35.71	41.94	31	EXTERIOR
			35.43	48.52	439	average injured occupant
			33.96	36.00	25	INSTRUMENT PANEL
			32.08	46.15	13	OTHER OCCS, SELF, CARGO
FAR-SIDE IMPACT	NO BELT	ARM	35.43	48.52	439	average injured occupant
			31.27	45.45	11	EXTERIOR
			28.68	52.63	19	SIDE INTERIOR SURFACE

**3+ standard deviations higher than average injured occupant (p < .002).
*1.96-3 standard deviations higher than average injured occupant (p < .05).

IMPACT TYPE	OCCUPANT PROTECTION	BODY REGION	AVERAGE AGE	PERCENT FEMALE	N	INJURY SOURCE
FAR-SIDE IMPACT	NO BELT	LEG	35.43	48.52	439	average injured occupant
			29.89	43.18	44	SIDE INTERIOR SURFACE
			28.88	57.69	26	INSTRUMENT PANEL
			27.91	36.36	11	SIDE HARDWARE OR ARMREST
ROLLOVER FIRST/WORST EVENT	BELTED	HEAD	38.84	55.26	38	ROOF & ITS COMPONENTS
			37.43	51.11	180	average injured occupant
			31.40	40.00	10	EXTERIOR
ROLLOVER FIRST/WORST EVENT	BELTED	NECK	37.43	51.11	180	average injured occupant
			37.22	40.74	27	ROOF & ITS COMPONENTS
ROLLOVER FIRST/WORST EVENT	BELTED	TORSO	39.96	70.37	27	SIDE INTERIOR SURFACE
			37.43	51.11	180	average injured occupant
			35.46	53.85	13	BELT SYSTEM
ROLLOVER FIRST/WORST EVENT	NO BELT	HEAD	31.90	42.60	331	average injured occupant
			30.03	29.03	31	ROOF & ITS COMPONENTS
			28.28	42.68	82	EXTERIOR
ROLLOVER FIRST/WORST EVENT	NO BELT	NECK	32.80	30.00	10	ROOF & ITS COMPONENTS
			31.90	42.60	331	average injured occupant
			28.33	40.00	15	EXTERIOR
ROLLOVER FIRST/WORST EVENT	NO BELT	TORSO	32.95	52.38	21	SIDE INTERIOR SURFACE
			32.33	33.33	15	INSTRUMENT PANEL
			31.90	42.60	331	average injured occupant
			28.95	41.28	109	EXTERIOR
			26.83	41.67	12	ROOF & ITS COMPONENTS
ROLLOVER FIRST/WORST EVENT	NO BELT	ARM	35.31	38.46	13	EXTERIOR
			31.90	42.60	331	average injured occupant
ROLLOVER FIRST/WORST EVENT	NO BELT	LEG	31.90	42.60	331	average injured occupant
			31.33	50.00	18	EXTERIOR
REAR & OTHER	BELTED	TORSO	44.50	50.00	10	SEAT & ITS COMPONENTS
			42.86	62.75	51	average injured occupant
REAR & OTHER	NO BELT	TORSO	40.00	63.64	11	SEAT & ITS COMPONENTS
			37.16	56.76	37	average injured occupant

3+ standard deviations higher than average injured occupant (p < .002).
*1.96-3 standard deviations higher than average injured occupant (p < .05).

TABLE E-6: AVERAGE AGE OF RIGHT-FRONT PASSENGERS AGE 18-96 WITH AIS 4+ INJURIES, CDS 1988-2010

IMPACT TYPE	OCCUPANT PROTECTION	BODY REGION	AVERAGE AGE	PERCENT FEMALE	N	INJURY SOURCE
FRONTAL	BELTED, AIR BAG	HEAD	* 69.20	100.00	5	AIR BAG
			52.00	42.86	7	A-PILLAR
			47.91	61.74	115	average injured occupant
			43.17	50.00	6	EXTERIOR
			38.13	50.00	8	INSTRUMENT PANEL
			38.13	50.00	8	ROOF & ITS COMPONENTS
			32.83	50.00	6	B-PILLAR
FRONTAL	BELTED, AIR BAG	TORSO	52.56	68.29	41	BELT SYSTEM
			49.60	60.00	5	SIDE HARDWARE OR ARMREST
			47.91	61.74	115	average injured occupant
			45.57	57.14	7	SIDE INTERIOR SURFACE
			37.57	57.14	14	INSTRUMENT PANEL
FRONTAL	BELTED, NO AIR BAG	HEAD	52.99	60.93	151	average injured occupant
			47.00	30.77	13	INSTRUMENT PANEL
			39.56	66.67	9	ROOF & ITS COMPONENTS
			31.83	50.00	6	EXTERIOR
FRONTAL	BELTED, NO AIR BAG	TORSO	54.75	64.91	57	BELT SYSTEM
			52.99	60.93	151	average injured occupant
			52.14	57.14	21	INSTRUMENT PANEL
			44.29	28.57	7	SIDE INTERIOR SURFACE
FRONTAL	NO BELT, AIR BAG	HEAD	41.00	33.33	6	INSTRUMENT PANEL
			39.70	49.24	132	average injured occupant
			37.83	33.33	6	OTHER FRONTAL COMPONENT
			34.60	20.00	10	ROOF & ITS COMPONENTS
			32.63	43.75	16	EXTERIOR
FRONTAL	NO BELT, AIR BAG	TORSO	51.00	40.00	5	SIDE INTERIOR SURFACE
			39.70	49.24	132	average injured occupant
			37.95	55.81	43	INSTRUMENT PANEL
			30.70	30.00	10	EXTERIOR
FRONTAL	NO BELT, NO AIR BAG	HEAD	41.93	50.88	57	OTHER FRONTAL COMPONENT
			40.51	45.27	349	average injured occupant
			37.18	41.18	17	ROOF & ITS COMPONENTS
			36.04	34.62	26	INSTRUMENT PANEL
			32.10	30.00	20	EXTERIOR
			28.60	20.00	5	A-PILLAR
FRONTAL	NO BELT, NO AIR BAG	TORSO	43.38	50.35	141	INSTRUMENT PANEL
			40.51	45.27	349	average injured occupant
			39.17	50.00	6	SIDE INTERIOR SURFACE
			33.80	0.00	5	EXTERIOR
NEARSIDE IMPACT	BELTED	HEAD	51.83	83.33	6	B-PILLAR
			49.90	60.98	82	average injured occupant
			37.00	57.14	7	SEAT & ITS COMPONENTS
			29.50	25.00	8	ROOF & ITS COMPONENTS
NEARSIDE IMPACT	BELTED	TORSO	65.09	81.82	11	BELT SYSTEM
			60.60	60.00	5	FLOOR & ITS COMPONENTS
			52.00	63.64	11	OTHER OCCS, SELF, CARGO
			49.90	60.98	82	average injured occupant
			43.27	63.64	11	SEAT & ITS COMPONENTS
NEARSIDE IMPACT	NO BELT	HEAD	44.60	40.00	5	INSTRUMENT PANEL
			34.95	35.71	126	average injured occupant
			34.00	42.86	14	EXTERIOR
			32.40	20.00	5	OTHER FRONTAL COMPONENT
			29.82	45.45	11	ROOF & ITS COMPONENTS
			21.20	40.00	5	OTHER
NEARSIDE IMPACT	NO BELT	TORSO	39.43	42.86	7	STEERING ASSEMBLY
			36.57	28.57	7	INSTRUMENT PANEL
			34.95	35.71	126	average injured occupant
			32.00	40.00	5	SIDE INTERIOR SURFACE
			31.89	11.11	9	OTHER OCCS, SELF, CARGO
			28.64	27.27	11	EXTERIOR
			28.50	0.00	6	SEAT & ITS COMPONENTS

**3+ standard deviations higher than average injured occupant (p < .002).
*1.96-3 standard deviations higher than average injured occupant (p < .05).

280

IMPACT TYPE	OCCUPANT PROTECTION	BODY REGION	AVERAGE AGE	PERCENT FEMALE	N	INJURY SOURCE
FAR-SIDE IMPACT	BELTED	HEAD	62.43	85.71	7	OTHER SIDE COMPONENT
			48.18	62.58	310	average injured occupant
			47.56	66.67	36	B-PILLAR
			44.78	66.67	27	EXTERIOR
			41.00	22.22	9	A-PILLAR
			40.50	41.67	12	SIDE INTERIOR SURFACE
			40.00	53.33	15	ROOF & ITS COMPONENTS
			29.14	14.29	7	OTHER
FAR-SIDE IMPACT	BELTED	NECK	54.43	85.71	7	B-PILLAR
			48.18	62.58	310	average injured occupant
FAR-SIDE IMPACT	BELTED	TORSO	51.41	59.46	148	SIDE INTERIOR SURFACE
			50.36	64.29	14	B-PILLAR
			48.18	62.58	310	average injured occupant
			41.29	67.74	31	SIDE HARDWARE OR ARMREST
FAR-SIDE IMPACT	NO BELT	HEAD	39.57	42.86	7	OTHER SIDE COMPONENT
			36.35	46.64	298	average injured occupant
			35.00	42.11	38	EXTERIOR
			33.36	31.82	22	A-PILLAR
			33.00	20.00	5	SIDE INTERIOR SURFACE
			32.60	48.00	25	B-PILLAR
			31.21	14.29	14	ROOF & ITS COMPONENTS
			25.92	25.00	12	OTHER
FAR-SIDE IMPACT	NO BELT	TORSO	40.06	51.00	100	SIDE INTERIOR SURFACE
			39.69	53.85	13	B-PILLAR
			39.58	42.31	26	SIDE HARDWARE OR ARMREST
			39.17	83.33	6	OTHER OCCS, SELF, CARGO
			36.50	38.89	18	EXTERIOR
			36.35	46.64	298	average injured occupant
			35.67	33.33	9	INSTRUMENT PANEL
ROLLOVER FIRST/WORST EVENT	BELTED	HEAD	38.71	49.51	103	average injured occupant
			38.58	45.83	24	ROOF & ITS COMPONENTS
			30.33	33.33	9	EXTERIOR
ROLLOVER FIRST/WORST EVENT	BELTED	NECK	38.71	49.51	103	average injured occupant
			28.60	20.00	5	ROOF & ITS COMPONENTS
ROLLOVER FIRST/WORST EVENT	BELTED	TORSO	42.00	61.54	13	SIDE INTERIOR SURFACE
			38.71	49.51	103	average injured occupant
			29.25	50.00	8	BELT SYSTEM
ROLLOVER FIRST/WORST EVENT	NO BELT	HEAD	32.34	42.62	237	average injured occupant
			30.23	31.82	22	ROOF & ITS COMPONENTS
			28.16	39.34	61	EXTERIOR
ROLLOVER FIRST/WORST EVENT	NO BELT	NECK	32.34	42.62	237	average injured occupant
			31.60	20.00	5	EXTERIOR
ROLLOVER FIRST/WORST EVENT	NO BELT	TORSO	32.34	42.62	237	average injured occupant
			29.94	41.27	63	EXTERIOR
			29.88	37.50	8	INSTRUMENT PANEL
			29.22	55.56	9	SIDE INTERIOR SURFACE
			25.00	37.50	8	ROOF & ITS COMPONENTS
REAR & OTHER	BELTED	TORSO	52.00	60.00	5	SEAT & ITS COMPONENTS
			42.08	62.50	24	average injured occupant

3+ standard deviations higher than average injured occupant (p < .002).
*1.96-3 standard deviations higher than average injured occupant (p < .05).

TABLE E-7: AVERAGE AGE OF BACKSEAT OUTBOARD PASSENGERS AGE 18-96 WITH AIS 2+ INJURIES, CDS 1988-2010

IMPACT TYPE	OCCUPANT PROTECTION	BODY REGION	AVERAGE AGE	PERCENT FEMALE	N	INJURY SOURCE
FRONTAL	BELTED	TORSO	* 50.11	75.38	65	BELT SYSTEM
			42.79	66.17	133	average injured occupant
FRONTAL	NO BELT	HEAD	37.27	53.33	15	OTHER FRONTAL COMPONENT
			35.51	48.32	358	average injured occupant
			34.18	47.89	71	SEAT & ITS COMPONENTS
			31.89	61.11	18	ROOF & ITS COMPONENTS
			27.18	35.29	17	B-PILLAR
FRONTAL	NO BELT	NECK	38.20	66.67	15	SEAT & ITS COMPONENTS
			35.51	48.32	358	average injured occupant
FRONTAL	NO BELT	TORSO	38.66	39.81	103	SEAT & ITS COMPONENTS
			35.51	48.32	358	average injured occupant
			35.38	50.00	16	SIDE INTERIOR SURFACE
FRONTAL	NO BELT	ARM	35.51	48.32	358	average injured occupant
			35.02	47.37	57	SEAT & ITS COMPONENTS
FRONTAL	NO BELT	LEG	** 43.71	68.57	70	SEAT & ITS COMPONENTS
			35.51	48.32	358	average injured occupant
			34.93	60.00	15	FLOOR & ITS COMPONENTS
NEARSIDE IMPACT	BELTED	TORSO	47.45	63.64	22	SIDE INTERIOR SURFACE
			47.26	66.04	53	average injured occupant
NEARSIDE IMPACT	NO BELT	HEAD	34.25	48.80	166	average injured occupant
			32.50	40.00	20	ROOF & ITS COMPONENTS
			30.82	41.18	17	EXTERIOR
			30.47	46.67	15	OTHER SIDE COMPONENT
NEARSIDE IMPACT	NO BELT	TORSO	39.89	61.11	18	SIDE HARDWARE OR ARMREST
			38.12	49.02	51	SIDE INTERIOR SURFACE
			35.75	55.00	20	EXTERIOR
			34.25	48.80	166	average injured occupant
FAR-SIDE IMPACT	BELTED	TORSO	* 51.27	65.38	26	SIDE INTERIOR SURFACE
			40.03	56.76	74	average injured occupant
FAR-SIDE IMPACT	NO BELT	HEAD	37.33	33.33	18	B-PILLAR
			36.63	57.89	19	EXTERIOR
			33.91	42.70	185	average injured occupant
			33.50	55.56	18	OTHER SIDE COMPONENT
			30.95	36.84	19	ROOF & ITS COMPONENTS
FAR-SIDE IMPACT	NO BELT	TORSO	37.61	40.63	64	SIDE INTERIOR SURFACE
			37.13	43.48	23	SEAT & ITS COMPONENTS
			37.00	52.17	23	SIDE HARDWARE OR ARMREST
			34.63	62.50	16	EXTERIOR
			33.91	42.70	185	average injured occupant
FAR-SIDE IMPACT	NO BELT	ARM	41.80	46.67	15	SIDE INTERIOR SURFACE
			33.91	42.70	185	average injured occupant
FAR-SIDE IMPACT	NO BELT	LEG	33.91	42.70	185	average injured occupant
			29.79	29.41	17	SIDE INTERIOR SURFACE
ROLLOVER FIRST/WORST EVENT	NO BELT	HEAD	29.99	46.79	156	average injured occupant
			29.60	40.00	20	ROOF & ITS COMPONENTS
			29.41	57.14	49	EXTERIOR
ROLLOVER FIRST/WORST EVENT	NO BELT	NECK	29.99	46.79	156	average injured occupant
			28.06	58.82	17	EXTERIOR
ROLLOVER FIRST/WORST EVENT	NO BELT	TORSO	29.99	46.79	156	average injured occupant
			28.00	56.60	53	EXTERIOR
REAR & OTHER	NO BELT	TORSO	42.09	72.73	22	SEAT & ITS COMPONENTS
			35.42	70.83	48	average injured occupant

**3+ standard deviations higher than average injured occupant (p < .002).
*1.96-3 standard deviations higher than average injured occupant (p < .05).

282

TABLE E-8: AVERAGE AGE OF BACKSEAT OUTBOARD PASSENGERS AGE 18-96 WITH AIS 3+ INJURIES, CDS 1988-2010

IMPACT TYPE	OCCUPANT PROTECTION	BODY REGION	AVERAGE AGE	PERCENT FEMALE	N	INJURY SOURCE
FRONTAL	BELTED	TORSO	49.18	73.68	38	BELT SYSTEM
			45.39	68.18	66	average injured occupant
FRONTAL	NO BELT	HEAD	36.10	47.76	201	average injured occupant
			35.82	45.45	22	SEAT & ITS COMPONENTS
			28.15	15.38	13	EXTERIOR
FRONTAL	NO BELT	TORSO	40.64	28.89	45	SEAT & ITS COMPONENTS
			36.10	47.76	201	average injured occupant
			32.20	50.00	10	SIDE INTERIOR SURFACE
			32.00	45.45	11	EXTERIOR
FRONTAL	NO BELT	ARM	36.10	47.76	201	average injured occupant
			31.96	46.15	26	SEAT & ITS COMPONENTS
FRONTAL	NO BELT	LEG	* 45.53	67.44	43	SEAT & ITS COMPONENTS
			36.10	47.76	201	average injured occupant
NEARSIDE IMPACT	BELTED	TORSO	49.13	70.00	30	average injured occupant
			43.91	54.55	11	SIDE INTERIOR SURFACE
NEARSIDE IMPACT	NO BELT	HEAD	35.42	45.16	124	average injured occupant
			33.17	41.67	12	ROOF & ITS COMPONENTS
			28.54	46.15	13	OTHER SIDE COMPONENT
NEARSIDE IMPACT	NO BELT	TORSO	39.51	48.78	41	SIDE INTERIOR SURFACE
			35.42	45.16	124	average injured occupant
			31.18	45.45	11	EXTERIOR
FAR-SIDE IMPACT	BELTED	TORSO	59.38	75.00	16	SIDE INTERIOR SURFACE
			45.26	57.14	42	average injured occupant
FAR-SIDE IMPACT	NO BELT	HEAD	34.40	40.00	125	average injured occupant
			31.71	57.14	14	EXTERIOR
			25.27	18.18	11	ROOF & ITS COMPONENTS
FAR-SIDE IMPACT	NO BELT	TORSO	38.17	50.00	12	SIDE HARDWARE OR ARMREST
			37.25	35.00	40	SIDE INTERIOR SURFACE
			36.18	72.73	11	EXTERIOR
			34.40	40.00	125	average injured occupant
FAR-SIDE IMPACT	NO BELT	LEG	38.30	30.00	10	SEAT & ITS COMPONENTS
			34.40	40.00	125	average injured occupant
			30.33	33.33	15	SIDE INTERIOR SURFACE
ROLLOVER FIRST/WORST EVENT	NO BELT	HEAD	31.32	52.83	106	average injured occupant
			28.75	58.10	90	EXTERIOR
ROLLOVER FIRST/WORST EVENT	NO BELT	NECK	31.32	52.83	106	average injured occupant
			27.58	41.67	12	EXTERIOR
ROLLOVER FIRST/WORST EVENT	NO BELT	TORSO	31.32	52.83	106	average injured occupant
			28.83	60.98	41	EXTERIOR
ROLLOVER FIRST/WORST EVENT	NO BELT	LEG	31.80	60.00	10	EXTERIOR
			31.32	52.83	106	average injured occupant
REAR & OTHER	NO BELT	TORSO	42.94	81.25	16	SEAT & ITS COMPONENTS
			39.83	72.41	29	average injured occupant

**3+ standard deviations higher than average injured occupant (p < .002).
*1.96-3 standard deviations higher than average injured occupant (p < .05).

TABLE E-9: AVERAGE AGE OF BACKSEAT OUTBOARD PASSENGERS AGE 18-96 WITH AIS 4+ INJURIES, CDS 1988-2010

IMPACT TYPE	OCCUPANT PROTECTION	BODY REGION	AVERAGE AGE	PERCENT FEMALE	N	INJURY SOURCE
FRONTAL	BELTED	TORSO	57.31	76.92	13	BELT SYSTEM
			47.84	60.00	25	average injured occupant
FRONTAL	NO BELT	HEAD	39.36	36.36	11	SEAT & ITS COMPONENTS
			37.83	50.00	6	ROOF & ITS COMPONENTS
			37.66	38.04	92	average injured occupant
			30.56	22.22	9	EXTERIOR
FRONTAL	NO BELT	NECK	38.00	80.00	5	SEAT & ITS COMPONENTS
			37.66	38.04	92	average injured occupant
FRONTAL	NO BELT	TORSO	43.70	35.00	20	SEAT & ITS COMPONENTS
			37.66	38.04	92	average injured occupant
			35.00	33.33	6	EXTERIOR
NEARSIDE IMPACT	BELTED	TORSO	48.00	68.75	16	average injured occupant
			45.43	71.43	7	SIDE INTERIOR SURFACE
NEARSIDE IMPACT	NO BELT	HEAD	42.29	42.86	7	ROOF & ITS COMPONENTS
			37.36	45.78	83	average injured occupant
			31.11	55.56	9	OTHER SIDE COMPONENT
			26.00	40.00	5	EXTERIOR
NEARSIDE IMPACT	NO BELT	TORSO	47.00	40.00	5	SIDE HARDWARE OR ARMREST
			42.83	50.00	24	SIDE INTERIOR SURFACE
			37.36	45.78	83	average injured occupant
			27.83	50.00	6	EXTERIOR
FAR-SIDE IMPACT	BELTED	TORSO	57.00	87.50	8	SIDE INTERIOR SURFACE
			51.91	63.64	22	average injured occupant
FAR-SIDE IMPACT	NO BELT	HEAD	48.80	60.00	5	B-PILLAR
			35.22	42.86	77	average injured occupant
			33.29	71.43	7	EXTERIOR
			28.29	28.57	7	ROOF & ITS COMPONENTS
			21.20	0.00	5	OTHER
FAR-SIDE IMPACT	NO BELT	TORSO	44.00	57.14	7	SIDE HARDWARE OR ARMREST
			40.13	37.50	16	SIDE INTERIOR SURFACE
			35.22	42.86	77	average injured occupant
			27.71	57.14	7	EXTERIOR
ROLLOVER FIRST/WORST EVENT	NO BELT	HEAD	30.97	51.35	74	average injured occupant
			29.59	50.00	22	EXTERIOR
			23.20	20.00	5	ROOF & ITS COMPONENTS
ROLLOVER FIRST/WORST EVENT	NO BELT	TORSO	30.97	51.35	74	average injured occupant
			28.89	62.96	27	EXTERIOR
REAR & OTHER	BELTED	TORSO	46.40	40.00	5	SEAT & ITS COMPONENTS
			42.33	33.33	6	average injured occupant
REAR & OTHER	NO BELT	HEAD	41.04	69.57	23	average injured occupant
			37.00	60.00	5	ROOF & ITS COMPONENTS
REAR & OTHER	NO BELT	TORSO	49.09	72.73	11	SEAT & ITS COMPONENTS
			41.04	69.57	23	average injured occupant

**3+ standard deviations higher than average injured occupant (p < .002).
*1.96-3 standard deviations higher than average injured occupant (p < .05).

284

IMPACT TYPE	OCCUPANT PROTECTION	BODY REGION	PERCENT FEMALE	AVERAGE AGE	N	INJURY SOURCE
FRONTAL	BELTED, AIR BAG	HEAD	54.55	37.86	22	OTHER SIDE COMPONENT
			54.02	40.52	174	NON-CONTACT
			51.30	43.37	4255	average injured occupant
			47.37	46.63	19	BELT SYSTEM
			46.15	45.65	26	INSTRUMENT PANEL
			45.61	40.68	57	B-PILLAR
			45.06	39.27	324	STEERING ASSEMBLY
			43.08	42.62	130	AIR BAG
			42.68	39.71	82	A-PILLAR
			38.89	43.00	18	SEAT & ITS COMPONENTS
			37.65	39.43	170	ROOF & ITS COMPONENTS
			35.85	40.25	53	OTHER FRONTAL COMPONENT
			35.29	43.29	34	EXTERIOR
			26.67	39.40	15	OTHER OCCS, SELF, CARGO
			16.00	38.64	25	OTHER
FRONTAL	BELTED, AIR BAG	NECK	66.67	56.00	21	BELT SYSTEM
			64.58	52.00	48	NON-CONTACT
			52.63	51.53	19	A-PILLAR
			51.30	43.37	4255	average injured occupant
			47.37	55.11	19	SEAT & ITS COMPONENTS
			46.81	49.26	47	STEERING ASSEMBLY
			37.25	45.82	51	ROOF & ITS COMPONENTS
FRONTAL	BELTED, AIR BAG	TORSO	60.00	62.37	60	AIR BAG
			51.30	43.37	4255	average injured occupant
			48.18	47.09	110	SIDE HARDWARE OR ARMREST
			48.02	48.63	683	BELT SYSTEM
			46.15	42.51	234	INSTRUMENT PANEL
			45.83	51.71	48	NON-CONTACT
			42.86	44.80	133	SIDE INTERIOR SURFACE
			42.55	45.69	463	STEERING ASSEMBLY
			41.38	41.96	116	SEAT & ITS COMPONENTS
			40.91	44.23	22	B-PILLAR
			35.82	40.42	67	FLOOR & ITS COMPONENTS
FRONTAL	BELTED, AIR BAG	ARM	** 80.00	44.83	150	AIR BAG
			** 69.88	43.33	83	A-PILLAR
			** 62.40	46.08	258	STEERING ASSEMBLY
			59.26	41.98	54	ROOF & ITS COMPONENTS
			52.53	44.02	158	SIDE INTERIOR SURFACE
			51.65	43.85	333	INSTRUMENT PANEL
			51.30	43.37	4255	average injured occupant
			50.00	46.64	98	OTHER FRONTAL COMPONENT
			47.37	42.42	19	SIDE HARDWARE OR ARMREST
			37.93	45.59	29	EXTERIOR
FRONTAL	BELTED, AIR BAG	LEG	** 63.36	43.89	1209	FLOOR & ITS COMPONENTS
			51.30	43.37	4255	average injured occupant
			49.55	42.46	1112	INSTRUMENT PANEL
			47.62	43.43	21	SIDE HARDWARE OR ARMREST
			33.87	35.26	62	STEERING ASSEMBLY
			27.78	42.28	54	SIDE INTERIOR SURFACE
FRONTAL	BELTED, NO AIR BAG	HEAD	52.94	39.76	17	B-PILLAR
			46.67	39.23	30	EXTERIOR
			43.76	38.91	713	STEERING ASSEMBLY
			43.32	38.94	247	OTHER FRONTAL COMPONENT
			43.18	42.17	3175	average injured occupant
			41.67	39.61	36	OTHER SIDE COMPONENT
			40.74	38.15	27	NON-CONTACT
			35.29	39.98	102	A-PILLAR
			35.00	36.33	60	INSTRUMENT PANEL
			31.58	41.75	95	ROOF & ITS COMPONENTS
FRONTAL	BELTED, NO AIR BAG	NECK	46.03	48.13	63	STEERING ASSEMBLY
			45.00	52.55	20	NON-CONTACT
			43.18	42.17	3175	average injured occupant
FRONTAL	BELTED, NO AIR BAG	TORSO	53.66	47.54	41	NON-CONTACT
			45.12	45.96	492	BELT SYSTEM
			43.18	42.17	3175	average injured occupant
			42.86	37.10	21	FLOOR & ITS COMPONENTS
			41.06	47.24	699	STEERING ASSEMBLY
			38.68	42.38	106	INSTRUMENT PANEL
			33.33	41.09	33	SIDE HARDWARE OR ARMREST
			30.23	47.02	43	SEAT & ITS COMPONENTS
			26.32	45.66	76	SIDE INTERIOR SURFACE

**3+ standard deviations higher than average injured occupant (p < .002).
*1.96-3 standard deviations higher than average injured occupant (p < .05).

285

IMPACT TYPE	OCCUPANT PROTECTION	BODY REGION	PERCENT FEMALE	AVERAGE AGE	N	INJURY SOURCE
FRONTAL	BELTED, NO AIR BAG	ARM	** 53.45	43.08	232	STEERING ASSEMBLY
			43.18	42.17	3175	average injured occupant
			42.86	41.84	259	INSTRUMENT PANEL
			42.22	46.87	45	SIDE INTERIOR SURFACE
			31.25	48.13	16	OTHER FRONTAL COMPONENT
FRONTAL	BELTED, NO AIR BAG	LEG	** 54.21	41.34	629	FLOOR & ITS COMPONENTS
			43.77	40.16	658	INSTRUMENT PANEL
			43.18	42.17	3175	average injured occupant
			38.89	38.28	36	SIDE INTERIOR SURFACE
			28.07	38.18	57	STEERING ASSEMBLY
FRONTAL	NO BELT, AIR BAG	HEAD	** 58.33	37.55	60	AIR BAG
			35.90	39.85	39	NON-CONTACT
			32.63	37.24	2400	average injured occupant
			32.57	35.32	175	STEERING ASSEMBLY
			28.00	32.52	25	OTHER
			25.24	38.78	103	A-PILLAR
			22.87	35.36	188	ROOF & ITS COMPONENTS
			20.93	33.81	43	INSTRUMENT PANEL
			20.87	31.77	115	EXTERIOR
			19.93	36.95	301	OTHER FRONTAL COMPONENT
FRONTAL	NO BELT, AIR BAG	NECK	39.13	43.00	46	STEERING ASSEMBLY
			32.63	37.24	2400	average injured occupant
			26.19	42.19	42	OTHER FRONTAL COMPONENT
			25.00	33.69	36	EXTERIOR
			22.22	39.78	18	A-PILLAR
			17.65	39.35	17	INSTRUMENT PANEL
			16.67	38.52	42	ROOF & ITS COMPONENTS
FRONTAL	NO BELT, AIR BAG	TORSO	42.86	44.91	35	AIR BAG
			35.71	35.21	42	SIDE HARDWARE OR ARMREST
			33.33	40.83	36	FLOOR & ITS COMPONENTS
			32.63	37.24	2400	average injured occupant
			31.76	40.22	677	STEERING ASSEMBLY
			27.55	38.66	98	SIDE INTERIOR SURFACE
			27.06	37.13	255	INSTRUMENT PANEL
			26.09	30.96	23	ROOF & ITS COMPONENTS
			25.00	33.20	84	EXTERIOR
			16.67	38.06	36	SEAT & ITS COMPONENTS
FRONTAL	NO BELT, AIR BAG	ARM	** 68.42	46.84	19	AIR BAG
			45.45	38.05	22	A-PILLAR
			* 42.86	35.87	91	STEERING ASSEMBLY
			41.18	36.50	34	OTHER FRONTAL COMPONENT
			33.33	33.83	18	ROOF & ITS COMPONENTS
			33.33	31.17	30	EXTERIOR
			32.63	37.24	2400	average injured occupant
			31.91	40.19	47	SIDE INTERIOR SURFACE
			28.64	40.15	206	INSTRUMENT PANEL
FRONTAL	NO BELT, AIR BAG	LEG	** 46.57	37.58	523	FLOOR & ITS COMPONENTS
			32.63	37.24	2400	average injured occupant
			32.22	37.32	568	INSTRUMENT PANEL
			26.17	36.64	107	STEERING ASSEMBLY
			25.00	32.08	24	EXTERIOR
			23.08	34.58	26	SIDE INTERIOR SURFACE
FRONTAL	NO BELT, NO AIR BAG	HEAD	38.00	36.58	50	OTHER SIDE COMPONENT
			* 35.27	36.91	601	STEERING ASSEMBLY
			34.68	37.93	173	INSTRUMENT PANEL
			30.55	37.61	5081	average injured occupant
			28.68	35.81	1440	OTHER FRONTAL COMPONENT
			28.64	34.55	206	A-PILLAR
			26.67	40.47	15	SIDE INTERIOR SURFACE
			23.53	36.65	17	NON-CONTACT
			18.85	37.36	191	ROOF & ITS COMPONENTS
			18.75	35.25	160	EXTERIOR
			15.79	31.16	19	OTHER
FRONTAL	NO BELT, NO AIR BAG	NECK	40.00	38.60	20	INSTRUMENT PANEL
			32.53	39.99	83	STEERING ASSEMBLY
			32.17	44.10	115	OTHER FRONTAL COMPONENT
			30.55	37.61	5081	average injured occupant
			20.83	35.96	24	A-PILLAR
			16.67	37.50	36	EXTERIOR
			14.29	44.29	42	ROOF & ITS COMPONENTS

**3+ standard deviations higher than average injured occupant (p < .002).
*1.96-3 standard deviations higher than average injured occupant (p < .05).

IMPACT TYPE	OCCUPANT PROTECTION	BODY REGION	PERCENT FEMALE	AVERAGE AGE	N	INJURY SOURCE
FRONTAL	NO BELT, NO AIR BAG	TORSO	38.89	50.56	18	NON-CONTACT
			34.21	41.50	38	FLOOR & ITS COMPONENTS
			33.33	39.52	27	OTHER FRONTAL COMPONENT
			32.00	35.04	50	SEAT & ITS COMPONENTS
			31.58	38.79	19	SIDE HARDWARE OR ARMREST
			30.55	37.61	5081	average injured occupant
			28.57	42.39	119	SIDE INTERIOR SURFACE
			28.08	41.88	1631	STEERING ASSEMBLY
			25.58	33.26	86	EXTERIOR
			22.28	40.53	359	INSTRUMENT PANEL
			15.00	35.95	20	ROOF & ITS COMPONENTS
FRONTAL	NO BELT, NO AIR BAG	ARM	** 41.26	39.15	288	STEERING ASSEMBLY
			38.78	37.65	49	OTHER FRONTAL COMPONENT
			35.00	39.50	20	A-PILLAR
			33.69	40.91	374	INSTRUMENT PANEL
			31.25	38.14	64	SIDE INTERIOR SURFACE
			30.55	37.61	5081	average injured occupant
			17.14	34.40	35	EXTERIOR
FRONTAL	NO BELT, NO AIR BAG	LEG	44.00	33.68	25	EXTERIOR
			** 43.90	38.15	779	FLOOR & ITS COMPONENTS
			36.99	36.90	146	STEERING ASSEMBLY
			* 33.69	37.83	1021	INSTRUMENT PANEL
			30.55	37.61	5081	average injured occupant
			10.53	39.16	38	SIDE INTERIOR SURFACE
NEARSIDE IMPACT	BELTED	HEAD	* 64.06	43.59	64	SIDE INTERIOR SURFACE
			61.90	39.57	21	AIR BAG
			60.00	34.87	30	BELT SYSTEM
			53.85	34.54	26	OTHER FRONTAL COMPONENT
			53.19	43.77	47	STEERING ASSEMBLY
			52.46	43.75	183	OTHER SIDE COMPONENT
			50.00	42.88	140	A-PILLAR
			49.32	42.34	148	EXTERIOR
			48.96	40.82	96	NON-CONTACT
			48.72	43.94	2974	average injured occupant
			47.08	42.83	308	B-PILLAR
			38.89	39.06	18	OTHER OCCS, SELF, CARGO
			34.39	37.48	221	ROOF & ITS COMPONENTS
			31.82	33.92	66	OTHER
NEARSIDE IMPACT	BELTED	NECK	58.97	47.44	39	NON-CONTACT
			53.13	50.05	64	B-PILLAR
			52.00	41.16	25	SIDE INTERIOR SURFACE
			48.72	43.94	2974	average injured occupant
			36.36	44.45	33	EXTERIOR
			34.38	39.97	32	ROOF & ITS COMPONENTS
NEARSIDE IMPACT	BELTED	TORSO	59.26	44.63	27	INSTRUMENT PANEL
			58.82	37.82	17	OTHER OCCS, SELF, CARGO
			* 52.77	46.10	741	SIDE HARDWARE OR ARMREST
			50.00	41.39	18	OTHER
			49.45	40.33	91	FLOOR & ITS COMPONENTS
			49.33	40.88	75	SEAT & ITS COMPONENTS
			48.72	43.94	2974	average injured occupant
			47.20	44.29	125	STEERING ASSEMBLY
			46.53	46.48	1255	SIDE INTERIOR SURFACE
			45.10	47.40	204	BELT SYSTEM
			41.04	47.70	212	B-PILLAR
NEARSIDE IMPACT	BELTED	ARM	* 63.24	48.04	68	STEERING ASSEMBLY
			56.25	41.00	16	B-PILLAR
			55.10	44.39	49	INSTRUMENT PANEL
			48.72	43.94	2974	average injured occupant
			46.82	42.02	173	SIDE INTERIOR SURFACE
			32.14	38.93	28	SIDE HARDWARE OR ARMREST
NEARSIDE IMPACT	BELTED	LEG	* 59.05	46.91	105	FLOOR & ITS COMPONENTS
			51.19	38.81	84	SIDE HARDWARE OR ARMREST
			50.75	43.10	134	INSTRUMENT PANEL
			48.72	43.94	2974	average injured occupant
			43.56	40.04	163	SIDE INTERIOR SURFACE
			33.33	32.93	15	STEERING ASSEMBLY

**3+ standard deviations higher than average injured occupant (p < .002).
*1.96-3 standard deviations higher than average injured occupant (p < .05).

IMPACT TYPE	OCCUPANT PROTECTION	BODY REGION	PERCENT FEMALE	AVERAGE AGE	N	INJURY SOURCE
NEARSIDE IMPACT	NO BELT	HEAD	* 56.67	44.17	30	SIDE INTERIOR SURFACE
			46.88	37.94	32	NON-CONTACT
			39.29	26.29	28	STEERING ASSEMBLY
			37.62	39.41	101	OTHER SIDE COMPONENT
			34.78	34.50	207	EXTERIOR
			34.33	36.10	67	OTHER FRONTAL COMPONENT
			33.71	37.83	1771	average injured occupant
			32.58	37.16	132	A-PILLAR
			32.43	38.63	111	B-PILLAR
			26.09	33.32	161	ROOF & ITS COMPONENTS
			17.74	30.56	62	OTHER
NEARSIDE IMPACT	NO BELT	NECK	40.00	45.80	15	SIDE INTERIOR SURFACE
			33.71	37.83	1771	average injured occupant
			31.58	41.21	19	B-PILLAR
			26.09	39.35	46	EXTERIOR
			18.42	35.55	38	ROOF & ITS COMPONENTS
NEARSIDE IMPACT	NO BELT	TORSO	45.45	38.03	33	SEAT & ITS COMPONENTS
			** 43.46	40.47	306	SIDE HARDWARE OR ARMREST
			41.18	40.35	17	A-PILLAR
			37.61	40.42	694	SIDE INTERIOR SURFACE
			37.50	37.63	48	FLOOR & ITS COMPONENTS
			35.00	37.99	160	STEERING ASSEMBLY
			33.71	37.83	1771	average injured occupant
			33.33	39.54	81	B-PILLAR
			26.88	35.85	93	EXTERIOR
			21.05	31.79	19	OTHER OCCS, SELF, CARGO
			10.00	40.50	20	INSTRUMENT PANEL
NEARSIDE IMPACT	NO BELT	ARM	48.57	40.71	35	STEERING ASSEMBLY
			36.13	37.97	119	SIDE INTERIOR SURFACE
			33.71	37.83	1771	average injured occupant
			32.56	32.47	43	EXTERIOR
			26.09	39.61	23	INSTRUMENT PANEL
NEARSIDE IMPACT	NO BELT	LEG	33.78	41.34	74	FLOOR & ITS COMPONENTS
			33.71	37.83	1771	average injured occupant
			33.58	34.78	134	SIDE INTERIOR SURFACE
			25.30	36.89	83	INSTRUMENT PANEL
			22.64	36.79	53	SIDE HARDWARE OR ARMREST
			22.58	34.55	31	EXTERIOR
FAR-SIDE IMPACT	BELTED	HEAD	64.71	34.53	17	AIR BAG
			52.38	42.65	63	STEERING ASSEMBLY
			49.58	37.09	119	SIDE INTERIOR SURFACE
			48.48	43.12	33	INSTRUMENT PANEL
			47.40	43.05	1310	average injured occupant
			45.28	40.62	53	B-PILLAR
			42.86	45.47	49	OTHER FRONTAL COMPONENT
			41.86	36.26	43	NON-CONTACT
			41.67	37.96	24	OTHER SIDE COMPONENT
			38.89	47.72	18	EXTERIOR
			36.11	38.33	36	OTHER OCCS, SELF, CARGO
			28.89	36.16	90	ROOF & ITS COMPONENTS
			28.57	39.14	91	SEAT & ITS COMPONENTS
FAR-SIDE IMPACT	BELTED	NECK	60.00	42.32	25	NON-CONTACT
			56.00	37.92	25	SIDE INTERIOR SURFACE
			47.40	43.05	1310	average injured occupant
			29.63	37.07	27	ROOF & ITS COMPONENTS
FAR-SIDE IMPACT	BELTED	TORSO	* 55.17	51.16	203	BELT SYSTEM
			52.71	44.53	129	FLOOR & ITS COMPONENTS
			51.79	53.00	56	STEERING ASSEMBLY
			50.00	52.58	24	INSTRUMENT PANEL
			49.60	40.70	125	SIDE INTERIOR SURFACE
			47.40	43.05	1310	average injured occupant
			44.44	47.50	18	NON-CONTACT
			43.59	49.87	39	OTHER
			39.47	39.39	38	SIDE HARDWARE OR ARMREST
			39.00	39.82	100	SEAT & ITS COMPONENTS
			30.00	35.80	20	B-PILLAR
			26.83	38.78	41	OTHER OCCS, SELF, CARGO
FAR-SIDE IMPACT	BELTED	ARM	** 72.00	50.94	50	STEERING ASSEMBLY
			55.26	51.08	38	INSTRUMENT PANEL
			48.78	36.66	41	SIDE INTERIOR SURFACE
			47.40	43.05	1310	average injured occupant

**3+ standard deviations higher than average injured occupant (p < .002).
*1.96-3 standard deviations higher than average injured occupant (p < .05).

288

IMPACT TYPE	OCCUPANT PROTECTION	BODY REGION	PERCENT FEMALE	AVERAGE AGE	N	INJURY SOURCE
FAR-SIDE IMPACT	BELTED	LEG	* 64.20	51.51	81	FLOOR & ITS COMPONENTS
			58.02	47.25	81	INSTRUMENT PANEL
			47.40	43.05	1310	average injured occupant
FAR-SIDE IMPACT	NO BELT	HEAD	38.46	41.42	52	OTHER SIDE COMPONENT
			36.24	38.99	149	SIDE INTERIOR SURFACE
			35.53	39.65	197	OTHER FRONTAL COMPONENT
			33.33	44.17	18	SIDE HARDWARE OR ARMREST
			32.65	35.78	49	B-PILLAR
			32.08	42.94	53	INSTRUMENT PANEL
			31.21	38.60	1535	average injured occupant
			30.00	34.07	30	STEERING ASSEMBLY
			29.59	42.37	98	A-PILLAR
			24.14	35.54	145	ROOF & ITS COMPONENTS
			22.22	34.83	144	EXTERIOR
			16.13	33.74	31	OTHER
FAR-SIDE IMPACT	NO BELT	NECK	36.36	46.84	44	SIDE INTERIOR SURFACE
			31.21	38.60	1535	average injured occupant
			30.95	41.33	42	OTHER FRONTAL COMPONENT
			23.81	51.57	21	A-PILLAR
			22.22	37.24	45	EXTERIOR
			18.42	33.74	38	ROOF & ITS COMPONENTS
FAR-SIDE IMPACT	NO BELT	TORSO	37.04	38.96	27	OTHER OCCS, SELF, CARGO
			32.75	40.77	284	SIDE INTERIOR SURFACE
			31.21	38.60	1535	average injured occupant
			30.59	40.38	85	STEERING ASSEMBLY
			29.85	36.31	67	FLOOR & ITS COMPONENTS
			29.59	34.81	98	EXTERIOR
			28.07	42.27	114	INSTRUMENT PANEL
			26.09	44.43	23	OTHER
			25.42	39.46	59	SEAT & ITS COMPONENTS
			25.42	36.39	59	SIDE HARDWARE OR ARMREST
			13.33	44.60	15	B-PILLAR
			10.00	33.00	20	ROOF & ITS COMPONENTS
FAR-SIDE IMPACT	NO BELT	ARM	* 51.85	38.67	27	STEERING ASSEMBLY
			38.30	45.94	47	INSTRUMENT PANEL
			36.54	41.40	52	SIDE INTERIOR SURFACE
			31.21	38.60	1535	average injured occupant
			25.71	38.31	35	EXTERIOR
FAR-SIDE IMPACT	NO BELT	LEG	* 50.00	47.09	56	FLOOR & ITS COMPONENTS
			* 46.27	43.13	67	INSTRUMENT PANEL
			31.21	38.60	1535	average injured occupant
			27.27	34.64	22	EXTERIOR
			21.74	34.52	23	SIDE INTERIOR SURFACE
ROLLOVER FIRST/WORST EVENT	BELTED	HEAD	50.00	36.77	22	STEERING ASSEMBLY
			50.00	29.50	18	NON-CONTACT
			48.15	38.70	27	B-PILLAR
			46.15	34.62	26	OTHER FRONTAL COMPONENT
			38.24	35.35	34	EXTERIOR
			37.75	38.00	1224	average injured occupant
			36.00	37.62	400	ROOF & ITS COMPONENTS
ROLLOVER FIRST/WORST EVENT	BELTED	NECK	37.75	38.00	1224	average injured occupant
			27.59	42.01	145	ROOF & ITS COMPONENTS
ROLLOVER FIRST/WORST EVENT	BELTED	TORSO	45.65	42.70	46	SIDE HARDWARE OR ARMREST
			41.94	39.17	124	SIDE INTERIOR SURFACE
			38.94	42.15	113	BELT SYSTEM
			38.46	42.83	65	SEAT & ITS COMPONENTS
			37.75	38.00	1224	average injured occupant
			30.88	39.29	68	ROOF & ITS COMPONENTS
			26.32	41.37	19	FLOOR & ITS COMPONENTS
			26.32	36.07	76	STEERING ASSEMBLY
			26.09	43.59	46	B-PILLAR
ROLLOVER FIRST/WORST EVENT	BELTED	ARM	* 50.68	41.21	73	ROOF & ITS COMPONENTS
			47.50	39.08	40	SIDE INTERIOR SURFACE
			45.00	28.25	20	STEERING ASSEMBLY
			43.94	39.86	66	EXTERIOR
			42.11	45.21	19	INSTRUMENT PANEL
			37.75	38.00	1224	average injured occupant

3+ standard deviations higher than average injured occupant (p < .002).
*1.96-3 standard deviations higher than average injured occupant (p < .05).

IMPACT TYPE	OCCUPANT PROTECTION	BODY REGION	PERCENT FEMALE	AVERAGE AGE	N	INJURY SOURCE
ROLLOVER FIRST/WORST EVENT	BELTED	LEG	52.50	35.83	40	FLOOR & ITS COMPONENTS
			47.06	34.24	17	STEERING ASSEMBLY
			39.22	35.41	51	INSTRUMENT PANEL
			37.75	38.00	1224	average injured occupant
ROLLOVER FIRST/WORST EVENT	NO BELT	HEAD	30.00	27.00	20	STEERING ASSEMBLY
			29.41	32.82	34	A-PILLAR
			29.41	31.88	17	OTHER SIDE COMPONENT
			26.70	31.66	442	EXTERIOR
			26.25	30.86	80	OTHER FRONTAL COMPONENT
			24.02	32.53	1815	average injured occupant
			17.65	32.04	306	ROOF & ITS COMPONENTS
ROLLOVER FIRST/WORST EVENT	NO BELT	NECK	* 35.38	33.06	130	EXTERIOR
			24.02	32.53	1815	average injured occupant
			18.18	35.43	99	ROOF & ITS COMPONENTS
ROLLOVER FIRST/WORST EVENT	NO BELT	TORSO	* 40.74	33.11	27	SIDE HARDWARE OR ARMREST
			* 29.22	31.46	503	EXTERIOR
			27.78	38.61	18	FLOOR & ITS COMPONENTS
			26.56	32.27	128	SIDE INTERIOR SURFACE
			24.02	32.53	1815	average injured occupant
			20.83	34.25	24	INSTRUMENT PANEL
			17.65	33.86	119	ROOF & ITS COMPONENTS
			17.56	35.47	131	STEERING ASSEMBLY
			9.38	39.19	32	SEAT & ITS COMPONENTS
			6.25	29.19	16	OTHER SIDE COMPONENT
ROLLOVER FIRST/WORST EVENT	NO BELT	ARM	42.11	38.53	19	INSTRUMENT PANEL
			36.84	31.74	19	SIDE INTERIOR SURFACE
			29.92	32.80	127	EXTERIOR
			24.02	32.53	1815	average injured occupant
			20.59	32.62	34	ROOF & ITS COMPONENTS
ROLLOVER FIRST/WORST EVENT	NO BELT	LEG	** 38.89	32.77	122	EXTERIOR
			31.71	33.37	41	FLOOR & ITS COMPONENTS
			26.09	33.52	46	INSTRUMENT PANEL
			24.02	32.53	1815	average injured occupant
			6.67	26.80	15	STEERING ASSEMBLY
REAR & OTHER	BELTED	HEAD	49.37	43.41	79	SEAT & ITS COMPONENTS
			47.22	44.81	36	B-PILLAR
			44.63	42.43	475	average injured occupant
			40.00	39.15	20	NON-CONTACT
			38.89	41.56	36	STEERING ASSEMBLY
			31.25	46.06	16	OTHER
			27.27	40.39	33	ROOF & ITS COMPONENTS
REAR & OTHER	BELTED	NECK	44.63	42.43	475	average injured occupant
			34.29	47.57	35	SEAT & ITS COMPONENTS
REAR & OTHER	BELTED	TORSO	50.53	45.27	95	SEAT & ITS COMPONENTS
			48.39	49.03	31	BELT SYSTEM
			46.67	38.27	15	SIDE INTERIOR SURFACE
			44.63	42.43	475	average injured occupant
			37.93	44.41	29	STEERING ASSEMBLY
			30.00	43.30	20	B-PILLAR
REAR & OTHER	BELTED	LEG	* 63.33	40.83	30	INSTRUMENT PANEL
			55.00	43.10	20	FLOOR & ITS COMPONENTS
			44.63	42.43	475	average injured occupant
REAR & OTHER	NO BELT	HEAD	42.86	38.89	35	SEAT & ITS COMPONENTS
			28.51	36.55	242	average injured occupant
			18.75	40.94	16	B-PILLAR
			13.79	30.28	29	ROOF & ITS COMPONENTS
			13.64	38.27	22	EXTERIOR
REAR & OTHER	NO BELT	NECK	28.51	36.55	242	average injured occupant
			20.00	40.80	20	SEAT & ITS COMPONENTS
REAR & OTHER	NO BELT	TORSO	28.51	36.55	242	average injured occupant
			21.74	39.57	23	EXTERIOR
			21.43	40.12	42	SEAT & ITS COMPONENTS
			14.29	36.54	28	STEERING ASSEMBLY

**3+ standard deviations higher than average injured occupant (p < .002).
*1.96-3 standard deviations higher than average injured occupant (p < .05).

TABLE E-11: PERCENT FEMALE, DRIVERS AGE 18-96 WITH AIS 3+ INJURIES, CDS 1988-2010

IMPACT TYPE	OCCUPANT PROTECTION	BODY REGION	PERCENT FEMALE	AVERAGE AGE	N	INJURY SOURCE
FRONTAL	BELTED, AIR BAG	HEAD	60.00	50.47	15	INSTRUMENT PANEL
			54.35	39.04	46	A-PILLAR
			50.91	43.30	110	STEERING ASSEMBLY
			48.73	44.45	2048	average injured occupant
			46.67	46.87	30	NON-CONTACT
			45.45	54.91	11	AIR BAG
			33.33	43.43	30	EXTERIOR
			31.65	40.96	79	ROOF & ITS COMPONENTS
			28.57	43.57	14	OTHER FRONTAL COMPONENT
			27.27	39.23	22	B-PILLAR
			15.79	36.89	19	OTHER
FRONTAL	BELTED, AIR BAG	NECK	* 76.19	53.33	21	NON-CONTACT
			48.73	44.45	2048	average injured occupant
			41.67	50.42	24	ROOF & ITS COMPONENTS
			40.00	50.85	20	STEERING ASSEMBLY
FRONTAL	BELTED, AIR BAG	TORSO	* 69.70	64.45	33	AIR BAG
			50.20	47.17	245	BELT SYSTEM
			48.73	44.45	2048	average injured occupant
			47.73	47.66	44	SIDE HARDWARE OR ARMREST
			47.44	44.76	78	SIDE INTERIOR SURFACE
			46.36	44.97	110	INSTRUMENT PANEL
			43.48	39.57	23	FLOOR & ITS COMPONENTS
			42.57	46.26	343	STEERING ASSEMBLY
			38.46	52.23	13	NON-CONTACT
			36.11	44.22	36	SEAT & ITS COMPONENTS
FRONTAL	BELTED, AIR BAG	ARM	** 86.84	49.25	78	AIR BAG
			** 71.59	44.32	88	STEERING ASSEMBLY
			* 67.57	43.68	37	A-PILLAR
			60.00	47.64	25	OTHER FRONTAL COMPONENT
			57.14	43.60	35	ROOF & ITS COMPONENTS
			53.44	45.97	131	INSTRUMENT PANEL
			48.73	44.45	2048	average injured occupant
			46.58	41.05	73	SIDE INTERIOR SURFACE
			33.33	45.33	15	EXTERIOR
FRONTAL	BELTED, AIR BAG	LEG	** 64.45	45.37	256	FLOOR & ITS COMPONENTS
			48.73	44.45	2048	average injured occupant
			45.16	42.07	589	INSTRUMENT PANEL
			45.00	43.20	20	SIDE HARDWARE OR ARMREST
			32.00	33.40	50	STEERING ASSEMBLY
			23.68	39.32	38	SIDE INTERIOR SURFACE
FRONTAL	BELTED, NO AIR BAG	HEAD	42.55	41.11	47	A-PILLAR
			42.13	41.66	216	STEERING ASSEMBLY
			41.97	43.81	1413	average injured occupant
			38.89	38.39	18	EXTERIOR
			38.46	41.48	52	OTHER FRONTAL COMPONENT
			28.13	39.69	32	ROOF & ITS COMPONENTS
			15.00	39.50	20	INSTRUMENT PANEL
FRONTAL	BELTED, NO AIR BAG	NECK	48.15	50.22	27	STEERING ASSEMBLY
			41.97	43.81	1413	average injured occupant
FRONTAL	BELTED, NO AIR BAG	TORSO	50.00	39.25	12	SIDE HARDWARE OR ARMREST
			43.75	48.14	144	BELT SYSTEM
			41.97	43.81	1413	average injured occupant
			38.07	47.44	415	STEERING ASSEMBLY
			30.00	39.80	10	SEAT & ITS COMPONENTS
			29.17	42.58	48	INSTRUMENT PANEL
			25.71	44.74	35	SIDE INTERIOR SURFACE
FRONTAL	BELTED, NO AIR BAG	ARM	* 54.22	46.57	83	STEERING ASSEMBLY
			47.13	41.67	87	INSTRUMENT PANEL
			42.11	43.26	19	SIDE INTERIOR SURFACE
			41.97	43.81	1413	average injured occupant
FRONTAL	BELTED, NO AIR BAG	LEG	** 59.09	42.50	132	FLOOR & ITS COMPONENTS
			41.97	43.81	1413	average injured occupant
			39.62	39.78	318	INSTRUMENT PANEL
			30.43	37.48	23	SIDE INTERIOR SURFACE
			28.21	36.77	39	STEERING ASSEMBLY

**3+ standard deviations higher than average injured occupant (p < .002).
*1.96-3 standard deviations higher than average injured occupant (p < .05).

291

IMPACT TYPE	OCCUPANT PROTECTION	BODY REGION	PERCENT FEMALE	AVERAGE AGE	N	INJURY SOURCE
FRONTAL	NO BELT, AIR BAG	HEAD	* 63.64	49.55	11	AIR BAG
			33.33	31.52	21	OTHER
			29.71	38.15	1518	average injured occupant
			25.00	37.25	24	INSTRUMENT PANEL
			23.73	37.85	59	STEERING ASSEMBLY
			18.56	36.37	97	ROOF & ITS COMPONENTS
			18.52	39.44	81	OTHER FRONTAL COMPONENT
			16.05	31.99	81	EXTERIOR
			15.87	41.11	63	A-PILLAR
FRONTAL	NO BELT, AIR BAG	NECK	50.00	40.50	16	STEERING ASSEMBLY
			29.71	38.15	1518	average injured occupant
			20.00	41.00	20	OTHER FRONTAL COMPONENT
			17.39	33.61	23	EXTERIOR
			14.29	39.71	14	A-PILLAR
			10.53	37.68	19	ROOF & ITS COMPONENTS
FRONTAL	NO BELT, AIR BAG	TORSO	* 52.63	47.11	19	AIR BAG
			32.06	40.58	496	STEERING ASSEMBLY
			30.00	41.60	10	FLOOR & ITS COMPONENTS
			29.71	38.15	1518	average injured occupant
			28.33	36.98	60	SIDE INTERIOR SURFACE
			26.36	38.97	129	INSTRUMENT PANEL
			23.21	33.38	56	EXTERIOR
			20.00	37.40	15	SIDE HARDWARE OR ARMREST
			16.67	37.25	12	SEAT & ITS COMPONENTS
FRONTAL	NO BELT, AIR BAG	ARM	* 63.64	41.09	11	OTHER FRONTAL COMPONENT
			* 55.56	42.06	18	SIDE INTERIOR SURFACE
			42.86	36.71	14	A-PILLAR
			36.11	34.22	36	STEERING ASSEMBLY
			30.00	38.50	10	OTHER
			29.71	38.15	1518	average injured occupant
			22.97	40.08	74	INSTRUMENT PANEL
			21.43	33.00	14	EXTERIOR
FRONTAL	NO BELT, AIR BAG	LEG	** 47.86	40.55	117	FLOOR & ITS COMPONENTS
			29.71	38.15	1518	average injured occupant
			28.76	36.75	372	INSTRUMENT PANEL
			27.72	36.38	101	STEERING ASSEMBLY
			16.67	32.33	12	EXTERIOR
			12.50	31.13	16	SIDE INTERIOR SURFACE
FRONTAL	NO BELT, NO AIR BAG	HEAD	* 42.19	39.72	64	INSTRUMENT PANEL
			29.07	39.22	172	STEERING ASSEMBLY
			28.26	39.66	2788	average injured occupant
			27.74	39.58	310	OTHER FRONTAL COMPONENT
			27.62	35.04	105	A-PILLAR
			25.00	38.13	16	OTHER SIDE COMPONENT
			19.78	40.41	91	ROOF & ITS COMPONENTS
			18.64	36.47	118	EXTERIOR
			13.33	30.07	15	OTHER
FRONTAL	NO BELT, NO AIR BAG	NECK	37.50	46.52	48	OTHER FRONTAL COMPONENT
			33.33	39.67	39	STEERING ASSEMBLY
			30.00	31.30	10	A-PILLAR
			28.26	39.66	2788	average injured occupant
			25.00	37.92	12	EXTERIOR
			14.81	41.22	27	ROOF & ITS COMPONENTS
FRONTAL	NO BELT, NO AIR BAG	TORSO	35.00	48.10	20	FLOOR & ITS COMPONENTS
			34.62	33.10	52	EXTERIOR
			29.51	43.13	61	SIDE INTERIOR SURFACE
			28.26	39.66	2788	average injured occupant
			25.29	42.83	1103	STEERING ASSEMBLY
			21.88	40.83	192	INSTRUMENT PANEL
			21.43	38.43	14	SEAT & ITS COMPONENTS
			18.18	37.55	11	OTHER FRONTAL COMPONENT
			0.00	32.30	10	ROOF & ITS COMPONENTS
FRONTAL	NO BELT, NO AIR BAG	ARM	** 42.27	41.05	97	STEERING ASSEMBLY
			* 37.40	40.82	131	INSTRUMENT PANEL
			36.36	44.27	11	A-PILLAR
			28.26	39.66	2788	average injured occupant
			26.92	37.58	26	SIDE INTERIOR SURFACE
			14.29	30.93	14	EXTERIOR

**3+ standard deviations higher than average injured occupant (p < .002).
*1.96-3 standard deviations higher than average injured occupant (p < .05).

292

IMPACT TYPE	OCCUPANT PROTECTION	BODY REGION	PERCENT FEMALE	AVERAGE AGE	N	INJURY SOURCE
FRONTAL	NO BELT, NO AIR BAG	LEG	** 47.19	37.66	178	FLOOR & ITS COMPONENTS
			40.00	31.07	15	EXTERIOR
			31.97	36.97	122	STEERING ASSEMBLY
			28.41	37.04	542	INSTRUMENT PANEL
			28.26	39.66	2788	average injured occupant
			9.68	41.06	31	SIDE INTERIOR SURFACE
FRONTAL	NO BELT, NO AIR BAG	BURN	28.26	39.66	2788	average injured occupant
			20.00	30.70	10	FIRE
NEARSIDE IMPACT	BELTED	HEAD	* 63.64	43.61	33	SIDE INTERIOR SURFACE
			57.38	45.08	61	A-PILLAR
			53.85	45.04	26	NON-CONTACT
			52.31	43.64	130	EXTERIOR
			50.00	56.00	14	STEERING ASSEMBLY
			46.55	44.51	174	B-PILLAR
			46.34	44.77	1830	average injured occupant
			43.59	48.41	39	OTHER SIDE COMPONENT
			31.52	38.64	92	ROOF & ITS COMPONENTS
			28.33	34.77	60	OTHER
NEARSIDE IMPACT	BELTED	NECK	50.00	53.04	26	B-PILLAR
			50.00	49.25	16	NON-CONTACT
			46.34	44.77	1830	average injured occupant
			33.33	40.92	12	EXTERIOR
			31.25	44.88	16	ROOF & ITS COMPONENTS
NEARSIDE IMPACT	BELTED	TORSO	61.54	43.15	13	INSTRUMENT PANEL
			57.69	37.27	26	SEAT & ITS COMPONENTS
			55.88	43.03	34	FLOOR & ITS COMPONENTS
			* 52.66	45.84	395	SIDE HARDWARE OR ARMREST
			46.34	44.77	1830	average injured occupant
			46.15	45.66	65	STEERING ASSEMBLY
			45.94	46.46	849	SIDE INTERIOR SURFACE
			41.05	45.79	95	B-PILLAR
			35.56	43.89	45	BELT SYSTEM
NEARSIDE IMPACT	BELTED	ARM	54.55	46.27	22	STEERING ASSEMBLY
			52.00	47.44	25	INSTRUMENT PANEL
			46.34	44.77	1830	average injured occupant
			42.86	42.79	14	SIDE HARDWARE OR ARMREST
			40.54	40.43	74	SIDE INTERIOR SURFACE
NEARSIDE IMPACT	BELTED	LEG	50.82	41.97	61	INSTRUMENT PANEL
			49.33	38.47	75	SIDE HARDWARE OR ARMREST
			46.34	44.77	1830	average injured occupant
			35.29	41.53	17	FLOOR & ITS COMPONENTS
			35.11	39.59	94	SIDE INTERIOR SURFACE
NEARSIDE IMPACT	NO BELT	HEAD	** 75.00	44.31	16	SIDE INTERIOR SURFACE
			41.67	40.33	24	OTHER SIDE COMPONENT
			37.50	39.80	56	B-PILLAR
			37.01	36.00	154	EXTERIOR
			31.95	38.39	1280	average injured occupant
			29.63	36.81	27	OTHER FRONTAL COMPONENT
			28.38	36.58	74	A-PILLAR
			27.17	32.74	92	ROOF & ITS COMPONENTS
			16.98	30.09	53	OTHER
NEARSIDE IMPACT	NO BELT	NECK	31.95	38.39	1280	average injured occupant
			22.22	40.44	18	EXTERIOR
			11.11	41.44	18	ROOF & ITS COMPONENTS
NEARSIDE IMPACT	NO BELT	TORSO	47.37	39.16	19	FLOOR & ITS COMPONENTS
			* 39.88	40.53	173	SIDE HARDWARE OR ARMREST
			36.05	40.79	516	SIDE INTERIOR SURFACE
			34.88	42.53	43	B-PILLAR
			31.95	38.39	1280	average injured occupant
			29.79	34.63	94	STEERING ASSEMBLY
			29.23	36.54	65	EXTERIOR
			0.00	38.50	10	INSTRUMENT PANEL
NEARSIDE IMPACT	NO BELT	ARM	* 58.82	43.35	17	STEERING ASSEMBLY
			40.00	30.00	15	EXTERIOR
			31.95	38.39	1280	average injured occupant
			31.11	38.44	45	SIDE INTERIOR SURFACE

**3+ standard deviations higher than average injured occupant (p < .002).
*1.96-3 standard deviations higher than average injured occupant (p < .05).

293

IMPACT TYPE	OCCUPANT PROTECTION	BODY REGION	PERCENT FEMALE	AVERAGE AGE	N	INJURY SOURCE
NEARSIDE IMPACT	NO BELT	LEG	33.33	32.96	90	SIDE INTERIOR SURFACE
			31.95	38.39	1280	average injured occupant
			21.05	41.53	19	FLOOR & ITS COMPONENTS
			18.75	37.06	48	INSTRUMENT PANEL
			15.91	35.52	44	SIDE HARDWARE OR ARMREST
			15.38	34.31	26	EXTERIOR
FAR-SIDE IMPACT	BELTED	HEAD	60.00	39.60	10	OTHER SIDE COMPONENT
			48.72	40.45	78	SIDE INTERIOR SURFACE
			42.75	44.05	648	average injured occupant
			41.94	41.23	31	B-PILLAR
			35.71	53.64	14	EXTERIOR
			33.33	48.13	15	INSTRUMENT PANEL
			30.00	46.60	10	OTHER OCCS, SELF, CARGO
			30.00	33.70	10	NON-CONTACT
			28.00	37.52	50	ROOF & ITS COMPONENTS
FAR-SIDE IMPACT	BELTED	NECK	42.75	44.05	648	average injured occupant
			18.18	34.09	11	ROOF & ITS COMPONENTS
FAR-SIDE IMPACT	BELTED	TORSO	55.56	51.11	27	STEERING ASSEMBLY
			50.00	44.80	10	INSTRUMENT PANEL
			47.67	39.38	86	SIDE INTERIOR SURFACE
			47.06	46.00	68	FLOOR & ITS COMPONENTS
			44.87	55.05	78	BELT SYSTEM
			42.75	44.05	648	average injured occupant
			42.11	40.95	19	SIDE HARDWARE OR ARMREST
			38.10	51.00	21	OTHER
			34.55	36.56	55	SEAT & ITS COMPONENTS
			28.57	37.75	28	OTHER OCCS, SELF, CARGO
FAR-SIDE IMPACT	BELTED	ARM	** 83.33	54.08	18	STEERING ASSEMBLY
			* 75.00	38.75	16	SIDE INTERIOR SURFACE
			42.75	44.05	648	average injured occupant
			38.46	47.38	13	INSTRUMENT PANEL
FAR-SIDE IMPACT	BELTED	LEG	* 66.67	53.33	18	FLOOR & ITS COMPONENTS
			42.75	44.05	648	average injured occupant
			42.31	50.54	26	INSTRUMENT PANEL
FAR-SIDE IMPACT	NO BELT	HEAD	37.08	40.15	89	SIDE INTERIOR SURFACE
			33.33	35.30	27	B-PILLAR
			29.25	39.74	988	average injured occupant
			29.03	41.76	62	A-PILLAR
			28.81	41.83	59	OTHER FRONTAL COMPONENT
			28.00	37.16	25	OTHER SIDE COMPONENT
			22.73	51.14	22	INSTRUMENT PANEL
			22.45	34.81	98	EXTERIOR
			21.11	36.61	90	ROOF & ITS COMPONENTS
			18.18	30.09	11	STEERING ASSEMBLY
			13.64	32.59	22	OTHER
FAR-SIDE IMPACT	NO BELT	NECK	31.82	46.41	22	OTHER FRONTAL COMPONENT
			29.25	39.74	988	average injured occupant
			27.78	43.61	18	SIDE INTERIOR SURFACE
			25.00	45.42	12	A-PILLAR
			25.00	36.35	20	EXTERIOR
			18.18	35.05	22	ROOF & ITS COMPONENTS
FAR-SIDE IMPACT	NO BELT	TORSO	35.94	35.09	64	EXTERIOR
			35.71	42.07	14	OTHER OCCS, SELF, CARGO
			32.65	39.12	49	STEERING ASSEMBLY
			30.77	48.23	13	OTHER
			29.90	41.12	204	SIDE INTERIOR SURFACE
			29.25	39.74	988	average injured occupant
			27.03	44.74	74	INSTRUMENT PANEL
			25.71	37.97	35	SEAT & ITS COMPONENTS
			24.14	40.34	29	FLOOR & ITS COMPONENTS
			24.14	38.83	29	SIDE HARDWARE OR ARMREST
FAR-SIDE IMPACT	NO BELT	ARM	42.86	39.95	21	SIDE INTERIOR SURFACE
			40.00	43.67	15	INSTRUMENT PANEL
			33.33	39.13	15	EXTERIOR
			29.25	39.74	988	average injured occupant

**3+ standard deviations higher than average injured occupant (p < .002).
*1.96-3 standard deviations higher than average injured occupant (p < .05).

IMPACT TYPE	OCCUPANT PROTECTION	BODY REGION	PERCENT FEMALE	AVERAGE AGE	N	INJURY SOURCE
FAR-SIDE IMPACT	NO BELT	LEG	50.00	33.50	12	EXTERIOR
			42.86	43.66	35	INSTRUMENT PANEL
			41.67	52.33	12	STEERING ASSEMBLY
			29.25	39.74	988	average injured occupant
			26.67	43.53	15	FLOOR & ITS COMPONENTS
			16.67	32.22	18	SIDE INTERIOR SURFACE
ROLLOVER FIRST/WORST EVENT	BELTED	HEAD	45.45	39.55	11	B-PILLAR
			37.64	37.32	178	ROOF & ITS COMPONENTS
			36.67	36.10	30	EXTERIOR
			36.32	39.13	669	average injured occupant
ROLLOVER FIRST/WORST EVENT	BELTED	NECK	36.32	39.13	669	average injured occupant
			29.27	41.94	82	ROOF & ITS COMPONENTS
ROLLOVER FIRST/WORST EVENT	BELTED	TORSO	42.86	41.62	21	SEAT & ITS COMPONENTS
			41.76	38.56	91	SIDE INTERIOR SURFACE
			40.00	44.60	10	FLOOR & ITS COMPONENTS
			37.04	41.59	27	SIDE HARDWARE OR ARMREST
			37.04	35.07	27	BELT SYSTEM
			36.32	39.13	669	average injured occupant
			29.17	35.83	24	ROOF & ITS COMPONENTS
			27.27	48.55	11	B-PILLAR
			22.45	35.45	49	STEERING ASSEMBLY
ROLLOVER FIRST/WORST EVENT	BELTED	ARM	54.55	44.36	22	SIDE INTERIOR SURFACE
			50.00	51.00	12	INSTRUMENT PANEL
			48.28	40.97	29	ROOF & ITS COMPONENTS
			42.42	41.79	33	EXTERIOR
			36.32	39.13	669	average injured occupant
ROLLOVER FIRST/WORST EVENT	BELTED	LEG	36.32	39.13	669	average injured occupant
			33.33	38.10	21	INSTRUMENT PANEL
			30.00	35.50	10	FLOOR & ITS COMPONENTS
ROLLOVER FIRST/WORST EVENT	NO BELT	HEAD	36.36	30.14	22	OTHER FRONTAL COMPONENT
			23.99	32.37	296	EXTERIOR
			23.67	33.20	1293	average injured occupant
			20.00	34.00	15	A-PILLAR
			18.98	33.18	137	ROOF & ITS COMPONENTS
ROLLOVER FIRST/WORST EVENT	NO BELT	NECK	26.79	33.16	56	EXTERIOR
			23.67	33.20	1293	average injured occupant
			13.95	35.53	43	ROOF & ITS COMPONENTS
ROLLOVER FIRST/WORST EVENT	NO BELT	TORSO	30.26	32.18	76	SIDE INTERIOR SURFACE
			28.53	32.43	354	EXTERIOR
			23.67	33.20	1293	average injured occupant
			17.50	34.65	40	ROOF & ITS COMPONENTS
			16.25	35.61	80	STEERING ASSEMBLY
			9.09	35.18	11	INSTRUMENT PANEL
			0.00	41.11	18	SEAT & ITS COMPONENTS
ROLLOVER FIRST/WORST EVENT	NO BELT	ARM	31.91	33.17	47	EXTERIOR
			23.67	33.20	1293	average injured occupant
			13.33	31.80	15	ROOF & ITS COMPONENTS
ROLLOVER FIRST/WORST EVENT	NO BELT	LEG	* 36.00	33.07	75	EXTERIOR
			23.67	33.20	1293	average injured occupant
			16.67	34.92	24	INSTRUMENT PANEL
			10.00	26.80	10	STEERING ASSEMBLY
REAR & OTHER	BELTED	HEAD	54.55	56.00	11	B-PILLAR
			46.88	45.44	32	SEAT & ITS COMPONENTS
			45.74	44.24	188	average injured occupant
			40.00	39.07	15	ROOF & ITS COMPONENTS
REAR & OTHER	BELTED	NECK	45.74	44.24	188	average injured occupant
			30.77	54.62	13	SEAT & ITS COMPONENTS
REAR & OTHER	BELTED	TORSO	57.69	46.58	52	SEAT & ITS COMPONENTS
			45.74	44.24	188	average injured occupant
			38.89	43.33	18	STEERING ASSEMBLY
REAR & OTHER	NO BELT	HEAD	26.32	40.89	19	SEAT & ITS COMPONENTS
			20.00	40.10	10	B-PILLAR
			17.48	36.65	143	average injured occupant
			8.33	31.42	12	EXTERIOR
			6.25	28.69	16	ROOF & ITS COMPONENTS

3+ standard deviations higher than average injured occupant (p < .002).
*1.96-3 standard deviations higher than average injured occupant (p < .05).

IMPACT TYPE	OCCUPANT PROTECTION	BODY REGION	PERCENT FEMALE	AVERAGE AGE	N	INJURY SOURCE
REAR & OTHER	NO BELT	NECK	23.08	36.62	13	SEAT & ITS COMPONENTS
			17.48	36.65	143	average injured occupant
REAR & OTHER	NO BELT	TORSO	21.43	43.14	14	EXTERIOR
			20.00	38.00	10	SIDE INTERIOR SURFACE
			17.48	36.65	143	average injured occupant
			13.64	39.45	22	SEAT & ITS COMPONENTS
			9.52	35.86	21	STEERING ASSEMBLY

**3+ standard deviations higher than average injured occupant (p < .002).
*1.96-3 standard deviations higher than average injured occupant (p < .05).

TABLE E-12: PERCENT FEMALE, DRIVERS AGE 18-96 WITH AIS 4+ INJURIES, CDS 1988-2010

IMPACT TYPE	OCCUPANT PROTECTION	BODY REGION	PERCENT FEMALE	AVERAGE AGE	N	INJURY SOURCE
FRONTAL	BELTED, AIR BAG	HEAD	* 62.96	36.81	27	A-PILLAR
			* 57.63	43.41	59	STEERING ASSEMBLY
			47.37	49.47	19	NON-CONTACT
			44.44	49.00	9	INSTRUMENT PANEL
			42.86	59.57	7	AIR BAG
			40.68	47.79	649	average injured occupant
			40.00	46.60	5	SEAT & ITS COMPONENTS
			37.50	45.63	8	OTHER FRONTAL COMPONENT
			29.63	39.87	54	ROOF & ITS COMPONENTS
			28.00	44.00	25	EXTERIOR
			26.67	40.93	15	B-PILLAR
			20.00	47.20	5	OTHER OCCS, SELF, CARGO
			7.69	34.46	13	OTHER
FRONTAL	BELTED, AIR BAG	NECK	42.86	47.29	7	ROOF & ITS COMPONENTS
			40.68	47.79	649	average injured occupant
			33.33	54.56	9	STEERING ASSEMBLY
FRONTAL	BELTED, AIR BAG	TORSO	* 68.75	71.00	16	AIR BAG
			50.68	50.40	73	BELT SYSTEM
			50.00	44.00	28	SIDE INTERIOR SURFACE
			50.00	37.00	6	INSTRUMENT PANEL
			43.01	47.11	186	STEERING ASSEMBLY
			40.68	47.79	649	average injured occupant
			35.71	55.71	14	SIDE HARDWARE OR ARMREST
			20.00	40.80	5	NON-CONTACT
FRONTAL	BELTED, AIR BAG	BURN	40.68	47.79	649	average injured occupant
			0.00	45.86	7	FIRE
FRONTAL	BELTED, NO AIR BAG	HEAD	41.18	42.62	34	A-PILLAR
			40.00	42.60	5	B-PILLAR
			38.46	37.92	13	EXTERIOR
			37.73	46.05	546	average injured occupant
			35.29	42.54	85	STEERING ASSEMBLY
			33.33	42.95	21	OTHER FRONTAL COMPONENT
			33.33	38.50	6	OTHER
			25.00	42.40	20	ROOF & ITS COMPONENTS
			9.09	42.36	11	INSTRUMENT PANEL
FRONTAL	BELTED, NO AIR BAG	NECK	50.00	61.50	6	STEERING ASSEMBLY
			37.73	46.05	546	average injured occupant
FRONTAL	BELTED, NO AIR BAG	TORSO	50.00	49.86	42	BELT SYSTEM
			38.66	48.47	194	STEERING ASSEMBLY
			37.73	46.05	546	average injured occupant
			10.00	45.20	10	SIDE INTERIOR SURFACE
FRONTAL	NO BELT, AIR BAG	HEAD	33.33	38.36	36	STEERING ASSEMBLY
			28.57	35.86	14	INSTRUMENT PANEL
			28.57	31.93	14	OTHER
			25.20	38.94	742	average injured occupant
			20.00	42.43	35	A-PILLAR
			16.67	37.00	54	ROOF & ITS COMPONENTS
			16.42	31.01	67	EXTERIOR
			15.56	38.09	45	OTHER FRONTAL COMPONENT
FRONTAL	NO BELT, AIR BAG	NECK	* 80.00	43.20	5	STEERING ASSEMBLY
			25.20	38.94	742	average injured occupant
			0.00	37.71	7	EXTERIOR
			0.00	36.88	8	ROOF & ITS COMPONENTS
FRONTAL	NO BELT, AIR BAG	TORSO	42.86	49.86	7	AIR BAG
			29.69	41.23	256	STEERING ASSEMBLY
			28.57	37.29	7	SIDE HARDWARE OR ARMREST
			25.20	38.94	742	average injured occupant
			20.00	37.24	25	INSTRUMENT PANEL
			20.00	34.84	25	SIDE INTERIOR SURFACE
			18.92	35.49	37	EXTERIOR
			16.67	28.33	6	ROOF & ITS COMPONENTS
FRONTAL	NO BELT, AIR BAG	BURN	25.20	38.94	742	average injured occupant
			20.00	29.00	5	FIRE

**3+ standard deviations higher than average injured occupant (p < .002).
*1.96-3 standard deviations higher than average injured occupant (p < .05).

IMPACT TYPE	OCCUPANT PROTECTION	BODY REGION	PERCENT FEMALE	AVERAGE AGE	N	INJURY SOURCE
FRONTAL	NO BELT, NO AIR BAG	HEAD	* 43.24	35.05	37	INSTRUMENT PANEL
			33.33	41.83	6	SIDE INTERIOR SURFACE
			32.50	41.03	80	STEERING ASSEMBLY
			29.85	35.93	67	A-PILLAR
			28.37	40.94	141	OTHER FRONTAL COMPONENT
			25.02	41.46	1431	average injured occupant
			22.22	36.44	9	OTHER SIDE COMPONENT
			20.00	37.00	5	B-PILLAR
			17.86	35.81	84	EXTERIOR
			15.25	40.08	59	ROOF & ITS COMPONENTS
			11.11	29.78	9	OTHER
FRONTAL	NO BELT, NO AIR BAG	NECK	45.45	45.55	11	STEERING ASSEMBLY
			33.33	50.50	6	ROOF & ITS COMPONENTS
			31.58	45.63	19	OTHER FRONTAL COMPONENT
			25.02	41.46	1431	average injured occupant
FRONTAL	NO BELT, NO AIR BAG	TORSO	* 40.63	33.69	32	EXTERIOR
			36.00	41.88	25	SIDE INTERIOR SURFACE
			25.04	43.78	603	STEERING ASSEMBLY
			25.02	41.46	1431	average injured occupant
			20.00	38.60	30	INSTRUMENT PANEL
FRONTAL	NO BELT, NO AIR BAG	BURN	25.02	41.46	1431	average injured occupant
			22.22	32.00	9	FIRE
NEARSIDE IMPACT	BELTED	HEAD	* 70.00	43.05	20	SIDE INTERIOR SURFACE
			* 61.11	43.44	36	A-PILLAR
			54.55	45.91	22	NON-CONTACT
			51.04	42.90	96	EXTERIOR
			47.62	49.81	21	OTHER SIDE COMPONENT
			46.96	43.90	115	B-PILLAR
			43.34	45.92	1006	average injured occupant
			33.33	55.00	6	STEERING ASSEMBLY
			30.91	40.65	55	ROOF & ITS COMPONENTS
			29.41	36.20	51	OTHER
NEARSIDE IMPACT	BELTED	NECK	62.50	42.50	8	B-PILLAR
			43.34	45.92	1006	average injured occupant
			28.57	41.29	7	EXTERIOR
			20.00	44.80	5	NON-CONTACT
NEARSIDE IMPACT	BELTED	TORSO	51.61	47.87	31	STEERING ASSEMBLY
			43.34	45.92	1006	average injured occupant
			42.00	47.18	400	SIDE INTERIOR SURFACE
			40.00	45.20	10	BELT SYSTEM
			40.00	43.30	10	FLOOR & ITS COMPONENTS
			39.62	45.81	53	B-PILLAR
			38.79	48.80	116	SIDE HARDWARE OR ARMREST
			20.00	36.40	10	SEAT & ITS COMPONENTS
			0.00	33.80	5	EXTERIOR
NEARSIDE IMPACT	NO BELT	HEAD	* 66.67	51.33	9	SIDE INTERIOR SURFACE
			40.00	29.50	10	OTHER FRONTAL COMPONENT
			37.84	41.46	37	B-PILLAR
			37.60	36.88	8	NON-CONTACT
			35.71	30.36	14	OTHER SIDE COMPONENT
			32.38	36.35	105	EXTERIOR
			30.61	36.24	49	A-PILLAR
			30.52	39.00	842	average injured occupant
			29.41	33.12	51	ROOF & ITS COMPONENTS
			16.28	29.26	43	OTHER
NEARSIDE IMPACT	NO BELT	NECK	30.52	39.00	842	average injured occupant
			22.22	29.78	9	EXTERIOR
NEARSIDE IMPACT	NO BELT	TORSO	50.00	31.83	6	FLOOR & ITS COMPONENTS
			34.53	42.02	278	SIDE INTERIOR SURFACE
			32.50	36.05	40	EXTERIOR
			30.52	39.00	842	average injured occupant
			29.82	42.16	57	SIDE HARDWARE OR ARMREST
			29.55	34.18	44	STEERING ASSEMBLY
			24.00	45.52	25	B-PILLAR
			0.00	42.40	5	ROOF & ITS COMPONENTS

**3+ standard deviations higher than average injured occupant (p < .002).
*1.96-3 standard deviations higher than average injured occupant (p < .05).

IMPACT TYPE	OCCUPANT PROTECTION	BODY REGION	PERCENT FEMALE	AVERAGE AGE	N	INJURY SOURCE
FAR-SIDE IMPACT	BELTED	HEAD	60.00	39.60	10	OTHER SIDE COMPONENT
			49.18	40.31	61	SIDE INTERIOR SURFACE
			42.31	40.04	26	B-PILLAR
			37.01	42.52	354	average injured occupant
			36.36	50.18	11	INSTRUMENT PANEL
			33.33	54.33	12	EXTERIOR
			33.33	27.17	6	NON-CONTACT
			28.57	52.71	7	OTHER OCCS, SELF, CARGO
			27.03	38.22	37	ROOF & ITS COMPONENTS
			16.67	42.83	6	SEAT & ITS COMPONENTS
			12.50	39.50	8	A-PILLAR
FAR-SIDE IMPACT	BELTED	NECK	37.01	42.52	354	average injured occupant
			20.00	35.00	5	ROOF & ITS COMPONENTS
FAR-SIDE IMPACT	BELTED	TORSO	* 80.00	69.40	5	STEERING ASSEMBLY
			50.00	57.17	6	OTHER
			50.00	49.40	20	BELT SYSTEM
			50.00	44.58	12	SIDE HARDWARE OR ARMREST
			43.18	39.25	44	SIDE INTERIOR SURFACE
			40.00	34.80	5	EXTERIOR
			37.01	42.52	354	average injured occupant
			33.33	38.73	15	OTHER OCCS, SELF, CARGO
			31.58	37.47	19	SEAT & ITS COMPONENTS
			23.81	51.52	21	FLOOR & ITS COMPONENTS
FAR-SIDE IMPACT	NO BELT	HEAD	35.48	40.23	62	SIDE INTERIOR SURFACE
			33.33	54.33	6	SIDE HARDWARE OR ARMREST
			27.91	41.84	43	A-PILLAR
			27.78	34.67	18	B-PILLAR
			25.69	39.34	654	average injured occupant
			24.14	44.00	29	OTHER FRONTAL COMPONENT
			22.22	38.06	18	OTHER SIDE COMPONENT
			21.43	36.34	70	ROOF & ITS COMPONENTS
			20.59	34.29	68	EXTERIOR
			20.00	35.40	5	SEAT & ITS COMPONENTS
			16.67	32.28	18	OTHER
			0.00	44.33	12	INSTRUMENT PANEL
			0.00	29.40	5	STEERING ASSEMBLY
FAR-SIDE IMPACT	NO BELT	NECK	40.00	59.00	5	OTHER FRONTAL COMPONENT
			40.00	43.20	5	SIDE INTERIOR SURFACE
			40.00	40.80	5	A-PILLAR
			25.69	39.34	654	average injured occupant
			0.00	39.33	6	EXTERIOR
			0.00	38.00	5	ROOF & ITS COMPONENTS
FAR-SIDE IMPACT	NO BELT	TORSO	57.14	40.14	7	OTHER OCCS, SELF, CARGO
			34.78	35.57	23	STEERING ASSEMBLY
			27.64	41.41	123	SIDE INTERIOR SURFACE
			25.69	39.34	654	average injured occupant
			25.00	45.97	32	INSTRUMENT PANEL
			25.00	42.06	32	EXTERIOR
			25.00	41.75	8	OTHER
			21.43	36.71	14	FLOOR & ITS COMPONENTS
			7.14	41.57	14	SEAT & ITS COMPONENTS
			0.00	39.29	7	SIDE HARDWARE OR ARMREST
ROLLOVER FIRST/WORST EVENT	BELTED	HEAD	40.00	27.20	5	OTHER
			36.84	36.68	19	EXTERIOR
			31.40	35.76	121	ROOF & ITS COMPONENTS
			29.91	38.29	351	average injured occupant
ROLLOVER FIRST/WORST EVENT	BELTED	NECK	33.33	44.06	18	ROOF & ITS COMPONENTS
			29.91	38.29	351	average injured occupant
ROLLOVER FIRST/WORST EVENT	BELTED	TORSO	40.48	40.10	42	SIDE INTERIOR SURFACE
			33.33	43.50	6	EXTERIOR
			33.33	41.50	18	STEERING ASSEMBLY
			33.33	30.50	6	SEAT & ITS COMPONENTS
			29.91	38.29	351	average injured occupant
			27.27	35.36	11	BELT SYSTEM
			25.00	38.33	12	SIDE HARDWARE OR ARMREST
			23.53	30.88	17	ROOF & ITS COMPONENTS

**3+ standard deviations higher than average injured occupant (p < .002).
*1.96-3 standard deviations higher than average injured occupant (p < .05).

IMPACT TYPE	OCCUPANT PROTECTION	BODY REGION	PERCENT FEMALE	AVERAGE AGE	N	INJURY SOURCE
ROLLOVER FIRST/WORST EVENT	NO BELT	HEAD	27.27	32.55	11	OTHER FRONTAL COMPONENT
			21.92	32.34	219	EXTERIOR
			21.29	34.05	902	average injured occupant
			19.35	32.72	93	ROOF & ITS COMPONENTS
			14.29	25.43	7	OTHER
			0.00	38.57	7	A-PILLAR
ROLLOVER FIRST/WORST EVENT	NO BELT	NECK	21.29	34.05	902	average injured occupant
			15.38	41.15	13	EXTERIOR
			15.38	35.38	13	ROOF & ITS COMPONENTS
ROLLOVER FIRST/WORST EVENT	NO BELT	TORSO	29.03	32.74	31	SIDE INTERIOR SURFACE
			25.77	33.10	194	EXTERIOR
			21.29	34.05	902	average injured occupant
			17.14	37.57	35	STEERING ASSEMBLY
			13.04	33.57	23	ROOF & ITS COMPONENTS
			0.00	35.73	11	SEAT & ITS COMPONENTS
			0.00	28.83	6	OTHER SIDE COMPONENT
REAR & OTHER	BELTED	HEAD	50.00	55.50	6	B-PILLAR
			46.53	46.68	101	average injured occupant
			44.44	43.00	18	SEAT & ITS COMPONENTS
			41.67	39.42	12	ROOF & ITS COMPONENTS
REAR & OTHER	BELTED	NECK	46.53	46.68	101	average injured occupant
			40.00	53.80	5	SEAT & ITS COMPONENTS
REAR & OTHER	BELTED	TORSO	60.00	52.73	30	SEAT & ITS COMPONENTS
			46.53	46.68	101	average injured occupant
			27.27	46.18	11	STEERING ASSEMBLY
REAR & OTHER	NO BELT	HEAD	33.33	40.33	6	B-PILLAR
			25.00	40.33	12	SEAT & ITS COMPONENTS
			14.89	36.41	94	average injured occupant
			8.33	28.83	12	ROOF & ITS COMPONENTS
			0.00	30.33	9	EXTERIOR
REAR & OTHER	NO BELT	NECK	14.89	36.41	94	average injured occupant
			0.00	36.60	5	SEAT & ITS COMPONENTS
REAR & OTHER	NO BELT	TORSO	15.38	42.15	13	SEAT & ITS COMPONENTS
			14.89	36.41	94	average injured occupant
			10.00	39.20	10	STEERING ASSEMBLY

**3+ standard deviations higher than average injured occupant (p < .002).
*1.96-3 standard deviations higher than average injured occupant (p < .05).

TABLE E-13: PERCENT FEMALE, RIGHT-FRONT PASSENGERS AGE 18-96 WITH AIS 2+ INJURIES, CDS 1988-2010

IMPACT TYPE	OCCUPANT PROTECTION	BODY REGION	PERCENT FEMALE	AVERAGE AGE	N	INJURY SOURCE
FRONTAL	BELTED, AIR BAG	HEAD	77.27	45.36	22	NON-CONTACT
			75.86	52.41	29	AIR BAG
			70.00	48.10	20	B-PILLAR
			69.37	45.76	790	average injured occupant
			51.52	40.48	33	ROOF & ITS COMPONENTS
			50.00	33.62	26	INSTRUMENT PANEL
			40.00	41.07	15	A-PILLAR
FRONTAL	BELTED, AIR BAG	TORSO	77.78	49.56	36	SIDE INTERIOR SURFACE
			76.47	42.53	17	FLOOR & ITS COMPONENTS
			* 76.37	51.75	237	BELT SYSTEM
			70.00	55.75	20	AIR BAG
			69.37	45.76	790	average injured occupant
			59.57	38.26	47	INSTRUMENT PANEL
			59.09	33.77	22	SEAT & ITS COMPONENTS
			52.38	50.71	21	SIDE HARDWARE OR ARMREST
FRONTAL	BELTED, AIR BAG	ARM	87.50	57.13	16	OTHER FRONTAL COMPONENT
			78.57	51.46	28	SIDE INTERIOR SURFACE
			76.19	48.71	63	AIR BAG
			70.27	47.20	74	INSTRUMENT PANEL
			69.37	45.76	790	average injured occupant
FRONTAL	BELTED, AIR BAG	LEG	* 80.30	45.83	132	FLOOR & ITS COMPONENTS
			71.88	43.79	160	INSTRUMENT PANEL
			69.37	45.76	790	average injured occupant
FRONTAL	BELTED, NO AIR BAG	HEAD	70.46	45.93	765	average injured occupant
			64.00	41.36	25	ROOF & ITS COMPONENTS
			51.72	37.81	58	OTHER FRONTAL COMPONENT
			49.25	40.00	67	INSTRUMENT PANEL
			46.67	35.93	15	A-PILLAR
FRONTAL	BELTED, NO AIR BAG	TORSO	74.07	45.41	27	SEAT & ITS COMPONENTS
			73.49	51.76	298	BELT SYSTEM
			70.77	51.20	65	INSTRUMENT PANEL
			70.46	45.93	765	average injured occupant
			50.00	46.16	32	SIDE INTERIOR SURFACE
FRONTAL	BELTED, NO AIR BAG	ARM	73.95	51.46	119	INSTRUMENT PANEL
			70.46	45.93	765	average injured occupant
FRONTAL	BELTED, NO AIR BAG	LEG	72.99	47.31	137	INSTRUMENT PANEL
			70.46	45.93	765	average injured occupant
			70.10	40.89	97	FLOOR & ITS COMPONENTS
FRONTAL	NO BELT, AIR BAG	HEAD	58.82	34.18	34	INSTRUMENT PANEL
			52.07	36.19	484	average injured occupant
			50.00	30.00	28	EXTERIOR
			44.64	34.98	56	OTHER FRONTAL COMPONENT
			37.14	32.00	35	ROOF & ITS COMPONENTS
FRONTAL	NO BELT, AIR BAG	TORSO	59.09	40.45	22	SIDE INTERIOR SURFACE
			54.47	39.05	123	INSTRUMENT PANEL
			52.07	36.19	484	average injured occupant
			45.00	33.20	20	EXTERIOR
FRONTAL	NO BELT, AIR BAG	ARM	62.12	40.30	66	INSTRUMENT PANEL
			52.07	36.19	484	average injured occupant
FRONTAL	NO BELT, AIR BAG	LEG	* 65.93	35.24	91	FLOOR & ITS COMPONENTS
			56.03	38.63	116	INSTRUMENT PANEL
			52.07	36.19	484	average injured occupant
FRONTAL	NO BELT, NO AIR BAG	HEAD	55.48	36.76	146	INSTRUMENT PANEL
			51.34	36.53	1498	average injured occupant
			51.04	35.58	480	OTHER FRONTAL COMPONENT
			37.14	29.11	35	A-PILLAR
			36.73	33.92	49	ROOF & ITS COMPONENTS
			34.88	31.81	43	EXTERIOR
FRONTAL	NO BELT, NO AIR BAG	NECK	62.96	48.78	27	INSTRUMENT PANEL
			51.34	36.53	1498	average injured occupant
			45.00	46.78	40	OTHER FRONTAL COMPONENT

**3+ standard deviations higher than average injured occupant (p < .002).
*1.96-3 standard deviations higher than average injured occupant (p < .05).

IMPACT TYPE	OCCUPANT PROTECTION	BODY REGION	PERCENT FEMALE	AVERAGE AGE	N	INJURY SOURCE
FRONTAL	NO BELT, NO AIR BAG	TORSO	52.40	40.99	479	INSTRUMENT PANEL
			52.38	38.24	21	SIDE INTERIOR SURFACE
			51.34	36.53	1498	average injured occupant
			41.67	36.96	24	SEAT & ITS COMPONENTS
			41.18	39.71	17	STEERING ASSEMBLY
			40.00	41.53	15	FLOOR & ITS COMPONENTS
			10.00	30.80	20	EXTERIOR
FRONTAL	NO BELT, NO AIR BAG	ARM	* 61.42	39.83	254	INSTRUMENT PANEL
			57.14	41.86	21	OTHER FRONTAL COMPONENT
			51.34	36.53	1498	average injured occupant
FRONTAL	NO BELT, NO AIR BAG	LEG	* 59.52	36.87	168	FLOOR & ITS COMPONENTS
			54.67	37.81	300	INSTRUMENT PANEL
			51.34	36.53	1498	average injured occupant
NEARSIDE IMPACT	BELTED	HEAD	83.33	46.77	30	OTHER OCCS, SELF, CARGO
			71.43	42.57	21	SEAT & ITS COMPONENTS
			68.73	47.45	339	average injured occupant
			60.00	49.27	15	B-PILLAR
			35.00	32.40	20	ROOF & ITS COMPONENTS
NEARSIDE IMPACT	BELTED	TORSO	83.33	52.33	30	FLOOR & ITS COMPONENTS
			79.31	53.55	29	OTHER OCCS, SELF, CARGO
			79.10	53.73	67	BELT SYSTEM
			71.88	48.13	32	SEAT & ITS COMPONENTS
			68.73	47.45	339	average injured occupant
			62.50	61.56	16	OTHER
NEARSIDE IMPACT	BELTED	LEG	80.00	51.60	15	FLOOR & ITS COMPONENTS
			69.23	39.35	26	INSTRUMENT PANEL
			68.73	47.45	339	average injured occupant
NEARSIDE IMPACT	NO BELT	HEAD	46.88	28.94	32	EXTERIOR
			44.48	34.49	353	average injured occupant
			42.86	38.33	21	INSTRUMENT PANEL
			38.24	32.94	34	ROOF & ITS COMPONENTS
			37.50	30.56	16	OTHER OCCS, SELF, CARGO
			33.33	29.50	30	OTHER FRONTAL COMPONENT
NEARSIDE IMPACT	NO BELT	TORSO	53.57	41.32	28	STEERING ASSEMBLY
			44.48	34.49	353	average injured occupant
			41.03	41.15	39	INSTRUMENT PANEL
			38.10	31.67	21	SEAT & ITS COMPONENTS
			36.36	29.18	22	EXTERIOR
			27.27	30.91	22	OTHER OCCS, SELF, CARGO
			25.00	30.19	16	SIDE INTERIOR SURFACE
NEARSIDE IMPACT	NO BELT	ARM	* 68.18	41.50	22	INSTRUMENT PANEL
			44.48	34.49	353	average injured occupant
NEARSIDE IMPACT	NO BELT	LEG	52.00	39.48	25	INSTRUMENT PANEL
			44.48	34.49	353	average injured occupant
FAR-SIDE IMPACT	BELTED	HEAD	87.50	35.50	16	NON-CONTACT
			77.00	40.10	10	OTHER SIDE COMPONENT
			71.29	45.47	101	B-PILLAR
			68.42	44.21	38	EXTERIOR
			66.49	45.55	749	average injured occupant
			56.60	35.49	53	ROOF & ITS COMPONENTS
			55.17	45.62	29	SIDE INTERIOR SURFACE
			45.83	40.58	24	A-PILLAR
			20.00	28.93	15	OTHER
FAR-SIDE IMPACT	BELTED	NECK	80.00	48.56	25	B-PILLAR
			66.49	45.55	749	average injured occupant
			41.18	45.71	17	ROOF & ITS COMPONENTS
FAR-SIDE IMPACT	BELTED	TORSO	88.24	38.18	17	SEAT & ITS COMPONENTS
			80.00	46.93	15	OTHER OCCS, SELF, CARGO
			69.39	47.68	196	SIDE HARDWARE OR ARMREST
			67.86	51.29	56	BELT SYSTEM
			66.49	45.55	749	average injured occupant
			65.92	47.34	355	SIDE INTERIOR SURFACE
			64.15	49.34	53	B-PILLAR
			60.87	47.65	23	FLOOR & ITS COMPONENTS

**3+ standard deviations higher than average injured occupant (p < .002).
*1.96-3 standard deviations higher than average injured occupant (p < .05).

IMPACT TYPE	OCCUPANT PROTECTION	BODY REGION	PERCENT FEMALE	AVERAGE AGE	N	INJURY SOURCE
FAR-SIDE IMPACT	BELTED	ARM	73.33	55.27	15	INSTRUMENT PANEL
			66.49	45.55	749	average injured occupant
			60.27	42.15	73	SIDE INTERIOR SURFACE
FAR-SIDE IMPACT	BELTED	LEG	76.32	44.16	38	FLOOR & ITS COMPONENTS
			66.49	45.55	749	average injured occupant
			61.76	39.96	68	SIDE INTERIOR SURFACE
			60.00	38.04	25	INSTRUMENT PANEL
			45.00	43.40	20	SIDE HARDWARE OR ARMREST
FAR-SIDE IMPACT	NO BELT	HEAD	54.29	34.63	35	OTHER FRONTAL COMPONENT
			53.33	36.92	60	B-PILLAR
			50.00	35.61	74	EXTERIOR
			50.00	32.24	34	OTHER SIDE COMPONENT
			48.42	34.46	632	average injured occupant
			42.86	33.93	56	A-PILLAR
			28.21	28.95	39	ROOF & ITS COMPONENTS
			25.00	27.06	16	SIDE INTERIOR SURFACE
			16.67	25.39	18	OTHER
FAR-SIDE IMPACT	NO BELT	NECK	48.42	34.46	632	average injured occupant
			47.06	31.00	17	EXTERIOR
FAR-SIDE IMPACT	NO BELT	TORSO	57.55	35.33	106	SIDE HARDWARE OR ARMREST
			53.85	37.58	26	OTHER OCCS, SELF, CARGO
			50.00	37.61	234	SIDE INTERIOR SURFACE
			50.00	37.29	34	B-PILLAR
			48.42	34.46	632	average injured occupant
			46.15	36.97	39	INSTRUMENT PANEL
			38.64	34.59	44	EXTERIOR
FAR-SIDE IMPACT	NO BELT	ARM	52.27	34.45	44	SIDE INTERIOR SURFACE
			52.00	36.40	25	EXTERIOR
			48.42	34.46	632	average injured occupant
FAR-SIDE IMPACT	NO BELT	LEG	56.10	31.17	41	INSTRUMENT PANEL
			48.42	34.46	632	average injured occupant
			44.64	32.54	56	SIDE INTERIOR SURFACE
			43.48	25.35	23	FLOOR & ITS COMPONENTS
ROLLOVER FIRST/WORST EVENT	BELTED	HEAD	55.41	36.66	74	ROOF & ITS COMPONENTS
			49.85	37.06	337	average injured occupant
ROLLOVER FIRST/WORST EVENT	BELTED	NECK	50.00	38.94	48	ROOF & ITS COMPONENTS
			49.85	37.06	337	average injured occupant
ROLLOVER FIRST/WORST EVENT	BELTED	TORSO	57.14	38.03	35	SIDE INTERIOR SURFACE
			56.25	36.81	16	SIDE HARDWARE OR ARMREST
			54.76	35.33	42	BELT SYSTEM
			49.85	37.06	337	average injured occupant
			40.91	35.09	22	SEAT & ITS COMPONENTS
			31.58	39.16	19	ROOF & ITS COMPONENTS
ROLLOVER FIRST/WORST EVENT	BELTED	ARM	52.17	37.35	23	ROOF & ITS COMPONENTS
			50.00	32.00	16	SIDE INTERIOR SURFACE
			49.85	37.06	337	average injured occupant
			47.06	29.53	17	EXTERIOR
ROLLOVER FIRST/WORST EVENT	BELTED	LEG	65.00	39.60	20	FLOOR & ITS COMPONENTS
			65.00	37.90	20	INSTRUMENT PANEL
			49.85	37.06	337	average injured occupant
ROLLOVER FIRST/WORST EVENT	NO BELT	HEAD	42.15	28.34	121	EXTERIOR
			40.92	31.06	501	average injured occupant
			34.67	30.44	75	ROOF & ITS COMPONENTS
ROLLOVER FIRST/WORST EVENT	NO BELT	NECK	40.92	31.06	501	average injured occupant
			40.00	33.36	25	ROOF & ITS COMPONENTS
			37.50	29.34	32	EXTERIOR
ROLLOVER FIRST/WORST EVENT	NO BELT	TORSO	48.48	32.52	33	SIDE INTERIOR SURFACE
			44.22	30.01	147	EXTERIOR
			40.92	31.06	501	average injured occupant
			36.84	32.16	38	ROOF & ITS COMPONENTS
			36.00	33.04	25	INSTRUMENT PANEL
ROLLOVER FIRST/WORST EVENT	NO BELT	ARM	51.52	32.70	33	EXTERIOR
			40.92	31.06	501	average injured occupant

**3+ standard deviations higher than average injured occupant (p < .002).
*1.96-3 standard deviations higher than average injured occupant (p < .05).

IMPACT TYPE	OCCUPANT PROTECTION	BODY REGION	PERCENT FEMALE	AVERAGE AGE	N	INJURY SOURCE
ROLLOVER FIRST/WORST EVENT	NO BELT	LEG	58.06	30.68	31	EXTERIOR
			40.92	31.06	501	average injured occupant
REAR & OTHER	BELTED	TORSO	61.11	48.17	18	SEAT & ITS COMPONENTS
			60.00	41.54	120	average injured occupant
REAR & OTHER	NO BELT	TORSO	73.33	38.00	15	SEAT & ITS COMPONENTS
			57.14	36.53	77	average injured occupant

**3+ standard deviations higher than average injured occupant (p < .002).
*1.96-3 standard deviations higher than average injured occupant (p < .05).

IMPACT TYPE	OCCUPANT PROTECTION	BODY REGION	PERCENT FEMALE	AVERAGE AGE	N	INJURY SOURCE
FRONTAL	BELTED, AIR BAG	HEAD	69.61	48.81	385	average injured occupant
			60.00	45.07	15	ROOF & ITS COMPONENTS
			54.55	46.73	11	B-PILLAR
			50.00	38.50	12	INSTRUMENT PANEL
			40.00	43.80	10	A-PILLAR
FRONTAL	BELTED, AIR BAG	TORSO	76.19	49.81	21	SIDE INTERIOR SURFACE
			72.16	53.51	97	BELT SYSTEM
			70.00	57.40	10	AIR BAG
			69.61	48.81	385	average injured occupant
			68.97	39.83	29	INSTRUMENT PANEL
			50.00	47.92	12	SIDE HARDWARE OR ARMREST
			45.45	32.64	11	SEAT & ITS COMPONENTS
FRONTAL	BELTED, AIR BAG	ARM	83.33	56.33	12	SIDE INTERIOR SURFACE
			81.82	55.86	22	AIR BAG
			69.61	48.81	385	average injured occupant
			69.44	49.94	36	INSTRUMENT PANEL
FRONTAL	BELTED, AIR BAG	LEG	80.00	49.12	25	FLOOR & ITS COMPONENTS
			71.11	45.97	90	INSTRUMENT PANEL
			69.61	48.81	385	average injured occupant
FRONTAL	BELTED, NO AIR BAG	HEAD	72.73	46.00	11	ROOF & ITS COMPONENTS
			68.00	49.86	375	average injured occupant
			42.31	48.38	26	INSTRUMENT PANEL
FRONTAL	BELTED, NO AIR BAG	TORSO	70.92	55.52	141	BELT SYSTEM
			69.05	50.17	42	INSTRUMENT PANEL
			68.00	49.86	375	average injured occupant
			38.10	45.24	21	SIDE INTERIOR SURFACE
FRONTAL	BELTED, NO AIR BAG	ARM	74.47	50.21	47	INSTRUMENT PANEL
			68.00	49.86	375	average injured occupant
FRONTAL	BELTED, NO AIR BAG	LEG	73.91	43.43	23	FLOOR & ITS COMPONENTS
			68.00	49.86	375	average injured occupant
			65.28	45.11	72	INSTRUMENT PANEL
FRONTAL	NO BELT, AIR BAG	HEAD	64.29	42.36	14	INSTRUMENT PANEL
			55.87	38.10	281	average injured occupant
			52.63	30.95	19	EXTERIOR
			37.50	33.50	16	OTHER FRONTAL COMPONENT
			36.36	32.41	22	ROOF & ITS COMPONENTS
			20.00	34.30	10	A-PILLAR
FRONTAL	NO BELT, AIR BAG	TORSO	62.50	39.19	16	SIDE INTERIOR SURFACE
			58.33	39.01	72	INSTRUMENT PANEL
			55.87	38.10	281	average injured occupant
			46.67	35.20	15	EXTERIOR
FRONTAL	NO BELT, AIR BAG	ARM	* 76.67	43.00	30	INSTRUMENT PANEL
			55.87	38.10	281	average injured occupant
FRONTAL	NO BELT, AIR BAG	LEG	55.87	38.10	281	average injured occupant
			53.09	38.14	81	INSTRUMENT PANEL
			52.63	32.95	19	FLOOR & ITS COMPONENTS
FRONTAL	NO BELT, NO AIR BAG	HEAD	52.85	40.82	123	OTHER FRONTAL COMPONENT
			50.65	38.82	768	average injured occupant
			48.15	38.76	54	INSTRUMENT PANEL
			44.83	36.52	29	ROOF & ITS COMPONENTS
			33.33	31.30	30	EXTERIOR
			28.57	30.57	14	A-PILLAR
FRONTAL	NO BELT, NO AIR BAG	NECK	72.73	48.73	11	INSTRUMENT PANEL
			50.65	38.82	768	average injured occupant
			46.67	47.20	15	OTHER FRONTAL COMPONENT
FRONTAL	NO BELT, NO AIR BAG	TORSO	57.14	40.93	14	SIDE INTERIOR SURFACE
			52.07	42.74	290	INSTRUMENT PANEL
			50.65	38.82	768	average injured occupant
			41.67	38.83	12	STEERING ASSEMBLY
			14.29	28.21	14	EXTERIOR
FRONTAL	NO BELT, NO AIR BAG	ARM	* 62.26	40.39	106	INSTRUMENT PANEL
			50.65	38.82	768	average injured occupant

**3+ standard deviations higher than average injured occupant (p < .002).
*1.96-3 standard deviations higher than average injured occupant (p < .05).

TABLE E-14 (CONTINUED): PERCENT FEMALE, RIGHT-FRONT PASSENGERS AGE 18-96 WITH AIS 3+ INJURIES, CDS 1988-2010

IMPACT TYPE	OCCUPANT PROTECTION	BODY REGION	PERCENT FEMALE	AVERAGE AGE	N	INJURY SOURCE
FRONTAL	NO BELT, NO AIR BAG	LEG	59.26	39.59	27	FLOOR & ITS COMPONENTS
			52.00	38.39	200	INSTRUMENT PANEL
			50.65	38.82	768	average injured occupant
NEARSIDE IMPACT	BELTED	HEAD	84.62	53.92	13	OTHER OCCS, SELF, CARGO
			80.00	51.20	10	B-PILLAR
			67.48	49.34	163	average injured occupant
			38.46	32.00	13	ROOF & ITS COMPONENTS
NEARSIDE IMPACT	BELTED	TORSO	81.25	52.44	16	FLOOR & ITS COMPONENTS
			77.78	60.81	27	BELT SYSTEM
			76.47	50.82	17	OTHER OCCS, SELF, CARGO
			75.00	44.85	20	SEAT & ITS COMPONENTS
			67.48	49.34	163	average injured occupant
NEARSIDE IMPACT	NO BELT	HEAD	50.00	34.22	18	ROOF & ITS COMPONENTS
			45.00	32.00	20	EXTERIOR
			43.06	35.45	209	average injured occupant
NEARSIDE IMPACT	NO BELT	TORSO	64.29	43.21	14	STEERING ASSEMBLY
			50.00	31.86	14	SEAT & ITS COMPONENTS
			45.00	43.10	20	INSTRUMENT PANEL
			43.06	35.45	209	average injured occupant
			37.50	30.94	16	EXTERIOR
			27.27	26.91	11	SIDE INTERIOR SURFACE
			8.33	30.67	12	OTHER OCCS, SELF, CARGO
NEARSIDE IMPACT	NO BELT	ARM	* 83.33	45.33	12	INSTRUMENT PANEL
			43.06	35.45	209	average injured occupant
NEARSIDE IMPACT	NO BELT	LEG	43.06	35.45	209	average injured occupant
			30.00	36.90	10	INSTRUMENT PANEL
FAR-SIDE IMPACT	BELTED	HEAD	81.82	59.27	11	OTHER SIDE COMPONENT
			70.59	44.94	34	EXTERIOR
			64.41	46.99	503	average injured occupant
			63.49	48.41	63	B-PILLAR
			57.14	48.00	21	SIDE INTERIOR SURFACE
			55.17	39.31	29	ROOF & ITS COMPONENTS
			25.00	42.92	12	A-PILLAR
			16.67	28.08	12	OTHER
FAR-SIDE IMPACT	BELTED	NECK	85.71	54.50	14	B-PILLAR
			64.41	46.99	503	average injured occupant
			50.00	42.90	10	ROOF & ITS COMPONENTS
FAR-SIDE IMPACT	BELTED	TORSO	66.98	48.72	106	SIDE HARDWARE OR ARMREST
			64.41	46.99	503	average injured occupant
			64.31	49.01	255	SIDE INTERIOR SURFACE
			61.54	44.92	13	BELT SYSTEM
			54.84	49.77	31	B-PILLAR
FAR-SIDE IMPACT	BELTED	ARM	75.00	42.10	20	SIDE INTERIOR SURFACE
			64.41	46.99	503	average injured occupant
FAR-SIDE IMPACT	BELTED	LEG	64.41	46.99	503	average injured occupant
			57.45	00.10	17	SIDE INTERIOR SURFACE
			47.37	44.21	19	SIDE HARDWARE OR ARMREST
FAR-SIDE IMPACT	NO BELT	HEAD	56.76	37.76	37	B-PILLAR
			48.52	35.43	439	average injured occupant
			45.45	35.09	55	EXTERIOR
			35.71	33.50	28	A-PILLAR
			20.00	26.60	15	OTHER
			18.18	28.36	11	SIDE INTERIOR SURFACE
			16.67	27.92	24	ROOF & ITS COMPONENTS
FAR-SIDE IMPACT	NO BELT	TORSO	55.56	36.30	54	SIDE HARDWARE OR ARMREST
			51.18	39.25	170	SIDE INTERIOR SURFACE
			48.52	35.43	439	average injured occupant
			46.15	32.08	13	OTHER OCCS, SELF, CARGO
			44.44	41.06	18	B-PILLAR
			41.94	35.71	31	EXTERIOR
			36.00	33.96	25	INSTRUMENT PANEL
FAR-SIDE IMPACT	NO BELT	ARM	52.63	28.68	19	SIDE INTERIOR SURFACE
			48.52	35.43	439	average injured occupant
			45.45	31.27	11	EXTERIOR

3+ standard deviations higher than average injured occupant (p < .002).
*1.96-3 standard deviations higher than average injured occupant (p < .03).

IMPACT TYPE	OCCUPANT PROTECTION	BODY REGION	PERCENT FEMALE	AVERAGE AGE	N	INJURY SOURCE
FAR-SIDE IMPACT	NO BELT	LEG	57.69	28.88	26	INSTRUMENT PANEL
			48.52	35.43	439	average injured occupant
			43.18	29.89	44	SIDE INTERIOR SURFACE
			36.36	27.91	11	SIDE HARDWARE OR ARMREST
ROLLOVER FIRST/WORST EVENT	BELTED	HEAD	55.26	38.84	38	ROOF & ITS COMPONENTS
			51.11	37.43	180	average injured occupant
			40.00	31.40	10	EXTERIOR
ROLLOVER FIRST/WORST EVENT	BELTED	NECK	51.11	37.43	180	average injured occupant
			40.74	37.22	27	ROOF & ITS COMPONENTS
ROLLOVER FIRST/WORST EVENT	BELTED	TORSO	70.37	39.96	27	SIDE INTERIOR SURFACE
			53.85	35.46	13	BELT SYSTEM
			51.11	37.43	180	average injured occupant
ROLLOVER FIRST/WORST EVENT	NO BELT	HEAD	42.68	28.28	82	EXTERIOR
			42.60	31.90	331	average injured occupant
			29.03	30.03	31	ROOF & ITS COMPONENTS
ROLLOVER FIRST/WORST EVENT	NO BELT	NECK	42.60	31.90	331	average injured occupant
			40.00	28.33	15	EXTERIOR
			30.00	32.80	10	ROOF & ITS COMPONENTS
ROLLOVER FIRST/WORST EVENT	NO BELT	TORSO	52.38	32.95	21	SIDE INTERIOR SURFACE
			42.60	31.90	331	average injured occupant
			41.67	26.83	12	ROOF & ITS COMPONENTS
			41.28	28.95	109	EXTERIOR
			33.33	32.33	15	INSTRUMENT PANEL
ROLLOVER FIRST/WORST EVENT	NO BELT	ARM	42.60	31.90	331	average injured occupant
			38.46	35.31	13	EXTERIOR
ROLLOVER FIRST/WORST EVENT	NO BELT	LEG	50.00	31.33	18	EXTERIOR
			42.60	31.90	331	average injured occupant
REAR & OTHER	BELTED	TORSO	62.75	42.86	51	average injured occupant
			50.00	44.50	10	SEAT & ITS COMPONENTS
REAR & OTHER	NO BELT	TORSO	63.64	40.00	11	SEAT & ITS COMPONENTS
			56.76	37.16	37	average injured occupant

**3+ standard deviations higher than average injured occupant (p < .002).
*1.96-3 standard deviations higher than average injured occupant (p < .05).

TABLE E-15: PERCENT FEMALE, RIGHT-FRONT PASSENGERS AGE 18-96 WITH AIS 4+ INJURIES, CDS 1988-2010

IMPACT TYPE	OCCUPANT PROTECTION	BODY REGION	PERCENT FEMALE	AVERAGE AGE	N	INJURY SOURCE
FRONTAL	BELTED, AIR BAG	HEAD	100.00	69.20	5	AIR BAG
			61.74	47.91	115	average injured occupant
			50.00	43.17	6	EXTERIOR
			50.00	38.13	8	INSTRUMENT PANEL
			50.00	38.13	8	ROOF & ITS COMPONENTS
			50.00	32.83	6	B-PILLAR
			42.86	52.00	7	A-PILLAR
FRONTAL	BELTED, AIR BAG	TORSO	68.29	52.56	41	BELT SYSTEM
			61.74	47.91	115	average injured occupant
			60.00	49.60	5	SIDE HARDWARE OR ARMREST
			57.14	45.57	7	SIDE INTERIOR SURFACE
			57.14	37.57	14	INSTRUMENT PANEL
FRONTAL	BELTED, NO AIR BAG	HEAD	66.67	39.56	9	ROOF & ITS COMPONENTS
			60.93	52.99	151	average injured occupant
			50.00	31.83	6	EXTERIOR
			30.77	47.00	13	INSTRUMENT PANEL
FRONTAL	BELTED, NO AIR BAG	TORSO	64.91	54.75	57	BELT SYSTEM
			60.93	52.99	151	average injured occupant
			57.14	52.14	21	INSTRUMENT PANEL
			28.57	44.29	7	SIDE INTERIOR SURFACE
FRONTAL	NO BELT, AIR BAG	HEAD	49.24	39.70	132	average injured occupant
			43.75	32.63	16	EXTERIOR
			33.33	41.00	6	INSTRUMENT PANEL
			33.33	37.83	6	OTHER FRONTAL COMPONENT
			20.00	34.60	10	ROOF & ITS COMPONENTS
FRONTAL	NO BELT, AIR BAG	TORSO	55.81	37.95	43	INSTRUMENT PANEL
			49.24	39.70	132	average injured occupant
			40.00	51.00	5	SIDE INTERIOR SURFACE
			30.00	30.70	10	EXTERIOR
FRONTAL	NO BELT, NO AIR BAG	HEAD	50.88	41.93	57	OTHER FRONTAL COMPONENT
			45.27	40.51	349	average injured occupant
			41.18	37.18	17	ROOF & ITS COMPONENTS
			34.62	36.04	26	INSTRUMENT PANEL
			30.00	32.10	20	EXTERIOR
			20.00	28.60	5	A-PILLAR
FRONTAL	NO BELT, NO AIR BAG	TORSO	50.35	43.38	141	INSTRUMENT PANEL
			50.00	39.17	6	SIDE INTERIOR SURFACE
			45.27	40.51	349	average injured occupant
			0.00	33.80	5	EXTERIOR
NEARSIDE IMPACT	BELTED	HEAD	83.33	51.83	6	B-PILLAR
			60.98	49.90	82	average injured occupant
			57.14	37.00	7	SEAT & ITS COMPONENTS
			25.00	29.50	8	ROOF & ITS COMPONENTS
NEARSIDE IMPACT	BELTED	TORSO	81.82	65.09	11	BELT SYSTEM
			63.64	52.00	11	OTHER OCCS, SELF, CARGO
			63.64	43.27	11	SEAT & ITS COMPONENTS
			60.98	49.90	82	average injured occupant
			00.00	00.00	8	FLOOR & ITS COMPONENTS
NEARSIDE IMPACT	NO BELT	HEAD	45.45	29.82	11	ROOF & ITS COMPONENTS
			42.86	34.00	14	EXTERIOR
			40.00	44.60	5	INSTRUMENT PANEL
			40.00	21.20	5	OTHER
			35.71	34.95	126	average injured occupant
			20.00	32.40	5	OTHER FRONTAL COMPONENT
NEARSIDE IMPACT	NO BELT	TORSO	42.86	39.43	7	STEERING ASSEMBLY
			40.00	32.00	5	SIDE INTERIOR SURFACE
			35.71	34.95	126	average injured occupant
			28.57	36.57	7	INSTRUMENT PANEL
			27.27	28.64	11	EXTERIOR
			11.11	31.89	9	OTHER OCCS, SELF, CARGO
			0.00	28.50	6	SEAT & ITS COMPONENTS

**3+ standard deviations higher than average injured occupant (p < .002).
*1.96-3 standard deviations higher than average injured occupant (p < .05).

308

IMPACT TYPE	OCCUPANT PROTECTION	BODY REGION	PERCENT FEMALE	AVERAGE AGE	N	INJURY SOURCE
FAR-SIDE IMPACT	BELTED	HEAD	85.71	62.43	7	OTHER SIDE COMPONENT
			66.67	47.56	36	B-PILLAR
			66.67	44.78	27	EXTERIOR
			62.58	48.18	310	average injured occupant
			53.33	40.00	15	ROOF & ITS COMPONENTS
			41.67	40.50	12	SIDE INTERIOR SURFACE
			22.22	41.00	9	A-PILLAR
			14.29	29.14	7	OTHER
FAR-SIDE IMPACT	BELTED	NECK	85.71	54.43	7	B-PILLAR
			62.58	48.18	310	average injured occupant
FAR-SIDE IMPACT	BELTED	TORSO	67.74	41.29	31	SIDE HARDWARE OR ARMREST
			64.29	50.36	14	B-PILLAR
			62.58	48.18	310	average injured occupant
			59.46	51.41	148	SIDE INTERIOR SURFACE
FAR-SIDE IMPACT	NO BELT	HEAD	48.00	32.60	25	B-PILLAR
			46.64	36.35	298	average injured occupant
			42.86	39.57	7	OTHER SIDE COMPONENT
			42.11	35.00	38	EXTERIOR
			31.82	33.36	22	A-PILLAR
			25.00	25.92	12	OTHER
			20.00	33.00	5	SIDE INTERIOR SURFACE
			14.29	31.21	14	ROOF & ITS COMPONENTS
FAR-SIDE IMPACT	NO BELT	TORSO	83.33	39.17	6	OTHER OCCS, SELF, CARGO
			53.85	39.69	13	B-PILLAR
			51.00	40.06	100	SIDE INTERIOR SURFACE
			46.64	36.35	298	average injured occupant
			42.31	39.58	26	SIDE HARDWARE OR ARMREST
			38.89	36.50	18	EXTERIOR
			33.33	35.67	9	INSTRUMENT PANEL
ROLLOVER FIRST/WORST EVENT	BELTED	HEAD	49.51	38.71	103	average injured occupant
			45.83	38.58	24	ROOF & ITS COMPONENTS
			33.33	30.33	9	EXTERIOR
ROLLOVER FIRST/WORST EVENT	BELTED	NECK	49.51	38.71	103	average injured occupant
			20.00	28.60	5	ROOF & ITS COMPONENTS
ROLLOVER FIRST/WORST EVENT	BELTED	TORSO	61.54	42.00	13	SIDE INTERIOR SURFACE
			50.00	29.25	8	BELT SYSTEM
			49.51	38.71	103	average injured occupant
ROLLOVER FIRST/WORST EVENT	NO BELT	HEAD	42.62	32.34	237	average injured occupant
			39.34	28.16	61	EXTERIOR
			31.82	30.23	22	ROOF & ITS COMPONENTS
ROLLOVER FIRST/WORST EVENT	NO BELT	NECK	42.62	32.34	237	average injured occupant
			20.00	31.60	5	EXTERIOR
ROLLOVER FIRST/WORST EVENT	NO BELT	TORSO	55.56	29.22	9	SIDE INTERIOR SURFACE
			42.62	32.34	237	average injured occupant
			41.27	29.94	63	EXTERIOR
			37.50	29.88	8	INSTRUMENT PANEL
			37.50	25.00	8	ROOF & ITS COMPONENTS
REAR & OTHER	BELTED	TORSO	62.50	42.08	24	average injured occupant
			60.00	52.00	5	SEAT & ITS COMPONENTS

**3+ standard deviations higher than average injured occupant (p < .002).
*1.96-3 standard deviations higher than average injured occupant (p < .05).

IMPACT TYPE	OCCUPANT PROTECTION	BODY REGION	PERCENT FEMALE	AVERAGE AGE	N	INJURY SOURCE
FRONTAL	BELTED	TORSO	75.38	50.11	65	BELT SYSTEM
			66.17	42.79	133	average injured occupant
FRONTAL	NO BELT	HEAD	61.11	31.89	18	ROOF & ITS COMPONENTS
			53.33	37.27	15	OTHER FRONTAL COMPONENT
			48.32	35.51	358	average injured occupant
			47.89	34.18	71	SEAT & ITS COMPONENTS
			35.29	27.18	17	B-PILLAR
FRONTAL	NO BELT	NECK	66.67	38.20	15	SEAT & ITS COMPONENTS
			48.32	35.51	358	average injured occupant
FRONTAL	NO BELT	TORSO	50.00	35.38	16	SIDE INTERIOR SURFACE
			48.32	35.51	358	average injured occupant
			39.81	38.66	103	SEAT & ITS COMPONENTS
FRONTAL	NO BELT	ARM	48.32	35.51	358	average injured occupant
			47.37	35.02	57	SEAT & ITS COMPONENTS
FRONTAL	NO BELT	LEG	** 88.57	43.71	70	SEAT & ITS COMPONENTS
			60.00	34.93	15	FLOOR & ITS COMPONENTS
			48.32	35.51	358	average injured occupant
NEARSIDE IMPACT	BELTED	TORSO	66.04	47.26	53	average injured occupant
			63.64	47.45	22	SIDE INTERIOR SURFACE
NEARSIDE IMPACT	NO BELT	HEAD	48.80	34.25	166	average injured occupant
			46.67	30.47	15	OTHER SIDE COMPONENT
			41.18	30.82	17	EXTERIOR
			40.00	32.50	20	ROOF & ITS COMPONENTS
NEARSIDE IMPACT	NO BELT	TORSO	61.11	39.89	18	SIDE HARDWARE OR ARMREST
			55.00	35.75	20	EXTERIOR
			49.02	38.12	51	SIDE INTERIOR SURFACE
			48.80	34.25	166	average injured occupant
FAR-SIDE IMPACT	BELTED	TORSO	65.38	51.27	26	SIDE INTERIOR SURFACE
			56.76	40.03	74	average injured occupant
FAR-SIDE IMPACT	NO BELT	HEAD	57.89	36.63	19	EXTERIOR
			55.56	33.50	18	OTHER SIDE COMPONENT
			42.70	33.91	185	average injured occupant
			36.84	30.95	19	ROOF & ITS COMPONENTS
			33.33	37.33	18	B-PILLAR
FAR-SIDE IMPACT	NO BELT	TORSO	62.50	34.63	16	EXTERIOR
			52.17	37.00	23	SIDE HARDWARE OR ARMREST
			43.48	37.13	23	SEAT & ITS COMPONENTS
			42.70	33.91	185	average injured occupant
			40.63	37.61	64	SIDE INTERIOR SURFACE
FAR-SIDE IMPACT	NO BELT	ARM	46.67	41.80	15	SIDE INTERIOR SURFACE
			42.70	33.91	185	average injured occupant
FAR-SIDE IMPACT	NO BELT	LEG	42.70	33.91	185	average injured occupant
			29.41	29.76	17	SIDE INTERIOR SURFACE
ROLLOVER FIRST/WORST EVENT	NO BELT	HEAD	57.14	29.41	49	EXTERIOR
			46.79	29.99	156	average injured occupant
			40.00	29.60	20	ROOF & ITS COMPONENTS
ROLLOVER FIRST/WORST EVENT	NO BELT	NECK	58.82	28.06	17	EXTERIOR
			46.79	29.99	156	average injured occupant
ROLLOVER FIRST/WORST EVENT	NO BELT	TORSO	56.60	28.00	53	EXTERIOR
			46.79	29.99	156	average injured occupant
REAR & OTHER	NO BELT	TORSO	72.73	42.09	22	SEAT & ITS COMPONENTS
			70.83	35.42	48	average injured occupant

**3+ standard deviations higher than average injured occupant (p < .002).
*1.96-3 standard deviations higher than average injured occupant (p < .03).

TABLE E-17: PERCENT FEMALE, BACKSEAT OUTBOARD PASSENGERS AGE 18-96 WITH AIS 3+ INJURIES, CDS 1988-2010

IMPACT TYPE	OCCUPANT PROTECTION	BODY REGION	PERCENT FEMALE	AVERAGE AGE	N	INJURY SOURCE
FRONTAL	BELTED	TORSO	73.68	49.18	38	BELT SYSTEM
			68.18	45.39	66	average injured occupant
FRONTAL	NO BELT	HEAD	47.76	36.10	201	average injured occupant
			45.45	35.82	22	SEAT & ITS COMPONENTS
			15.38	28.15	13	EXTERIOR
FRONTAL	NO BELT	TORSO	50.00	32.20	10	SIDE INTERIOR SURFACE
			47.76	36.10	201	average injured occupant
			45.45	32.00	11	EXTERIOR
			28.89	40.64	45	SEAT & ITS COMPONENTS
FRONTAL	NO BELT	ARM	47.76	36.10	201	average injured occupant
			46.15	31.96	26	SEAT & ITS COMPONENTS
FRONTAL	NO BELT	LEG	* 67.44	45.53	43	SEAT & ITS COMPONENTS
			47.76	36.10	201	average injured occupant
NEARSIDE IMPACT	BELTED	TORSO	70.00	49.13	30	average injured occupant
			54.55	43.91	11	SIDE INTERIOR SURFACE
NEARSIDE IMPACT	NO BELT	HEAD	46.15	28.54	13	OTHER SIDE COMPONENT
			45.16	35.42	124	average injured occupant
			41.67	33.17	12	ROOF & ITS COMPONENTS
NEARSIDE IMPACT	NO BELT	TORSO	48.78	39.51	41	SIDE INTERIOR SURFACE
			45.45	31.18	11	EXTERIOR
			45.16	35.42	124	average injured occupant
FAR-SIDE IMPACT	BELTED	TORSO	75.00	59.38	16	SIDE INTERIOR SURFACE
			57.14	45.26	42	average injured occupant
FAR-SIDE IMPACT	NO BELT	HEAD	57.14	31.71	14	EXTERIOR
			40.00	34.40	125	average injured occupant
			18.18	25.27	11	ROOF & ITS COMPONENTS
FAR-SIDE IMPACT	NO BELT	TORSO	* 72.73	36.18	11	EXTERIOR
			50.00	38.17	12	SIDE HARDWARE OR ARMREST
			40.00	34.40	125	average injured occupant
			35.00	37.25	40	SIDE INTERIOR SURFACE
FAR-SIDE IMPACT	NO BELT	LEG	40.00	34.40	125	average injured occupant
			33.33	30.33	15	SIDE INTERIOR SURFACE
			30.00	38.30	10	SEAT & ITS COMPONENTS
ROLLOVER FIRST/WORST EVENT	NO BELT	HEAD	53.13	28.78	32	EXTERIOR
			52.83	31.32	106	average injured occupant
ROLLOVER FIRST/WORST EVENT	NO BELT	NECK	52.83	31.32	106	average injured occupant
			41.67	27.58	12	EXTERIOR
ROLLOVER FIRST/WORST EVENT	NO BELT	TORSO	60.98	28.83	41	EXTERIOR
			52.83	31.32	106	average injured occupant
ROLLOVER FIRST/WORST EVENT	NO BELT	LEG	60.00	31.80	10	EXTERIOR
			52.83	31.32	106	average injured occupant
REAR & OTHER	NO BELT	TORSO	81.25	42.94	16	SEAT & ITS COMPONENTS
			72.41	39.83	29	average injured occupant

**3+ standard deviations higher than average injured occupant (p < .002).
*1.96-3 standard deviations higher than average injured occupant (p < .05).

TABLE E-18: PERCENT FEMALE, BACKSEAT OUTBOARD PASSENGERS AGE 18-96 WITH AIS 4+ INJURIES, CDS 1988-2010

IMPACT TYPE	OCCUPANT PROTECTION	BODY REGION	PERCENT FEMALE	AVERAGE AGE	N	INJURY SOURCE
FRONTAL	BELTED	TORSO	76.92	57.31	13	BELT SYSTEM
			60.00	47.84	25	average injured occupant
FRONTAL	NO BELT	HEAD	50.00	37.83	6	ROOF & ITS COMPONENTS
			38.04	37.66	92	average injured occupant
			36.36	39.36	11	SEAT & ITS COMPONENTS
			22.22	30.56	9	EXTERIOR
FRONTAL	NO BELT	NECK	80.00	38.00	5	SEAT & ITS COMPONENTS
			38.04	37.66	92	average injured occupant
FRONTAL	NO BELT	TORSO	38.04	37.66	92	average injured occupant
			35.00	43.70	20	SEAT & ITS COMPONENTS
			33.33	35.00	6	EXTERIOR
NEARSIDE IMPACT	BELTED	TORSO	71.43	45.43	7	SIDE INTERIOR SURFACE
			68.75	48.00	16	average injured occupant
NEARSIDE IMPACT	NO BELT	HEAD	55.56	31.11	9	OTHER SIDE COMPONENT
			45.78	37.36	83	average injured occupant
			42.86	42.29	7	ROOF & ITS COMPONENTS
			40.00	26.00	5	EXTERIOR
NEARSIDE IMPACT	NO BELT	TORSO	50.00	42.83	24	SIDE INTERIOR SURFACE
			50.00	27.83	6	EXTERIOR
			45.78	37.36	83	average injured occupant
			40.00	47.00	5	SIDE HARDWARE OR ARMREST
FAR-SIDE IMPACT	BELTED	TORSO	87.50	57.00	8	SIDE INTERIOR SURFACE
			63.64	51.91	22	average injured occupant
FAR-SIDE IMPACT	NO BELT	HEAD	71.43	33.29	7	EXTERIOR
			60.00	48.80	5	B-PILLAR
			42.86	35.22	77	average injured occupant
			28.57	28.29	7	ROOF & ITS COMPONENTS
			0.00	21.20	5	OTHER
FAR-SIDE IMPACT	NO BELT	TORSO	57.14	44.00	7	SIDE HARDWARE OR ARMREST
			57.14	27.71	7	EXTERIOR
			42.86	35.22	77	average injured occupant
			37.50	40.13	16	SIDE INTERIOR SURFACE
ROLLOVER FIRST/WORST EVENT	NO BELT	HEAD	51.35	30.97	74	average injured occupant
			50.00	29.59	22	EXTERIOR
			20.00	23.20	5	ROOF & ITS COMPONENTS
ROLLOVER FIRST/WORST EVENT	NO BELT	TORSO	62.96	28.89	27	EXTERIOR
			51.35	30.97	74	average injured occupant
REAR & OTHER	BELTED	TORSO	40.00	46.40	5	SEAT & ITS COMPONENTS
			33.33	42.33	6	average injured occupant
REAR & OTHER	NO BELT	HEAD	69.57	41.04	23	average injured occupant
			60.00	37.00	5	ROOF & ITS COMPONENTS
REAR & OTHER	NO BELT	TORSO	72.73	43.00	11	SEAT & ITS COMPONENTS
			69.57	41.04	23	average injured occupant

**3+ standard deviations higher than average injured occupant (p < .002).
*1.96-3 standard deviations higher than average injured occupant (p < .05).

APPENDIX F

NUMBER OF DATA POINTS IN THE REGRESSION ANALYSES

A "data point" is a driver-passenger pair occupying the same vehicle, both 21 to 96, and:

At least one or both were fatally injured (in the FARS analyses of Chapter 3)
At least one or both were fatally injured and also had an injury of the type addressed by the analysis (in the MCOD analyses of Chapter 4)
- At least one or both had an injury of the type addressed by the analysis (in the CDS analyses of Chapter 5)

Chapter 3:

Table 3-1

All cars + LTVs, MY 1960-2011	154,467
Cars only	113,397
LTVs only	41,070
Cars only, MY ≥ 2000	8,496
LTVs only, MY ≥ 2000	5,404
Cars + LTVs, MY 1960-1966	5,030
1967-1974	25,724
1975-1979	26,993
1980-1984	23,295
1985-1989	29,205
1990-1994	16,183
1995-1999	14,137
2000-2004	10,313
2005-2011	3,587

Table 3-2
Cars

Unbelted, pre-MY 1968, no EA columns	4,272
Unbelted occupants of MY 1969-1982 cars	43,593
Unbelted, MY 1983-1996, no air bags	18,001
3-pt. belted occupants of cars w/o air bags	17,454
2-pt. belted occupants of cars w/o air bags	4,929
Unbelted occupants of cars with dual air bags	6,072
Belted, dual air bags, no pretens/load lim	7,148
Belted, dual air bags, pretensioners, load lim	4,199
LTVs	
Unbelted occupants of LTVs w/o air bags	26,407
Belted occupants of LTVs w/o air bags	5,804
Unbelted, dual air bags (no on-off switches)	2,993
Belted, dual air bags (no on-off switches)	5,817

Table 3-3

Cars + LTVs	
Frontal impacts	74,630
Left-side impacts	21,370
Right-side impacts	26,951
First-event rollovers	19,828
Rear impacts & other crashes	11,688
Cars only	
Frontal impacts	55,400
Left-side impacts	17,673
Right-side impacts	22,709
First-event rollovers	9,546
Rear impacts & other crashes	8,069

Table 3-4 (cars only, except where noted)

Frontal impacts	
Unbelted occs of cars without air bags	31,494
3-pt. belted occs of cars w/o air bags	7,862
Unbelted occs of cars with dual air bags	2,750
Belted, dual air bags, no pretens/load lim	2,815
Belted, dual air bags, pretensioners, load lim	1,925
Left-side impacts	
Unbelted occs of cars w/o side/curtain bags	9,137
3-pt belted occs of cars w/o side/curtain bags	6,192
Right-side impacts	
Unbelted occs of cars w/o side/curtain bags	12,148
3-pt belted occs of cars w/o side/curtain bags	7,401
All side impacts	
Cars/LTVs w curtain+torso or combo bags	744
First-event rollovers	
Unbelted occupants	7,730
3-point belted occupants	1,464
Rear impacts & other crashes	
Unbelted occupants	5,695
3-point belted occupants	2,056

Table 3-5 (back-seat occupants of cars)

All back-seat occupants	25,502
Unbelted (cars w/o air bags)	15,181
Lap-belted (cars w/o air bags)	1,352
3-point belted (cars w/o air bags)	535
Unbelted (cars w driver/dual air bags)	2,274
3-point belted (cars w driver/dual air bags)	2,494

Chapter 4:

Table 4-1
Severe & unk severity head injury
    All occupants   33,998
    Unbelted   20,739
    3-point belted   11,831
Severe head injury
    All occupants   9,304
    Unbelted   5,666
    3-point belted   3,265
Severe & unk severity torso injury
    All occupants   30,378
        Chest injury   23,680
        Abdominal injury   7,820
    Unbelted   16,105
        Chest injury   12,665
        Abdominal injury   3,859
    3-point belted   12,063
        Chest injury   9,342
        Abdominal injury   3,213
Severe torso injury
    All occupants   12,248
        Chest injury   9,655
        Abdominal injury   3,542
    Unbelted   6,200
    3-point belted   5,072
Severe & unk severity neck injury
    All occupants   8,049
    Unbelted   4,391
    3-point belted   3,198
Severe neck injury
    All occupants   6,244
    Unbelted   3,586
    3-point belted   2,293

Table 4-2 (head injuries, severe or unk severity)
Frontal impacts
    All occupants   15,518
    No air bag, unbelted occupants   8,560
    No air bag, 3-pt belted occupants   3,173
    Dual air bags   3,002
Left-side impacts
    All occupants   4,567
    Unbelted occupants   2,327
    3-pt belted occupants   2,041

315

Right-side impacts
All occupants	5,448
Unbelted occupants	2,884
3-pt belted occupants	2,287

First-event rollovers
All occupants	6,002
Unbelted occupants	4,255
3-pt belted occupants	1,578

Rear impacts & other crashes
All occupants	2,463
Unbelted occupants	1,441
3-pt belted occupants	908

Table 4-3 (torso injuries, severe or unk severity)

Frontal impacts
All occupants	15,302
No air bag, unbelted occupants	7,426
No air bag, 3-pt belted occupants	3,709
Dual air bags	2,811

Left-side impacts
All occupants	4,835
Unbelted occupants	2,032
3-pt belted occupants	2,506

Right-side impacts
All occupants	5,835
Unbelted occupants	2,568
3-pt belted occupants	2,844

First-event rollovers
All occupants	2,868
Unbelted occupants	2,282
3-pt belted occupants	521

Rear impacts & other crashes
All occupants	1,538
Unbelted occupants	834
3-pt belted occupants	617

Table 4-4 (neck injuries, severe or unk severity)

Frontal impacts
All occupants	3,627
No air bag, unbelted occupants	1,833
No air bag, 3-pt belted occupants	805
Dual air bags	757
Left-side impacts	1,120
Right-side impacts	1,401
First-event rollovers	1,287
Rear impacts & other crashes	614

Chapter 5

Table 5-1
Fatal Injuries (CDS data)	1,326
Fatal Injuries (FARS data)	
All occupants	154,467
Unbelted occupants	104,372
3-pt belted occupants	43,887
MAIS ≥ 4	
All occupants	1,849
Unbelted occupants	942
3-pt belted occupants	758
MAIS ≥ 3	
All occupants	3,554
Unbelted occupants	1,605
3-pt belted occupants	1,692
MAIS ≥ 2	
All occupants	6,132
Unbelted occupants	2,548
3-pt belted occupants	3,173
MAIS 3 exactly	
All occupants	2,098
Unbelted occupants	867
3-pt belted occupants	1,092
MAIS 2 exactly	
All occupants	3,384
Unbelted occupants	1,335
3-pt belted occupants	1,833

Table 5-2
Head injuries	
AIS ≥ 4	807
AIS ≥ 3	1,324
AIS ≥ 2	3,096
Torso injuries	
AIS ≥ 4	1,132
AIS ≥ 3	2,219
AIS ≥ 2	3,739
Chest injuries	
AIS ≥ 4	999
AIS ≥ 3	1,884
AIS ≥ 2	2,512
AIS ≥ 2 abdominal injuries	1,208
AIS ≥ 2 neck injuries	785
AIS ≥ 2 arm injuries	1,599
AIS ≥ 2 leg injuries	2,072

DOT HS 811 766
May 2013

U.S. Department
of Transportation

**National Highway
Traffic Safety
Administration**

www.nhtsa.gov

9585-051013-v3a

www.ingramcontent.com/pod-product-compliance
Lightning Source LLC
Chambersburg PA
CBHW081431170526

45166CB00008B/2166